USO

D0080904

Proteins and Enzymes

J. ELLIS BELL

EVELYN T. BELL

Department of Biochemistry
University of Rochester

PRENTICE-HALL, INC.

Englewood Cliffs, New Jersey / 07632

Library of Congress Cataloging-in-Publication Data

Bell, J. Ellis (John Ellis)
  Proteins and enzymes.

  Bibliography: p.
  Includes Index.
  1. Proteins. 2. Enzymes. I. Bell, Evelyn.
  II. Title.   QP551.B33 1988
ISBN 0-13-731647-X   547.7′5.   87-7004

© 1988 by Prentice-Hall, Inc.
A Division of Simon & Schuster
Englewood Cliffs, New Jersey 07632

Printed in the United States of America

10 9 8 7 6 5 4 3 2 1

Prentice-Hall International (UK) Limited, *London*
Prentice-Hall of Australia Pty Limited, *Sydney*
Prentice-Hall of Canada Inc., *Toronto*
Prentice-Hall Hispanoamericana, S.A., *Mexico*
Prentice-Hall of India Private Limited, *New Delhi*
Prentice-Hall of Japan, Inc., *Tokyo*
Prentice-Hall of Southeast Asia Pte., Ltd., *Singapore*
Editora Prentice-Hall do Brasil, Ltda., *Rio de Janeiro*

to KEITH DALZIEL, Oxford University

Contents

Preface

This text was developed over a number of years to complement a course in Protein Chemistry and Enzymology taught to first-year graduate students and senior-year undergraduates in the Biochemistry Department at the University of Rochester. It has been written to detail the background and development of current concepts of protein structure and function. The book has two sections, the first describing in detail the chemistry of proteins and the techniques for purifying and characterizing them. Included in this section is a consideration of the conformation and three-dimensional structure of proteins together with basic concepts of how structure and function are related. In the second section the major emphasis is on advanced methods of examining enzymatic function and the regulation of enzymatic function.

Throughout the book experimental techniques and overall approaches to problems in protein chemistry and enzymology are emphasized. At the end of many of the chapters a number of examples from the biochemical literature are introduced to complement the material presented in the chapter. Rather than compile an extensive list of supplementary reading material for each chapter, we have, where appropriate, given the references to certain pieces of work that were chosen to illustrate some part of the chapter's content.

Acknowledgments

We wish to acknowledge the help and encouragement of our many colleagues and students during the preparation of the manuscript. In particular we owe much to the influence of Keith Dalziel, Robert L. Hill, and H. Richard Levy over the years. The students in the B.S. Biochemistry program at the University of Rochester are also acknowledged for their patience and comments while the material presented here was being developed. In particular, several graduate students in the Medical Center have been most helpful. Bob Burne assisted with the treatment of gene cloning and sequencing in Chapter 5 and with the description of site-directed mutagenesis in Chapter 6. Tom Smith and Lee Janson helped with several of the diagrams, and Cheryl Gilbert contributed to the writing of Chapter 11. Beth Gingell was invaluable in compiling the index. We would also like to thank Drs. C. Brown, D. Goldstein, and B. Goldstein for their help with the preparation of some of the figures involving secondary structure in Chapter 9. These figures were generated by the molecular graphics and modeling system MOFO, developed by Christopher S. Brown, David A. Goldstein, and Barry M. Goldstein, in the Department of Radiation Biology and Biophysics at the University of Rochester.

1

Basic Concepts of Protein Purification and Characterization

INTRODUCTION

Before examining the ways in which proteins may be purified, two points must first be covered that are linked in the concept of a "specific activity." Specific activity is defined and discussed in more detail later, but for the moment it can be regarded as activity per unit amount of protein. Of paramount importance therefore in the determination of a specific activity are the experimental determination of the concentration and the activity of a protein. Each of these is considered before discussing specific activity in more detail and examining sources of protein to be purified. It is the goal of purification to obtain protein having the maximum specific activity, and achievement of this is often monitored by obtaining a constant specific activity during several purification steps.

DETERMINATION OF PROTEIN CONCENTRATIONS

The accurate quantitation of protein concentrations is, of course, increasingly important as purity is attained since the accuracy of the final specific activity of a "pure" preparation of protein depends in large measure on the accuracy of its concentration determination. This determination would appear to be a simple operation; however, many of the methods that have been used suffer from interference by reagents such as the detergents or salts that are commonly used during purifications, or from contaminants such as the nucleic acids or nucleotides that are frequently associated with proteins. Much effort has been directed toward developing approaches to circumvent such problems.

Here we consider the pitfalls and advantages of several commonly used techniques for determining protein concentration. These are the biuret protein determination method, the Lowry method, the Bradford method, the Amidoschwarz method, and the 280-nm absorbance method.

Biuret Method

The biuret method is one of the earliest colorimetric methods for determining protein concentrations and is still a rapid, if rather insensitive approach for use during the early stages of a purification, especially when ammonium sulfate precipitation (Chap. 2) is employed since it is unaffected by the presence of this salt. In this method a colored complex involving the complexation of copper in alkaline solution with peptide bonds and tyrosine side chains is used. As with many of the methods discussed here, a calibration curve with known concentrations is used to allow the concentration of the unknown to be determined. This aspect leads to many of the inaccuracies of such methods since the choice of standard protein affects the calibration curve. In the biuret method the use of two different standard proteins having quite different tyrosine contents yields different calibration curves. The nature of the standard protein should always be indicated.

Lowry Method

The approach first described by Lowry et al. in 1951 is also based on the formation of a copper–protein complex under alkaline conditions that subsequently reduce the folin–phenol reagent at pH 10 with color development that can be quantitated spectrophotometrically at 750 nm. Protein concentration of an unknown sample is determined by comparison with a standard curve. The folin–phenol reagent is a phosphomolybdate-containing mix, and maximum color development usually takes about 30 minutes. A fundamental problem is that the color developed depends on the nature of the protein—not all proteins give the same color intensity. Some of the problems encountered using nonionic and cationic detergents (which under Lowry conditions tend to cause precipitate formation) have been overcome by the inclusion of 0.5% sodium dodecylsulfate in the alkali reagent, which prevents precipitate formation and does not affect color development. This adaptation is particularly useful in the estimation of protein concentrations in detergent solubilized membrane preparations.

Bradford Method

The observation that the dye Coomassie Brilliant Blue G-250 exists in two different color forms (a red and a blue) and that the red form is converted to the blue form upon binding to protein, led to the development of the Bradford dye binding method. The binding of the dye to protein causes a shift in the absorption maximum of the dye from 465 nm to 595 nm, and the resultant increase in absorption at 595 nm is monitored. The dye–protein complex has a high extinction coefficient, leading to excellent sensitivity, and the colored complex develops rapidly and has relatively good stability. Although the color intensity is quite pH dependent, falling off considerably at high pH, appropriate buffering leads to accurate protein estimation. Although cations and carbohydrates do not significantly interfere with the color, various detergents do cause problems. If the concentration of detergent is small, the interference is minimized by the inclusion of the same amount of detergent in the calibration samples. For the most accurate determinations, the buffers for the sample and the standard curve should be the same.

Because of its ease of use, its high sensitivity (the accurate determination of 0.1 μg of protein is possible), and its overcoming of many of the Lowry method problems, the Bradford method of protein estimation is often the method of choice.

For both the Lowry and Bradford methods we have indicated that a variety of small molecules interfere with color formation. This can lead to quite large errors in protein concentration determination and often, compounds that decrease the sensitivity of the Bradford assay (e.g., CsCl and guanidine HCl) increase the sensitivity of the Lowry assay. Other compounds, such as Tris and dithiothreitol, have a significant effect on the Lowry but not on the Bradford method.

Some of these problems involve the effects of solvent composition on the color of the various noncovalent complexes involved in color formation. Such effects can be overcome by employing covalent labeling with a fluorescent reagent such as O-phthalaldehyde, which reacts with amines to give an adduct that fluoresces under alkaline conditions (pH 9 to 11). The derivative is essentially insensitive to all common buffer components except Tris.

Amidoschwarz Method

The Amidoschwarz approach is particularly useful since it is not affected by the majority of reagents that interfere with the Lowry and the Bradford methods. Its basis is simple. Protein is precipitated by trichloroacetic acid, the precipitate collected by filtration onto filter paper and stained with Amidoschwarz 10B dye, and after suitable washing to remove excess dye the protein–dye complex is eluted and quantitated by absorbance measurements at 630 nm. As with the Lowry and Bradford methods, the protein concentration is estimated by comparison with a standard curve, but because of the precipitation step prior to staining, the color development is virtually independent of buffer salts, detergents, and so on, which interfere with the previously discussed methods. Furthermore, because quite large volumes can be used, concentrations as low as 0.5 μg/ml can be conveniently determined. The major practical drawback of this method is that it is somewhat tedious; however, under circumstances that interfere with the Lowry and Bradford assays, this method represents a convenient way to estimate protein concentrations accurately. The only condition so far shown to lead to problems is the estimation of low-molecular-weight proteins where the precipitates are incompletely retained during the filtration step. This represents a problem with proteins such as insulin (mol. wt. 5770) and α-lactalbumin (mol. wt. 14,000).

280-nm Absorbance Method

Most proteins contain one or more aromatic amino acids (tyrosine, tryptophan, and phenylalanine) that absorb light in the region 250 to 300 nm. The absorption spectrum of most proteins features a broad absorption usually centered around 275 to 285 nm, and spectrophotometric measurement at 280 nm is a frequently used method of estimating protein concentration. As will be discussed in more detail, a pure protein has a characteristic and defined extinction coefficient at 280 nm that can, in pure solutions, be used to quantitate concentration very accurately based on spectrophotometric measurements. However, the extinction coefficient at 280 nm varies from approximately 0.5 to 1.5 for a 1-mg/ml solution, making it difficult to obtain accurate estimates of total protein concentration for a mixture of proteins (which, of course, is the normal situation encountered during a purification). However an "average" value of the extinction coefficient of 1.0 for a 1-mg/ml solution is frequently employed for the purposes of estimating protein concentrations.

The prime advantage of absorbance measurements is their ease, but unfortunately, proteins are not the only species in solution that absorb at 280 nm. Particularly troublesome are nucleic acids, nucleotides, and detergents. Various detergents used in protein solubilization absorb at 280 nm, frequently giving quite high background absorbances and making accurate protein determinations difficult. Although the use of appropriate blanks can often alleviate detergent problems, nucleic acids and nucleotides represent a more difficult situation. Nucleotides absorb strongly at 260 nm with resultant absorbance at 280 nm from the tail of their spectrum. The average protein has a 280:260 absorbance ratio of approximately 2.0, and the 280:260 ratio of a protein containing solution is frequently used to give some estimate of possible contamination by nucleotides. In fact, this approach is often used with pure proteins that bind nucleotides tightly, as a way of estimating how much nucleotide is bound per protein molecule.

DETERMINATION OF EXTINCTION COEFFICIENTS
OF PURE PROTEINS

Clearly, for the quantitation of a pure protein solution by absorbance measurements at 280 nm, its extinction coefficient must be known. There are two commonly used methods to determine this for a protein: the dry weight method and the amino acid analysis method. In the first, the absorbance of a protein containing solution is first determined after thorough dialysis versus a buffer of known composition that is not volatile, to give an absorbance per milliliter. Next, volumes of protein containing solution and dialysate are evaporated to dryness, heated at 110°C for 24 hours and cooled in a vacuum desiccator over phosphorus pentoxide, weighed, and the heating and weighing cycle repeated until a constant weight is obtained. The weight of protein in the original solution is thus obtained, allowing direct calculation of the extinction coefficient.

Alternatively, from the known molecular weight of the protein and amino acid analysis, a theoretical extinction coefficient can be calculated from the extinction coefficients of tyrosine, tryptophan, and phenylalanine. Usually, extinction coefficients are somewhat lower than those from dry weight determinations, probably because of environmental effects on the absorption properties of the aromatic amino acids.

Because of the experimental vagaries of either of these methods, they are best used in conjunction with some method of active site titration, discussed in more detail in Chap. 4. Where an accurate molecular weight can be obtained, active site titration, in conjunction with absorbance measurements, gives increased reliability of extinction coefficient determinations.

Amino Acid Analysis

The amino acid analysis of a polypeptide chain is achieved through some type of hydrolysis followed by analysis of the resultant amino acids using either ion-exchange chromatography or, for much greater sensitivity, high-performance liquid chromatography (HPLC).

There are two basic approaches to the hydrolysis procedure: (1) chemical hydrolysis, and (2) enzymatic hydrolysis, both of which suffer from drawbacks. With chemical hydrolysis it is the fact that the usually harsh hydrolysis conditions lead to destruction or chemical modification of some of the amino acids. For example, under typical acid hydrolysis conditions (5.7 M HCl, 108°C for 22 hours), the amides (asparagine and glutamine) are hydrolyzed to aspartate and glutamate, up to 10% losses of threonine or serine occur, there are variable losses of tyrosine or tryptophan (protective agents are usually added to help prevent these), and cysteine and cystine may be interconverted, oxidized to cystine oxides, or in the case of cysteine, react with tryptophan.

Early procedures for amino acid analysis employed multiple time points of hydrolysis and specific types of hydrolysis to optimize determination of certain amino acids. However, a general procedure has evolved that allows reasonable amino acid analysis to be made from a single hydrolysis of a protein. This approach is comprised of the following stages: (1) alkylation of cysteine, (2) acid hydrolysis in the presence of an amino acid protectant, and (3) neutralization and formation of S-sulfocysteine prior to analysis.

Alkylation of Cysteine. The two most commonly used methods of alkylation for amino acid analysis involve iodoacetate (a) or 4-vinyl-pyridine (b).

(a) $P-CH_2-SH + ICH_2-COO^- \longrightarrow P-CH_2-S-CH_2-COO^- + HI$

(b) $P-CH_2-SH + CH_2-CH\theta(N) \longrightarrow P-CH_2-S-CH_2-CH_2\theta(N)$

These yield derivatives that are fairly stable and more polar than cysteine (which is usually eluted with proline): S-carboxymethylcysteine is more acid than cysteine and elutes before aspartate, while S-β-(4 pyridylethyl) cysteine is more basic than cysteine and elutes after ammonia.

Alkylations are usually performed with a 10-to 20-fold excess of alkylating reagent over thiol at alkaline pH values. The reactions can be done in the presence of up to 8 M urea or 5 M guanidine hydrochloride, which is subsequently removed by dialysis or gel filtration.

Hydrolysis. Although a variety of hydrolysis conditions have been used, one of the best is 4 M methanesulfonic acid at 115°C for 22 hours. Here the addition of protective reagents for unstable amino acids is essential, with the most common additives being tryptamine (to protect tryptophan) and phenol (to protect tyrosine). One of the major problems with hydrolysis conditions selected to minimize breakdown of products is that these "mild" conditions may be insufficent to achieve complete cleavage of more resistant peptide bonds, such as the hindered ones of Ile-Ile, Ile-Val, or Val-Val, where time periods of up to 96 hours may be necessary to achieve complete hydrolysis.

Glycoproteins may be of particular trouble in hydrolysis since some amino acids are destroyed by hydrolysis in the presence of large amounts of carbohydrate. While protective agents such as tryptamine do work, there is a correlation between yield and percent carbohydrate—a plot of log (tryptophan recovery) versus percent carbohydrate gives a straight-line relationship.

As one may gather, the estimation of tryptophan content is most problematical in acid hydrolysis and most subject to error. For the most accurate estimation of tryptophan from protein hydrolysis, alkaline hydrolysis rather than acid hydrolysis gives the best results. It is carried out in vacuo at either 110°C or 136°C in 4.2 N NaOH containing 40 mg/ml starch. Tryptophan is then estimated by ion-exchange chromatography at pH 5.4 to separate it from a lysine derivative N-(DL-2-amino-2-carboxyethyl)-L-lysine formed during alkaline hydrolysis. As discussed in Chap. 7, the tryptophan contents of proteins can also be obtained from various chemical modifications.

Because of problems with chemical hydrolysis, digestion with mixtures of proteolytic enzymes has been used as an alternative. Although this approach is successful in some cases, low yields of aspartate and glutamine often result, possibly due to the resistance of Asp-Asp bonds to proteolysis, the cyclization of glutamine residues or α-β rearrangement of peptide bonds involving aspartate. A further problem is presented by autolysis of the digestive enzymes and corrections must be made to accommodate such effects. Disulfide-containing proteins that may be particularly resistant to proteolysis must be reduced and alkylated prior to enzymatic digestion.

Analysis. While the analysis of amino acids by ion-exchange chromatography has been well documented, sensitivity in the picomole range is achievable by HPLC. The strategy for such analysis often involves a pre-column derivatization which must react with all amino acids, including the imino acids proline and hydroxyproline, in a reproducible and quantitative manner to give a stable derivative. A particularly useful reagent for this purpose is DABS-Cl (dimethylaminoazobenzenesulfonyl chloride), which in addition to fulfilling these requirements produces derivatives that are detectable at 420 nm in ethanol with very high sensitivity.

In an alternative approach, the amino acids obtained from hydrolysis are first separated and then detected in a post-column derivatization procedure.

DETERMINATION OF ACTIVITY

The determination of the "activity" of a protein can represent a difficult problem. While many proteins are enzymes, where some catalytic reaction can be followed, others have biological functions that do not fall into this category. Apart from the class described as enzymes, proteins can, in terms of the way one attempts to assay for them, be placed into two groups: binding and structural proteins. Each of these three classes is considered in turn.

Enzymes

The activity of an enzyme can often be conveniently measured by following either the production of a product or the removal of a substrate. With certain classes of enzymes (e.g., dehydrogenases) the natural substrates are chromophoric and exhibit spectral changes that can be followed directly. For example, alcohol dehydrogenase catalyzes the reduction of acetaldehyde by the coenzyme NADH:

$$\text{acetaldehyde} + \text{NADH} \rightleftharpoons \text{ethanol} + \text{NAD}$$

NADH has an absorption band centered at 340 nm with an extinction coefficient of $6.22 \times 10^3 \text{ cm}^{-1} M^{-1}$, while NAD has no absorbance at this wavelength. When alcohol dehydrogenase is added to a mixture of acetaldehyde and NADH, there is a time-dependent loss of absorbance at 340 nm, as seen in Fig. 1-1.

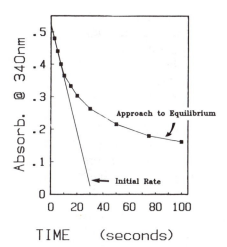

Figure 1-1 Complete time course of the reduction of acetaldehyde catalyzed by alcohol dehydrogenase using NADH as coenzyme.

If the reaction is allowed to proceed to equilibrium, the "rate" progressively slows until equilibrium is reached. Clearly, the reaction "rate" changes during the time course of the reaction as a consequence of both utilization of substrate and approach to equilibrium. To enable reproducible rate determinations, two aspects of the reaction are determined: (1) the initial rate, as shown in Fig. 1-1, and (2) the rate at saturating substrate concentrations. This rate (the "maximum rate") is calculated using concentrations of, in this case, acetaldehyde and NADH that give an experimentally determined maximum rate (Fig. 1-2).

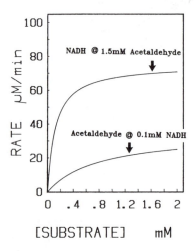

Figure 1-2   Dependence of the rate of the reaction catalyzed by alcohol dehydrogenase on the concentrations of NADH and acetaldehyde.

These substrate concentrations are used for two reasons, one pragmatic and the other theoretical. If substrate concentrations sufficient to give this maximum rate are used, any experimental error in making up the assay is minimized. From the theoretical standpoint, under these conditions the measured rate of the reaction is dependent only on the concentration of the enzyme, a situation necessary if the enzyme assay is used to determine the activity of the enzyme.

Our example has dealt with a very simple situation where the concentration of a substrate can be directly and continuously monitored, making it quite easy to determine an "initial rate." In cases where the product is the absorbing species, the rate can be followed by directly monitoring product formation. Many enzymes are not this simple to assay, and we must examine two types of assay procedures, direct and indirect.

Direct Assays. The previous example is a direct assay: The concentration of a particular substrate or product of the reaction is determined directly—in that case continuously. Consider another enzyme, galactosyl transferase, which catalyzes the reaction

UDP-galactose + *N*-acetylglucosamine ⇌ UDP + *N*-acetyllactosamine

None of these substrates or products can be conveniently monitored spectropho-tometrically; however, the reaction rate can be followed directly using radioactively labeled UDP-galactose (with the label in the galactose moiety). The reaction is initiated by addition of the enzyme and allowed to proceed for a particular time interval before termination [in this case by the addition of EDTA since galactosyl transferase is a Mn(II)-requiring enzyme]. The reaction mixture is then chromato-graphed on an ion-exchange column (which binds UDP-galactose), the flow-through collected, and the amount of radioactivity determined. From the specific activity of

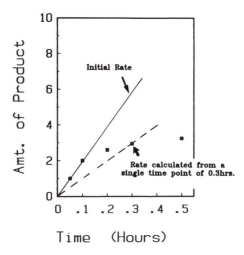

Figure 1-3   Time dependence of product accumulation during a discontinuous direct assay of enzyme activity.

the UDP-galactose the amount of N-acetyllactosamine produced in the time interval used can be measured and hence the reaction rate calculated. This rate is, however, not necessarily the initial rate discussed previously (see Fig. 1-3), and to establish that the initial rate is being determined the amount of product formed at several time intervals must be measured to show that at the particular time point chosen for routine assays the amount of product falls on a linear product versus time plot, proving that indeed the initial rate is being determined.

Such direct assays, where reaction mixtures are incubated, terminated, and then the amount of product or substrate measured after some type of separation procedure, are frequently employed, and provided that the separation procedure is effective, yield good results (usually this is controlled for by running blank reaction mixtures with no enzyme and subtracting "blanks" from values determined in the presence of enzyme).

Indirect Assays. Sometimes it is not convenient or possible to run direct assays, and indirect ones must be resorted to. In an indirect assay, product or substrate concentration is determined by linking the reaction to some second enzyme (or in some cases several enzymes) to give an easily measured variable. Very frequently, reactions are linked to dehydrogenases due to of the ease of measuring their reactions.

Consider the example just used, galactosyl transferase. Its activity can also be determined in a linked assay as follows. After reaction termination, in this case through the addition of EDTA to chelate the required metal ion of the enzyme, pyruvate kinase and lactate dehydrogenase are added, together with phospho-enolpyruvate (PEP) and NADH. Although pyruvate kinase normally uses ATP/ADP, it also functions with UDP and catalyzes the reaction

$$UDP + PEP \rightleftharpoons UTP + pyruvate$$

Lactate dehydrogenase catalyzes the reaction

$$NADH + pyruvate \rightleftharpoons NAD + lactate$$

which is monitored at 340 nm. The overall result of these reactions is that UDP production is linked to the utilization of quantitated NADH. Provided that conditions are such that the equilibria of the linking reactions lay completely to the right, the decrease in NADH concentration is directly proportional to the UDP concentration. Where relatively long incubation times are necessary for the linking enzyme, it is essential that the first reaction be completely terminated. This can usually be accomplished by brief periods of boiling, taking care to ensure that the product whose concentration is to be measured is not adversely affected.

Such linked reactions are in some cases used in a continuous manner; however, this is not recommended, as it is fraught with both practical and theoretical difficulties such as ensuring that the linking enzymes do not become rate limiting, or that the substrates of the linking reactions do not affect the rate of the reaction being followed.

At this point it is beneficial to make several comments about enzyme activity measurements and to work through some calculations for the rates of enzyme reactions. The use of saturating substrate concentrations in reaction mixtures to minimize experimental errors has been emphasized. It is also important that reaction rates be measured under conditions where a sufficiently small amount of substrate is utilized so that the rate does not change during the assay as a result of substrate depletion. Similarly, product buildup, which may lead to product inhibition, is to be avoided. In general, a convenient way to test that these factors do not become a problem is to measure activity at a series of protein concentrations: The rate should be directly proportional to the protein concentration, as in Fig. 1-4. Deviations below the line indicate that substrate depletion or product accumulation may be occurring. Deviations from linearity can also result from protein aggregation or subunit dissociation affecting the rate of the catalyzed reaction.

Calculations of Enzyme Activity. Consider again the example of alcohol dehydrogenase in Fig. 1-1. From the linear extrapolation of the initial rate it is found that an absorbance change of 0.53 occurs in 27 minutes. From this a "rate" of 0.0196 *A*/min is calculated. Given that the millimolar extinction coefficient of NADH is

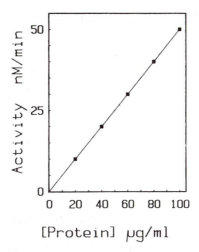

[Protein] µg/ml

Figure 1-4   Dependence of measured rate on protein concentration for a typical enzyme assay measured using a discontinuous system.

$6.22 \ \mathrm{cm}^{-1}$, the rate equals

$$0.0196/6.22 = 0.00316 \ \mathrm{m}M/\mathrm{min}$$

That is, in terms of *concentration*, the rate in the spectrophotometer cuvette is $0.00316 \ \mathrm{m}M/\mathrm{min}$. For convenience, assume that this measurement is made in a 1-ml cuvette. The rate in terms of *amount* of NADH utilized per minute is

$$0.00316 \ \mathrm{m}M = 0.00316 \ \mathrm{mmol/liter} = 3.16 \ \mu\mathrm{mol/liter} = 3.16 \ \mathrm{nmol/ml}$$

So the rate, in terms of amount, is 3.16 nmol/min *per milliliter* of reaction mixture.

If a radioactive substrate is employed, as in the example of the direct assay for galactosyl transferase, the concentration of product formed is calculated from the specific activity of the isotope. Assume that the UDP-galactose has a specific activity of 3.7×10^5 cpm/mol and that during the course of an activity measurement, as outlined in Fig. 1-3, an initial rate of 284 cpm/min is determined in a 100-µl reaction mixture.

The *rate* of the reaction in terms of µmol/min is

$$284/3.7^5 = 0.00077 \ \mu\mathrm{mol/liter} = 0.77 \ \mathrm{nmol/min}$$

If this rate is expressed in terms of *concentration*, the rate equals

$$0.77 \ \mathrm{nmol/min} \ \mathrm{per} \ 0.1 \ \mathrm{ml} = 7.7 \ \mathrm{nmol/min/ml} = 7.7 \ M/\mathrm{min}$$

Binding Proteins

The function of these proteins is usually assessed by their ability to bind a specific ligand or an analog of that ligand. Equilibrium binding of a ligand to a protein can be determined in various ways, some of which are dealt with from a theoretical standpoint in Chap. 17. For the present the various approaches are introduced briefly from the experimental standpoint.

Direct Methods of Studying Ligand Binding. The analogous parameter in ligand binding to the rate of an enzyme measured at saturating substrate concentrations is the maximum binding capacity—the amount of ligand bound by the protein in the presence of a large ligand excess. In terms of quantitating the amount of protein, we need not concern ourselves with the estimation of dissociation constants. Determination of the maximum binding capacity of a protein is not a simple matter, as is discussed thoroughly in Chap. 17. However, there are experimental difficulties as well as theoretical problems.

In the direct techniques for determining ligand binding, some method to separate physically bound ligand from free ligand is usually required. This may be achieved by equilibrium dialysis, gel filtration, or precipitation techniques.

Equilibrium Dialysis. In dialysis techniques, of which there are a variety, the protein-containing solution is separated from buffer by a semipermeable membrane that allows equilibration of the ligand. Ligand initially can be added to either side of the membrane, and after reaching equilibrium aliquots are withdrawn from both the free buffer and the protein-containing solution, and the concentration of ligand is determined (by radioactivity, spectral or chemical methods, or enzymic assay). The amount of bound ligand is obtained by subtracting the free ligand concentration from the total ligand concentration (bound + free) in the protein-containing solution. Equilibrium is ensured by running appropriate controls where ligand is added to only one of the compartments in the absence of any protein. The dialysis may be passive or forced by application of pressure to the protein-containing compartment in an apparatus such as the Amicon filter.

Gel Filtration. Separation of free ligand from protein-bound ligand can often be achieved rapidly and effectively by gel filtration using a matrix, where the free ligand penetrates the matrix but the protein and protein-bound ligand are excluded. The application of gel filtration techniques to ligand binding is usually achieved in one of two ways. Either the protein is passed through a gel filtration column equilibrated with a set concentration of ligand and the concentration of protein and ligand monitored in the effluent, or a small sample of protein plus ligand is applied to the top of a syringe or centrifuge tube column of matrix equilibrated in buffer containing no ligand. The column is then rapidly centrifuged, forcing the protein (and bound ligand) to be rapidly eluted, and the eluted protein and ligand concentration is determined. In the latter case the appropriate "blank" values must be subtracted.

Precipitation. In many instances the approaches already discussed cannot readily be applied. This is most often the case where the ligand is a protein or other large molecule. In such instances it may be possible to precipitate the complex between receptor and ligand but not the ligand free in solution. This approach may also be applied to small ligands where the complex can be precipitated without loss of affinity for the ligand and the precipitate washed to remove free ligand. This is frequently the case with proteins such as hormone receptors, which have very high affinity for their specific ligands.

Two points to bear in mind concerning these techniques need to be mentioned here as well as one experimental problem in determining the maximum binding capacity of a protein.

Nonspecific Binding. In many instances, especially with impure protein preparations, both specific (i.e., to the binding site of the desired protein) and nonspecific binding may occur. Specific binding, especially in the case of hormone receptors, is usually defined as being readily displaceable by a competing ligand, whereas nonspecific binding is usually nondisplaceable.

Chemically or Isotopically Altered Ligands. Often, the natural ligand for a protein cannot be used conveniently and an analog, either radioactively or spectrally labeled, is employed. Although this is usually not a problem, on occasion the alteration of the ligand may induce inhomogeneity: a mixture of two forms may arise where one is active in terms of binding and one inactive, leading to erroneous calculations of the total amount of ligand present in a particular compartment. It is therefore essential to determine that all of the ligand used can be bound by the protein. This is done by titrating a small amount of ligand with increasing amounts of protein to determine what proportion of the ligand can be bound. With small ligands this is not normally a problem unless the modification has produced racemization of an asymmetric carbon, but it is often encountered with protein or peptide ligands that have been radioactively labeled, where, due to differing extents of modification, some of the protein may have been inactivated. Where assays of biological function can be correlated with binding assays, additional confidence in the ligand is obtained.

Finally, we must consider an overall experimental problem in the determination of the maximum binding capacity of a protein. As with kinetic assays to determine activity, large excesses of ligand must be used to ensure saturation. This results in trying to measure the difference between two large numbers to assess the amount of bound ligand—obviously not the ideal situation, but one about which very little can be done except exercising experimental dexterity and using large numbers of replicates, the latter of which is usually inconvenient in the course of protein preparation.

Indirect Methods. As is discussed in considerable detail in Chap. 17, there are a wide variety of indirect methods for studying ligand binding, ranging from titrations employing spectral parameters of either the protein or ligand, to studies involving the ability of ligand to alter in some way an identifiable protein property. Although these approaches are frequently of use in the context of ligand binding studies, only those involving spectral parameters are of much use in the routine assay of protein binding activity. Briefly, these methods, whether they use absorbance, fluorescence, or magnetic properties, depend on the protein–ligand complex exhibiting some spectral difference from the free protein and ligand solutions that can be used to follow ligand binding. Although clearly such an approach cannot be used to quantitate the amount of protein when it is not pure, the method can be used to follow the protein activity and to calculate a "specific activity." If one makes assumptions about the number of binding sites for ligand, it can also be used to quantitate the protein.

Structural Proteins

The determination of the "activity" of structural proteins represents the most difficult aspect of protein purification: frequently, activity measurements are only possible by immunoassay using antibody to the purified material.

The problems may best be illustrated by reference to the purification of laminin, a glycoprotein obtained from basement membranes. In essence, laminin was discovered as an impurity in collagen preparations from tumor cells: It was purified by standard approaches to give a "pure" protein by protein chemistry criteria—antibodies were made, and, via immunofluorescence studies, an in vivo localization of the laminin was achieved. Subsequently, immunoassay techniques were used to follow the purification.

In other instances, assays of biological function may be employed to follow purification, as has been used for example with interleukin-II, a T-lymphocyte growth factor required for the stimulation and growth of responding T cells, which is produced by a particular subset of T cells.

Finally, it is sometimes possible to "purify" a protein based on some particular chemical property—a unique cofactor or unusual amino acid composition, for example, the hydroxyproline content of collagen. Clearly, there are no firm rules for the assay of structural proteins, and definition of activity has to be rather pragmatic.

SPECIFIC ACTIVITY

Finally, we come to a definition of specific activity. Depending on the state of knowledge concerning the molecular weight of the protein, the specific activity can be defined by one of the equations

$$\text{specific activity} = \frac{\text{activity}}{\text{mg protein}} \tag{1-1}$$

$$\text{specific activity} = \frac{\text{activity}}{\text{mol protein}} \tag{1-2}$$

In either instance, as the purity of the protein preparation increases, the apparent specific activity increases, reaching a defined value for the pure protein. As indicated earlier and considered in more detail in Chap. 2, it is often considered that a protein is pure when a constant specific activity is reached in a purification. If the specific activity of the pure protein is known, the apparent specific activity can be used to calculate the percent purity at a given point in a purification scheme.

Calculation of Specific Activities

From the example of alcohol dehydrogenase discussed earlier, a *specific activity* can be calculated if the concentration of enzyme or the amount of enzyme in the reaction mixture is known. In the example we calculated a rate of 0.00316 mM/min. If this rate was achieved with a protein concentration of, for instance 4 μg/ml, one can calculate the specific activity,

$$0.00316/4 = 0.00079 \text{ m}M/\mu\text{g/ml/min}$$

If the molecular weight of alcohol dehydrogenase is known, the molarity of the enzyme can be calculated and the specific activity expressed in terms of much simpler units. Assume that the molecular weight is 70,000; then 4 μg/ml gives a molecular concentration of

$$70,000 \text{ g/liter} = 1 \ M, \qquad 70,000 \text{ mg/ml} = 1 \ M, \qquad 1 \text{ mg/ml} = \tfrac{1}{7} \times 10^{-4} \ M$$

$$4 \ \mu\text{g/ml} = \tfrac{4}{7} \times 10^{-7} \ M = 5.7 \times 10^{-8} \ M$$

Since the rate is expressed in millimolar units, this becomes 5.7×10^{-5} mM for a 4-μg/ml solution. The *specific activity* is therefore

$$\frac{0.00316}{(5.7 \times 10^{-5}) \text{ min}^{-1}} = 55.4 \text{ min}^{-1}$$

In this type of calculation care must be taken to define the "molecule" of the enzyme. In these calculations a molecular weight of 70,000 was used: This is the *dimer* molecular weight (the dimer contains *two* active sites). The specific activity calculated on the basis of the overall molecular weight can be divided by the number of active sites per molecule to give the active site specific activity: in this example, $55.4/2 = 27.7 \text{ min}^{-1}$.

Finally, even when activities are measured with saturating substrate concentrations as described earlier, temperature and pH must be defined, and if buffer salts affect the enzyme activity, these must also be defined. Thus the active-site specific activity would be given as 27.7 min^{-1} at 25°C, pH 7.0 in phosphate buffer. If maximum-rate substrate concentrations are not used, these should also be given.

SOURCES OF PROTEIN

Before discussing fractionation procedures for proteins (Chap. 2), there are several generalities that can be made concerning sources of proteins for purification. Often the goal is to obtain, in pure form, a particular enzyme (or other activity) for further study, with little regard to the starting material. In such a case it is extremely helpful initially to screen a variety of species and tissues within a species to obtain a starting material with the highest initial specific activity. Once a starting tissue has been selected it is necessary to establish whether the protein is soluble (i.e., in the cytosol), membrane bound, or entrapped in a subcellular organelle. When the protein is cytostolic the cells must be ruptured to release the cytosol; this can be achieved by homogenization, sonic shock, or freeze-thawing.

Many proteins can be described as membrane bound, and it is often convenient as well as beneficial to isolate the appropriate membrane fraction prior to isolating the protein, as this can give a much higher apparent specific activity of the starting material relative to a whole-cell extract.

Initial stages of membrane preparation involve membrane disruption, usually by mechanical means such as homogenization. The homogenate is then centrifuged at low speed to remove nuclei and whole cells. Different subcellular organelles such as mitochondria and microsomes can then be obtained using differential centrifugation. Finally, discontinuous density gradient centrifugation (frequently using sucrose gradients) can be employed to yield various membrane fractions. A typical scheme is given in Table 1-1. Different membrane fractions are identified by marker enzymes, which are summarized in Table 1-2.

Membrane-bound enzymes usually need to be solubilized prior to further fractionation. Two common approaches are used for so-called peripheral and integral membrane proteins. Peripheral proteins can usually be solubilized by high ionic

TABLE 1-1   Preparation of membrane markers

Step	Procedure
1	Homogenize cells (Dounce homogenizer)
2	Centrifuge at 1000g to remove nuclei and whole cells
3	Differential centrifugation
	(a) 10,000g, 15 min → Mitochondrial preparation
	(b) Supernate from (a): 113,000g for 30 min → microsomal preparation
	Discontinuous sucrose gradient centrifugation, frequently with sucrose gradients ranging from 20 to 50%

TABLE 1-2   Membrane markers

Membrane	Marker enzyme
Mitochondrial	
Inner	Succinate–cytochrome c reductase
	Rotenone sensitive; NADH–cytochrome c reductase
Outer	Monoamine oxidase
	Rotenone-insensitive NADH–cytochrome c reductase
Endoplasmic reticulum	RNA and protein synthesis enzymes
	NADPH–cytochrome c reductase
Plasma	5-Nucleotidase
	Lectin binding
	Oxytocin (or hormone) binding
Golgi	Glycosyl transferases

strength, while integral proteins require detergent solubilization. In many instances differential solubilization, either of peripheral versus integral proteins or of different integral proteins, by different detergents, can be of considerable help in enhancing the specific activity of the starting extract.

Secreted Proteins

A number of proteins of considerable biological interest are secreted from cultured cells and can be harvested from culture filtrate. With bacterial cells this usually represents an excellent source of such proteins. With mammalian cells, however, a major problem is that the cultures frequently require quite high percentages of serum to grow, and thus secreted proteins may represent a small proportion of the extracellular proteins.

A similar problem is encountered with the secreted material from solid tissue cells such as hepatocytes, which often require a collagen-based matrix to grow. A particular problem with extracellular material from solid tissues such as the liver is encountered in the attempted isolation of glycocalyx material, a loose aggregate of glycoprotein and proteoglycan material that surrounds the plasma membrane and may be regarded as an extracellular organelle. Since solid tissue cells are often harvested by techniques such as collagenase perfusion to degrade the extracellular support of the cells, the existence and isolation of glycocalyx material were not recognized or achieved until gentler means of tissue disruption were introduced.

Protease Problem

In a number of instances the isolation of a particular protein is hindered by its susceptibility to various nonspecific (and in some cases specific) proteases which abound in the cell. Such protease activities may reside in lysosomes or be associated with membrane fractions and are often released or activated (e.g., by detergents) during the course of the initial stages of a protein purification. Their effects are frequently combated by the use of various protease inhibitors that can be added during the isolation procedures. Some of the more troublesome proteases and ways to inhibit them are outlined in Table 1-3.

TABLE 1-3   Proteases and their inhibitors

Protease	Specificity	Inhibitor
Chymotrypsin	Phe, Tyr, Trp	TPCK, DFP, PMSF
Trypsin	Lys, Arg	TLCK, DFP, SBTI
Pepsin	Phe, Trp, Tyr	p-Bromophenacylbromide, Diazo-p-bromoacetophenone
Papain	None	Iodoacetic acid, heavy metals
Thermolysin	Hydrophobics	EDTA, 1, 10-phenanthroline
Carboxypeptidase B		1, 10-phenanthroline, 2,2-dipyridyl
Cathepsin A		Antipain
Urokinase		ε-Aminocaproic acid
Kallikreins		α-2-macroglobulin, DFP

TABLE 1-4 Small-molecule substrates for proteases

Protease	Substrate	Monitored by:
Papain	L-Pyroglutamyl-L-Phe-L-Leu-*p*-nitroanilide	Liberation of *p*-nitroaniline monitored at 410 nm
Elastase	CBZ-Ala-*p*NP	Liberation of *p*-nitrophenol monitored at 400 nm
Chymotrypsin	Ac-Trp-*p*NP, *p*-nitrophenyl-ethyl carbonate	Liberation of *p*-nitrophenol monitored at 400 nm
Trypsin	CBZ-Lys-*p*NP	Liberation of *p*-nitrophenol monitored at 400 nm

It is important to consider the possibility of proteases in any purification scheme and it is useful to be able to assay for their activity in the initial extract as well as during subsequent stages in a purification. Protease assays are also helpful in determining the efficacy of added protease inhibitors. Two general types of protease assays are employed: one involving chromagen or radioactively labeled protein and the other small molecule protease substrates that release a chromophore after hydrolysis by a protease. In the former the labeled protein substrate is added and "soluble" label detected after precipitation of protein by trichloroacetic acid. In the latter type of assay, the release of chromophore as a result of the hydrolytic activity of the protease is monitored. A number of small-molecule protease substrates have been developed, which are summarized in Table 1-4.

Once a suitable starting material is obtained, an initial specific activity must be determined together with the total amount of the desired protein present in the starting material. When the specific activity of the pure protein is known, this amount can be calculated in terms of milligrams of protein. However, this is often not the case, and the total amount of active protein can be expressed as *units* of protein, where the definition of the *unit* must be made in terms of the activity being measured. During the subsequent purification stages (Chaps. 2 and 3) both the specific activity and overall yield (the percentage of the original amount of active protein) should be monitored. Such comparisons from step to step in the purification require that assays be performed under standard conditions of substrate concentration, pH, temperature, and buffer components, since each of these factors can affect the measured activity.

2

Protein Purification:
Classical Approaches

INTRODUCTION

In Chap. 1 we discussed various approaches to obtaining starting material for the purification of proteins. Now we consider some of the standard approaches in purification. While many proteins are now purified using various types of affinity chromatography (considered in Chap. 3), the classical approaches described in this chapter are still in everyday use, either alone, in systems where the appropriate affinity chromatographic approaches have not been worked out, or usually in conjunction with affinity chromatography.

Before examining these "classical" approaches, however, two important general considerations must be discussed. The aim in protein purification is self-evident—preparation of a "pure" protein. The achievement depends, however, on the definition of purity. With the increasingly sensitive methods of detecting proteins that have been developed in recent years (discussed in detail in Chap. 4 in the section on electrophoresis) it has become considerably more difficult to prepare a "pure" protein. The main question is, pure enough for what purpose? The purity required for the accurate determination of a molecular weight may be quite different from that required for structural studies of the sequence of the polypeptide chain or for enzyme kinetic or ligand binding studies. Thus a pragmatic approach to the question of protein purity must be used, and this is discussed in these and other contexts in subsequent chapters. Since the determination of purity is usually based on one or more of the various approaches used to establish the molecular weight of a protein, further discussion is contained in Chap. 4.

During the course of protein purification, the specific activity is followed from step to step and a frequently used criterion of purity is the achievement of a constant specific activity for several steps. This approach is particularly useful when constant specific activities are obtained for steps involving quite different physical bases for separation (such as molecular size and ionic properties). It is, however, advisable to determine the purity of the sample independently.

The second important consideration involves the yields from the purification scheme used and the amounts of "pure" protein that are required. For some purposes small amounts of highly purified material are desirable; on other occasions larger amounts may be required and judgments then have to be made as to whether particular steps in a purification scheme which may have low yields but good increases in specific activity are justified. As discussed previously, an informed choice based on a thorough understanding of the pitfalls of a particular experimental approach can allow the researcher to make such decisions.

PRELIMINARY FRACTIONATION PROCEDURES

The approaches that we consider in this section may yield only a few-fold purification; however, their use is not restricted to purification purposes alone. Early stages in most purification schemes have three other motives in addition to increased specific activity:

1. The rapid removal of proteolytic enzymes that might otherwise degrade the desired protein. Protease inhibitors may not always be sufficient to block the action of either specific or nonspecific proteases that may be present at the early stages of a purification or may be activated during a purification.

2. The concentration of starting material to more managable volumes. In many of the procedures used, large volumes of material are not desirable: some of the precipitation methods described here are useful for effective and rapid concentration of the starting material—with the added advantage that they yield a purification as well.

3. The removal of material that may interfere with subsequent stages of the purification. In various procedures the desired protein is adhered to an immobile phase to allow contaminating proteins to be washed away. Whether this immobilization is by specific affinity, as in affinity chromatography, or by the general characteristics of the protein, as in ion-exchange or hydrophobic chromatography, it is often necessary to remove as much nonspecific protein as possible first so as to prevent interference with the immobilization.

Ammonium Sulfate Precipitation

Differential precipitation of proteins by ammonium sulfate is one of the most widely used preliminary purification procedures. It is based on the differing solubility proteins have in ammonium sulfate solutions and can result in a two- to fivefold increase in specific activity (in the case of glyceraldehyde-3-phosphate dehydrogenase from rabbit muscle, essentially homogeneous protein can be prepared simply by using a three-step ammonium sulfate precipitation procedure). Provided that appropri-

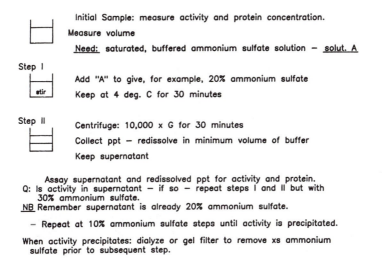

Initial Sample: measure activity and protein concentration.

Measure volume

<u>Need:</u> saturated, buffered ammonium sulfate solution — <u>solut. A</u>

Step I

stir

Add "A" to give, for example, 20% ammonium sulfate

Keep at 4 deg. C for 30 minutes

Step II

Centrifuge: 10,000 x G for 30 minutes

Collect ppt — redissolve in minimum volume of buffer

Keep supernatant

Assay supernatant and redissolved ppt for activity and protein.
Q: Is activity in supernatant — if so — repeat steps I and II but with
30% ammonium sulfate.
<u>NB</u> Remember supernatant is already 20% ammonium sulfate.

— Repeat at 10% ammonium sulfate steps until activity is precipitated.

When activity precipitates: dialyze or gel filter to remove xs ammonium
sulfate prior to subsequent step.

Figure 2-1 Outline of protocol for ammonium sulfate precipitation.

ately buffered ammonium sulfate solutions are used to protect the desired activity, recoveries approaching 100% can be expected. A typical protocol (as outlined in Fig. 2-1) consists of adding ammonium sulfate to give a certain percentage saturation, followed by a period of time for proteins to precipitate and a centrifugation step to collect the precipitate.

Once it is known in what range of ammonium sulfate concentrations the desired protein precipitates, the initial solution can be adjusted to a concentration sufficiently below this so that none (or very little—there are always judgments to be made) of the desired protein precipitates and the undesired protein can be removed by centrifugation. The ammonium sulfate concentration is then raised to a level sufficient to precipitate all (or most) of the desired protein while leaving in solution other undesirable proteins, and the precipitate retained for further purification. The appropriate concentration ranges are conveniently ascertained by screening a range of concentrations for small samples and determining the activity of the desired protein in the supernatant after centrifugation. Once this has been done, the appropriate concentration ranges can easily be chosen. It is important that when scaling up the total protein concentration in the sample is similar to that in the trial since the solubility of most proteins in ammonium sulfate is quite dependent on the total protein concentration.

Isoelectric Precipitation

Essentially similar in practice to ammonium sulfate precipitation, this approach is based on the fact that most proteins precipitate when there is no overall charge on the molecule—that is, at the isoelectric pH—since charge–charge repulsions tend to keep proteins in solution. Because proteins in general have fairly unique isoelectric points this procedure can give good, quick separation of unwanted proteins. In practice, the pH dependence of the stability of the desired protein can be a determining factor in the method's usefulness. Some limitations exist for the effective concentration of proteins depending on how readily the desired protein, once precipitated, can be redissolved. A variation of this procedure involves the pH denaturation of unwanted proteins and their removal by centrifugation, an approach that can be assisted by factors that affect (increase) the pH stability of the desired protein. Substrates or other ligands may increase stability to, for example, low pH, thereby allowing a lower pH to be used than would otherwise be the case. As with ammonium sulfate precipitation procedures, the appropriate conditions are established on a small scale.

Solvent Precipitation

As with isoelectric precipitation, solvent precipitation can be used in two basic ways. Ethanol or other organic reagents, by changing the dielectric constant of the solvent, frequently induce precipitation of proteins that can then be collected and treated in ways similar to those described for ammonium sulfate precipitation.

Polyethylene glycol, of a variety of polymer sizes, is commonly used in fractional precipitation procedures. Any of the readily soluble polyethylene glycols can be employed: Those of higher molecular weight are frequently useful in concentration schemes. The dilute protein solution is placed in dialysis tubing and surrounded with dry polyethylene glycol, which absorbs water through the semipermeable membrane and concentrates the dialysis tube contents.

In the second approach, unwanted proteins in a mixture might be specifically inactivated and denatured by an organic solvent, thus allowing the contaminating protein to be removed. During the purification of Jack Bean α-mannosidase, contaminating β-N-acetylhexosaminidase is removed by specific inactivation with pyridine followed by centrifugation of the precipitated contaminant.

Heat Precipitation

Finally, we consider precipitation of contaminating proteins by heat denaturation. Different proteins have different stabilities at elevated temperatures, and if the desired protein has a greater heat stability than contaminating proteins, incubation at elevated temperatures for periods of time varying from a few minutes to a few hours often precipitates unwanted proteins, which can then be removed by centrifugation. As with pH-induced denaturation, the stability of the desired protein at elevated temperatures may in some cases be enhanced by the presence of substrates or other specific ligands.

CHROMATOGRAPHY METHODS

Gel Filtration

A large variety of gel filtration media are available and all work primarily on the basis of an *exclusion limit*, which is generally defined as the protein size that cannot penetrate the bead space of the material and thus is excluded from the column matrix. Proteins larger than this size co-chromatograph through the column and elute in the *void volume* (V_0) of the column. Other material falls into two classes:

   1. Material smaller than the exclusion limit which does not physically interact with the matrix material. Such material can be considered as having "normal" gel filtration behavior, and its *elution volume* (V_e) depends on the size of the material relative to the pore size of the matrix.
   2. Material that interacts with the matrix material. Any physical interaction (causes of which are considered in the context of the nature of the matrix material) causes a retardation of the chromatographed material greater than what would be expected to occur simply by its ability to penetrate matrix space, and thus such material elutes at an anomalous elution volume. If the interaction with the matrix is of sufficient magnitude, the interacting material may elute at a volume larger than the normal *total elution volume* (V_t) of the column, which is the volume taken to elute molecules from the column having sizes similar to the bulk solvent volume.

Experimental Determination of Chromatography Parameters. The three parameters V_0, V_e, and V_t are used to describe the behavior of a particular molecule on a gel filtration column and must be determined experimentally. Three types of chromatography experiments can be envisaged:

   1. In the ideal case, the sample size loaded onto the column is very small compared to the volume of the packed matrix material. In this instance (Fig. 2-2A), the elution volume, V_e, is simply the volume of eluent collected from the *start* of loading the sample to the midpoint of the sample elution.

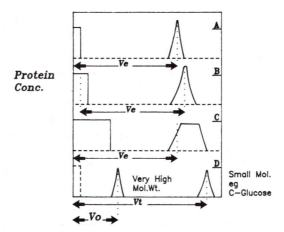

Figure 2-2   Elution profiles for gel filtration with sample sizes of 2% V_t (A), 10% V_t (B), and 25% V_t (C). (D) Summary of various gel filtration parameters.

2. When the sample size is not negligible compared to the bed volume of the column, the elution volume is usually calculated from the *midpoint* of the sample loading to the midpoint of the elution profile (Fig. 2-2B).

3. If the sample size is so large that a simple elution peak is not obtained (Fig. 2-2C), the elution volume is calculated from the *start* of the sample loading to the midpoint of the ascending side of the elution profile.

We should consider practical limits to the sample size that can be used in gel filtration chromatography. The *separation volume* (V_{sep}) between two peaks A and B can be defined as

$$V_{sep} = V_{e_B} - V_{e_A} \tag{2-1}$$

If a sample eluted from a column behaved ideally, the maximum sample size could be as great as V_{sep}. However, as the sample size is increased the size of the eluted peaks increases, and for resolution of the peaks the sample size should always be smaller than V_{sep}.

One problem is that under many experimental situations V_{sep} is not known, so as a general rule the sample size should be kept as small as is practical—in the range 2 to 5% of the column bed volume. For desalting applications, however, where the matrix has usually been chosen such that the desired protein elutes in the void volume while the elution volume of the "salt" approaches the total column elution volume, it is possible to use much larger sample sizes—in the region of 20% of V_t—and achieve effective desalting with minimal sample dilution.

The total volume, V_t, is usually obtained by loading a sample containing a small molecule (which does not physically interact with the matrix material) that can be conveniently monitored by absorbance or radioactivity and directly determining its elution position as described previously for V_e estimations.

The void volume of the column is determined similarly but by using a sample containing a macromolecule of sufficient size such that it is totally excluded from the matrix. In many instances blue dextran is used, although for columns employed to separate smaller molecules, a protein such as BSA is often convenient. As mentioned, the elution position of the void volume material is obtained as described for V_e determinations. Several other parameters can be defined and estimated once V_e, V_t, and V_0 are known.

The elution of a solute molecule in gel filtration chromatography can be characterized by a *distribution coefficient*, k_D:

$$k_D = \frac{V_e - V_0}{V_s} \tag{2-2}$$

where V_s is the volume of the stationary phase, which is the volume of solvent that can permeate the matrix and is accessible to small molecules (those which elute at V_t).

$$V_s = V_t - V_0 - V_{\text{gel matrix}} \tag{2-3}$$

In practice V_s is difficult to determine and is usually approximated by $V_t - V_0$, and k_D is replaced by K_{av},

$$K_{av} = \frac{V_e - V_0}{V_t - V_0} \tag{2-4}$$

and is *not* a true partition coefficient. These parameters are summarized in Fig. 2-2D.

Prior to considering the chemical nature of the various matrix materials available, we should discuss several other choices that have to be made concerning the practical setup of a gel filtration experiment. Many of the available matrices come in different particle sizes, from superfine to coarse. The smaller particles of the superfine grades give better physical packing of the matrix than does larger material, resulting in less zone broadening of peaks and consequentially, better resolution. The larger particle grades have considerably faster flow rates, however, which may be advantageous in working with unstable material or with rapid procedures such as desalting. The physical size of the column must also be chosen: Since the resolution of separated peaks increases as the square root of the column length, long columns in general give better separation than short columns but elute more slowly. The diameter of the column is important since narrow columns can hinder ideal passage of solvent through the column and wide columns give increased sample dilution. By far the most important choice regards the *sample viscosity*: High sample vicosity leads to distortion of elution peaks, which vary with the molecule size. The sample and buffer viscosity should not differ by more than a factor of 2, which, for most proteins, puts an upper limit for concentration of 50 to 70 mg/ml. It must be emphasized that many proteins undergo concentration-dependent aggregation, which can lead to anomalous gel filtration behavior not just due to viscosity problems, but also because of the molecular weight and size polydispersity that such a phenomenon can create.

Choice of Gel Filtration Matrix Material. Three basic types of matrix material have been used which differ somewhat in their physical and chemical properties. The most common are the cross-linked dextrans (e.g., Sephadex). This bead-formed gel is prepared by cross-linking dextran with epichlorohydrin. The resultant gel contains a large number of hydroxyl groups, which makes it quite hydrophilic and causes the gel to swell readily in water or electrolyte solutions. The porosity of the gel, and hence the useful fractionation range, is governed by the degree of cross-linking.

As discussed earlier, adsorption of material being chromatographed to the matrix leads to anomalous elution. Two principal types of adsorption must be considered: ionic and aromatic. With the cross-linked dextrans these effects are particularly noticeable on the highly cross-linked gels used to fractionate small molecules. The matrix material contains a low level of carboxyl groups, which at low ionic strength lead to the retardation of positively charged species and increased exclusion of negatively charged species. At ionic strengths above about 0.02, however, these effects become negligible with most proteins or peptides. A variety of aromatic compounds (such as purines, pyrimidines, dyes, and hydrophobic peptides) interact with the matrix material, causing additional retardation. These interactions can be suppressed by using urea or phenol–acetic acid–water buffer systems for elution. However, such interactions are not always undesirable. Frequently, fairly similar aromatic compounds can be separated by making use of their interactions with the matrix, which can be modulated by changing the composition of the elution buffer. The addition of methanol or ethanol tends to increase the strength of these interactions, while altered ionic strength or pH can be used to weaken them. In essence this is hydrophobic chromatography, which is discussed in more detail later in this section.

The second type of matrix material commonly used consists of allyl dextran cross-linked with *N,N'*-methylene bisacrylamide, which gives a quite rigid gel structure having well-defined porosity. Due to the rigidity of the matrix, this type of material (e.g., Sephacryl) can easily be used with organic solvents with a much smaller effect on pore size (and hence distortion of the fractionation range) than with the Sephadex-like matrices. In general, the Sephacryl-like resins give better flow rates for equivalent fractionation ranges, but are only available for the separation of larger molecules [20,000 − 10^6 daltons (Da)].

Because of its high matrix density (and consequent carboxyl group density) these matrices have more pronounced ionic adsorption properties than the simple dextrans. In general, higher-ionic-strength buffers are therefore used with this type of material to help suppress such effects.

Finally, various derivatives of agarose have been used. The gel structure of agarose-based gels is stabilized by hydrogen bonding rather than chemical cross-linking but is quite stable under most conditions. The porosity is governed by the concentration of agarose in the material. The open structure of the agarose-based matrix makes this type of material (e.g., Sepharose) most suitable for the fractionation of very large macromolecules, although matrices with high agarose contents (up to approximately 6%) can be used with proteins in the range 10,000 Da and upward.

Such resins do contain low levels of carboxyl and sulfate groups, which can cause retardation of basic proteins, although as discussed for the other resins, these effects can be minimized by using elution buffers of reasonable ionic strength ($I > 0.02$). The thermal and chemical stability of agarose gels can be increased (with negligible effect on porosity) by chemical cross-linking with 2,3-dibromopropanol. The enhanced stability of the resultant material allows alkaline hydrolysis (under reducing conditions) to be used to remove sulfate groups, giving a gel with a very low content of ionic groups and consequent elimination of most ionic adsorption effects. The basic structures of these various resins is given in Fig. 2-3.

Ion-Exchange Chromatography

Ion-exchange chromatography is based on the simple concept that at a given pH most proteins have a charge (either overall negative or positive, depending on the pI of the protein) and hence are attracted to (i.e., interact with) an opposite charge. Different proteins have differing amounts of charge and hence adhere more or less tightly to the opposite charge compared to other proteins. This interaction causes a retardation in chromatography provided that the matrix material has the

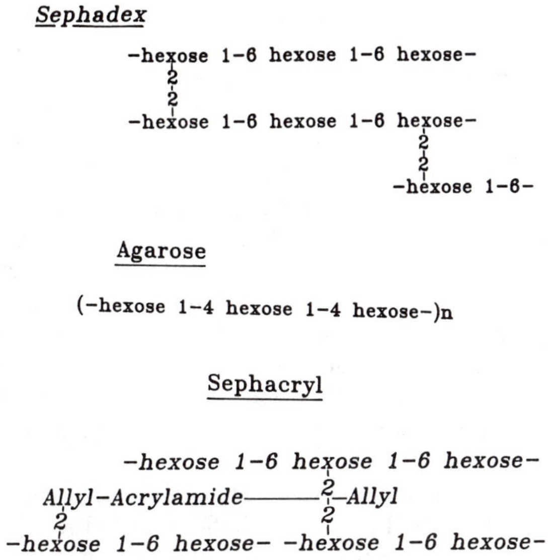

Sephadex

```
-hexose 1-6 hexose 1-6 hexose-
       2
       2
-hexose 1-6 hexose 1-6 hexose-
                        2
                        2
                 -hexose 1-6-
```

Agarose

(-hexose 1-4 hexose 1-4 hexose-)n

Sephacryl

```
        -hexose 1-6 hexose 1-6 hexose-
                           2
Allyl-Acrylamide───────────2-Allyl
     2                     2
-hexose 1-6 hexose- -hexose 1-6 hexose-
```

Figure 2-3 Schematic outline of the matrix structures of Sephadex, Sephacryl, and Agarose resins.

appropriate charge. In essence, the various matrices we have discussed for gel filtration chromatography are the basis of ion-exchange matrices: The matrix is derivatized to give it the desired anion- or cation-exchange properties. The commonly used functional groups are shown in Fig. 2-4. The basic properties of the support matrix are as discussed previously and should be selected based on the size of the proteins

Type	Functional Group	Counter-ion	Comment
DEAE	diethylamino ethyl $-O-CH_2CH_2N^+\!\!<^{C_2H_5}_{C_2H_5}$ with H	Cl^-	Weak anion exchange
QAE	diethyl-(2-hydroxypropyl)-amino ethyl $-O-CH_2CH_2N^+\!\!<^{C_2H_5}_{C_2H_5}$ with $CH_2CH(OH)CH$	Cl^-	Strong anion exchange
CM	Carboxymethyl $-O-CH_2COO^-$	Na^+	Weak cation exchange
SP	Sulfopropyl $-O-CH_2CH_2CH_2SO_3^-$	Na^+	Strong cation exchange

Figure 2-4 Structure and properties of ion-exchange groups.

to be fractionated. If it is necessary to use polar organic solvents, the matrix should be of the chemically cross-linked agarose type.

Once the appropriate resin has been chosen (more about this later) only the ionic strength and pH of the loading buffer need to be considered. Since the interaction of a protein with the matrix is through charge–charge, ionic strength of the loading buffer should be kept low to maximize interaction. The capacity of the column to bind the appropriately charged species is dependent on the number of oppositely charged groups available, which in turn depends on the pK values of the groups and the pH of the medium. Figure 2-5 shows titration curves for some of the commonly used ion exchangers. DEAE-based resins indicate the presence of multiple charged groups but have good capacity below a pH of about 8.5 (the pK of the normal DEAE group is about 9.5). If an anion-exchange resin is needed at higher pH, the QAE-type resins (pK around 12) can be used at significantly higher pH values. Similar considerations apply to the cation exchangers CM- (pK around 3.5) and SP- (pK around 2.0).

Elution of material from an ion-exchange matrix is generally achieved in one of two ways. The ionic strength of the elution buffer is raised to a level that decreases the charge–charge interaction of the chromatographed material with the matrix, or the pH of the eluent is changed so that the charge of the adhered protein is altered such that it no longer interacts with the matrix. The pH must be decreased with anion-exchange material but increased with cation-exchange material. In some cases a combination of these two effects is used. The change is usually produced by running a gradient of increased-ionic-strength buffer (or the appropriate pH gradient) through

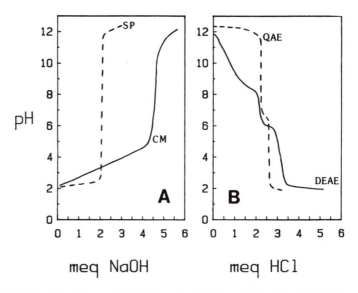

Figure 2-5   Titration curves: (A) CM (——) and SP (–––) Sephadex; (B) DEAE (——) and QAE-Sephadex (–––).

the column and monitoring the eluent for protein, activity, and so on, to locate the desired protein. Separation is achieved at two levels: First, not all proteins adhere to the column during the adsorption phase of the experiment. Second, as the elution gradient proceeds, different proteins elute based on the avidity of their interaction with the matrix; weakly bound proteins (i.e., those with the lowest charge density under the initial adsorption phase) are eluted first, while highly charged proteins require more drastic changes in pH or ionic strength.

Determination of Adsorption and Elution Conditions. During the initial stages of establishing a protein purification it is necessary to establish: (1) what type of ion exchanger should be used, (2) what conditions are necessary for adsorption, and (3) what conditions are necessary for elution. In general, conditions where the wanted protein adheres to the matrix should be established rather than conditions where other proteins adhere but not the wanted protein, since in the former case separation is achieved at both the loading and elution stages. In the absence of prior knowledge about the molecular properties of the protein it is convenient to screen a wide range of pH values rapidly with a particular resin type using the simple mixing and centrifugation procedure outlined in Fig. 2-6. Activity measurements on the supernatant allow one to establish adsorption (and, of course, elution) conditions rapidly.

An alternative approach for establishing optimal separation conditions for closely related molecules such as lactate dehydrogenase isoenzymes involves electrophoretic titration curves. This depends on the normal charge on the protein and its isoelectric

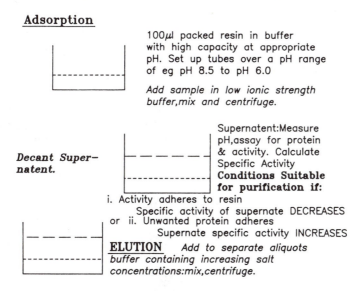

Figure 2-6 Experimental determination of conditions for adsorption and elution of material using ion-exchange resins.

CATHODE

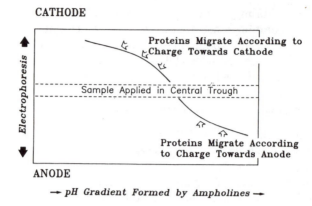

Figure 2-7   Outline of experimental determination of electrophoretic titration curves.

point (p*I*). Electrophoresis is carried out in a vertical plane using a large-pore gel matrix such as agarose or a low-percentage acrylamide which has a preformed *horizontal* pH gradient generated from the appropriate ampholines. As indicated in Fig. 2-7, the sample containing the mixture of proteins is added to a central horizontal well and electrophoresis is begun.

During electrophoresis the proteins move either toward the cathode or the anode or, if the pH is at their isoelectric point, they do not move at all. The rate of movement depends on the pH relative to the p*I* of the proteins. After electrophoresis is terminated the proteins are stained (for activity if appropriate; see later) and the titration curves examined. A typical set of curves for lactate dehydrogenase isoenzymes is shown schematically in Fig. 2-8. From the results the pH that gives the largest separation on the basis of charge can easily be evaluated. This pH gives optimal separation during elution from the appropriate ion-exchange resin.

In addition to being a suitable purification procedure for many proteins, ion-exchange chromatography has a number of other attributes that are outlined in Table 2-1. Particularly useful are the potential concentration of a wanted protein during a

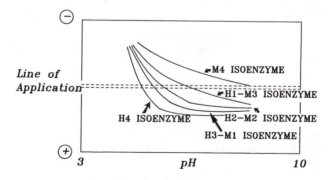

Figure 2-8   Electrophoretic titration curves for lactate dehydrogenase isoenzymes.

TABLE 2-1   Other uses of ion-exchange resins

Use	Resin of choice	Comments
Protein concentration	Any	Use steep salt gradient or stepwise elution
Demetallizing	Chelex	Dialyze protein vs. Chelex in buffer
Electrolyte removal	Mixed bed	
Peptide separation	Cation exchange	Gradient elution
Nucleotide separation	DEAE, QAE	Use gradient elution with QAE

purification procedure and the removal of metals from metalloproteins during the preparation of apoenzymes. In protein concentration it is often convenient to use a stepwise elution procedure rather than gradient elution.

Although we have discussed gel filtration and ion exchange in terms of column chromatography, both approaches are readily adaptable to thin-layer chromatography, which is particularly useful when a two-dimensional separation involving electrophoresis in addition to gel filtration (for example) is used. Ion-exchange methods are also particularly suitable for batchwise procedures since nonadsorbed material can easily be removed by washing and centrifugation prior to elution.

Hydrophobic Chromatography

Although the use of hydrophobic chromatography in protein purification has been popularized only recently, the idea owes its genesis both to gel filtration and affinity chromatography (Chap. 3). The matrix employed is usually based on agarose that has been derivatized in aprotic solvents with epoxides (which have relatively large alkyl chains). A generalized formula for the derivatives is

$$\text{agarose} - \text{O} - \text{CH}_2 - \underset{\underset{\text{OH}}{|}}{\text{CH}} - \text{CH}_2 - \text{O} - \text{R} \tag{2-5}$$

where R represents the alkyl chain and usually contains between 5 and 12 carbons. Any protein with some external hydrophobic characteristics tends to interact with such a matrix and be retarded relative to proteins lacking such characteristics. In general, the capacity of such columns for protein increases with increasing hydrophobicity of the substituent, with increasing degree of substitution, and with increasing ionic strength. The latter characteristic is quite distinct from the charge–charge interactions described earlier for ion-exchange chromatography, and leads to the principal method of elution from such a matrix: The ionic strength of the loading buffer is kept high and elution is achieved using a decreasing-ionic-strength gradient. Because the porosity of the matrix is decreased as the hydrophobicity of the substituent increases, generally a lower degree of substitution is employed, which is compensated for by using a higher initial ionic strength to maximize capacity and adsorption. In circumstances where adsorption is particularly tight (i.e., long alkyl chains, high degree of substitution, high ionic strength), complete desorption of adhered protein is sometimes difficult to achieve by decreased ionic strength alone. In such cases the addition of glycerol or ethylene glycol to the elution buffer tends to enhance desorption. In some instances increased desorption can be achieved by adding to the elution buffer ligands specific for the wanted protein that bind and change the conformation to one with lower external hydrophobicity. The converse of this situation—a ligand that upon binding leads to increased hydrophobicity—can be used to increase adsorption of the wanted protein to the matrix. In such an instance elution would be enhanced by omitting the ligand from the elution buffer.

High-Pressure Liquid Chromatography; High-Performance Liquid Chromatography

Both terms above are represented by "HPLC," and over the last five years these techniques have become increasingly useful in the isolation and characterization of molecules of biological interest whether, in the context of this book, they are proteins, peptides, or amino acids. HPLC is a philosophy rather than a particular technique, and in fact under the term "HPLC" fall each of the chromatographic techniques we have discussed so far, together with affinity chromatography. The fundamental principles remain the same whether used in conventional column chromatography or in HPLC methods and are not reiterated here, although some aspects of reverse-phase HPLC (which is derived from hydrophobic chromatography) are amplified since at present this is the most commonly used of the HPLC techniques.

In general, each of the approaches employs an immobile phase bonded onto a porous silica, which allows high flow rates to be used, and a mobile phase, whose composition is appropriate for the particular technique. We now briefly consider some of the characteristics of each of the HPLC techniques.

1. *Gel Filtration Chromatography.* A variety of bonded phases have been used to cover the cationic surface of silica and prevent nonpermeation effects. These include glycerylpropyl, diol, and *N*-acetylaminopropyl silane. Although a number of non-silica-based support materials have been used, most work has involved the silica-based material.

2. *Ion-Exchange Chromatography.* Again, silica supports with an associated immobile phase of, for example, polyethyleneimine have produced column packing with good stability and high (in the context of HPLC) capacity. A variety of organic polymer supports such as polystyrene have also been used, but primarily for low-molecular-weight molecules.

3. *Reverse-Phase Chromatography.* This technique uses reversed phases such as octadecylsilyl (C_{18}), octylsilyl (C_8), butylsilyl (C_4), and propysilyl (C_3) bonded to silica supports. RP-HPLC is essentially derived from hydrophobic chromatography and is probably the most widely used of the HPLC techniques, having found applications in both purification and characterization of proteins, peptides, and peptide mixtures such as might be obtained by proteolytic digestion of a protein. In general, the retardation of a molecule in RP-HPLC depends on no one parameter such as

size or charge, although there is an approximate correlation between retention time (i.e., elution time) and the percentage of hydrophobic residues in the protein or peptide, although conformational effects often distort this relationship. Two types of elution are frequently used in RP-HPLC.

(a) *Isocratic Elution.* The composition of the mobile phase is kept constant (this phase usually contains an organic solvent such as acetonitrile and an aqueous solvent such as trifluoroacetic acid or phosphoric acid). With isocratic elution the composition of the mobile phase must be predetermined since the retention time of a protein changes with its composition, especially in reference to the context of the organic solvent.

(b) *Gradient Elution.* Because of the sensitivity of retention time to the content of the organic solvent, proteins and peptides are usually eluted with an acidic mobile phase using a gradually increasing organic solvent content.

ELECTROPHORETIC METHODS

Two electrophoretic methods are available that can conveniently be used in general protein purification. These are native polyacrylamide gel electrophoresis (PAGE) and isoelectric focusing. Both depend on the movement of proteins through a matrix support on the basis of the charge of the native protein. In native PAGE, the rate of movement is governed by other factors, such as the porosity of the gel and the molecular weight and the shape of the protein, all of which are discussed in detail in Chap. 4 in the context of molecular weight determination. In isoelectric focusing the distance of movement is governed by the isoelectric point of the protein, and the rate of movement is less important since the experiments are continued until equilibrium is reached. In principle, both techniques are somewhat similar: a mixture of proteins is separated electrophoretically and the desired protein is identified on the electrophoretogram and eluted from the support material.

Detection of Active Material after Electrophoresis

There are three basic procedures that can be used for the detection of native material after electrophoresis: (1) measurement of enzymatic activity; (2) detection by specific ligand binding, using, for example, a fluorescent or radioactive ligand; and (3) detection by antibody binding if a suitable antibody to the native protein is available. These procedures can be performed directly in the separating matrix, or after the separated material has been transferred to a more suitable matrix for such detection, via a process known as "blotting." Although detection of active proteins directly in the separating matrix can in theory be achieved, a number of factors may result in such a procedure being ineffective or undesirable. Substrate or ligand diffusion through the matrix may limit the sensitivity or success of direct staining. The desired protein may have little or no direct activity while constrained within the rigid separating matrix. The problem of substrate or ligand diffusion can be overcome by using ultrathin support material; however, the latter problem remains a potential pitfall. These difficulties are largely overcome by the process of blotting. Blotting is gradually replacing the more laborious but still quite effective process of slicing the matrix material into small pieces, eluting protein with a suitable buffer for subsequent enzymic analysis, and then assaying the eluted material for its specific activity.

Nitrocellulose paper is the most widely used material for blotting since most proteins adhere to nitrocellulose, and such papers have reasonable capacity, making subsequent detection more facile. Under conditions for transfer both the nitrocellulose paper and the protein are probably negatively charged, and hydrophobic rather than ionic effects are probably involved in protein–paper binding. The major problem with nitrocellulose is that low-molecular-weight proteins may bind with low affinity and as a result be washed from the paper during subsequent handling. Alternative types of matrix involving covalent immobilization of blotted proteins can be useful; however, such a process may inactivate the protein.

Three principal ways exist for transferring proteins from the separation matrix to the detection matrix in blotting. In the simplest, the separation matrix is sandwiched between two sheets of nitrocellulose filter paper and the sandwich completed with appropriate support material, then placed into a chamber containing buffer and transfer by simple diffusion takes place. Although slow, such a procedure can be quite effective and provided that denaturation of the desired protein does not occur during the transfer, is simple, cheap, and effective. The second procedure is a variant of the first, where mass flow of solvent is induced through the gel and the nitrocellulose paper. The gel is placed in a buffer reservoir, the nitrocellulose paper placed on top, and a stack of absorbing material (such as paper towels) placed on top of the filter paper. This leads to buffer being drawn through the gel and filter paper, resulting in elution of the proteins from the gel and their immobilization on the nitrocellulose paper. The third method involves additional apparatus but can be quite fast and effective. It is based on the electroelution of the sample from the separating matrix onto the nitrocellulose paper. This is possible because proteins adhere to nitrocellulose even in low-ionic-strength buffers. The original gel and nitrocellulose paper are sandwiched together with porous support material and placed into a tank containing a transfer buffer and electrodes. In electroelution proteins of different charge and molecular weight "elute" at different rates, which can present a problem. As an attempt to counter this, PAGE is done with a reversible gel cross-linker such as N,N'-diallyltartardiamide in place of bisacrylamide. The gel is depolymerized, in this instance by incubation with 10 mM periodate for 30 minutes at 22°C, prior to being placed in the sandwich used in electroelution. The inclusion of low concentrations of detergent such as 0.1% SDS in the transfer buffer also facilitates electroelution and does not appear to affect the adherence of protein to nitrocellulose (it may, however, in some cases have adverse effects on protein activity or stability).

Detection of Specific Proteins after Electrophoresis

As indicated earlier, it is possible, with gel electrophoresis systems where the native structure of the protein is retained, to stain for specific proteins in situ in the electrophoresis matrix. This can be accomplished by one of three approaches.

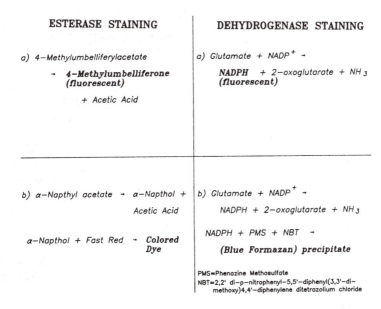

ESTERASE STAINING | DEHYDROGENASE STAINING

a) 4—Methylumbelliferylacetate

→ **4—Methylumbelliferone**
   (fluorescent)

   + Acetic Acid

a) Glutamate + NADP$^+$ →

   NADPH *+ 2—oxoglutarate + NH$_3$*
   (fluorescent)

b) α—Napthyl acetate → α—Napthol +

                              Acetic Acid

α—Napthol + Fast Red → **Colored**
                            Dye

b) Glutamate + NADP$^+$ →

   NADPH + 2—oxoglutarate + NH$_3$

   NADPH + PMS + NBT →

   (Blue Formazan) precipitate

PMS=Phenazine Methosulfate
NBT=2,2' di—p—nitrophenyl—5,5'—diphenyl(3,3'—di—
    methoxy)4,4'—diphenylene ditetrazolium chloride

Figure 2-9   Outline of procedures for activity staining of native gels.

Enzymatic Activity Stains. A variety of staining procedures making use of the catalytic reaction of the specific protein have been developed. There are two approaches: (1) either a fluorescent substrate or product is followed, and after reaction the loss or appearance of fluorescence is monitored, or (2) the enzymatic reaction product is coupled to a chemical reaction that produces a colored dye located in the gel at the site of the specific protein. Examples of these types of staining procedures are the detection of esterases or dehydrogenases, both shown schematically in Fig. 2-9.

Detection by Specific Ligand Binding. In a number of cases a specifically bound ligand, either fluorescent or radioactive, can be diffused into the gel and after washing to remove background labeling is located by fluorescence or autoradiography. An extension of this approach is the use of the fluorescent dye ANS to bind proteins in gels run in the presence of denaturants. ANS binds to most proteins with an enhancement of fluorescence, and protein bands can be detected by this. Background effects are minimized in the general gel matrix by the fact that the denaturants tend to quench the fluorescence of free ANS.

Detection by Binding Fluorescently Labeled Proteins to Specific Target. Any protein that shows a specific interaction with a desired target protein and can be fluorescently labeled can also be used to detect the specific protein in a gel matrix. Usually, the protein is labeled with fluorescein isothiocyanate (the method could also be used with radioactively labeled protein and autoradiography), which is then diffused into the gel matrix, and following destaining the specific interaction is detected by location of fluorescence after exposure to long-range ultraviolet (UV) light. A variety of such stains have been developed using fluorescently labeled antibodies of the appropriate specificity, as well as using fluorescently labeled lectins for the detection of specific glycoproteins. Lectins in particular can be quite useful in this type of approach, as general glycoproteins can be labeled using lectins with little sugar specificity such as concanavalin A, while glycoproteins with specific terminal sugar residues can be labeled using lectins having defined precise specificities (many of which are given in Chap. 3).

Finally, we must consider briefly pertinent aspects of native PAGE and preparative isoelectric focusing. In native PAGE the separating gel is usually topped with a spacer gel of increased porosity that is used to concentrate the sample at the top of the separating gel. They are usually run at fairly high pH (around 8.3), where most proteins carry a net negative charge; however, occasionally a protein might be encountered that electrophoreses away from the gel. Electrophoresis can still be achieved, either by changing the pH or reversing the polarity of the electrodes.

In preparative electrophoretic techniques the major difficulty after separation and detection of the separated material is the quantitative recovery of the desired protein from the matrix. In most cases this can be achieved rapidly and quantitatively by preparative-scale electroelution using an apparatus of the design shown in Fig. 2-10. Because of the multiple membrane construction of such a device it can be used to separate proteins from SDS and from native or isoelectric focusing gels, and quite effectively concentrates the eluted protein.

Although analytic isoelectrofocusing is usually run in either tube gels (infrequently) or in slab gels (more usually), preparative isoelectric focusing is conveniently

Electro–elution of Proteins from Polyacrylamide Gels

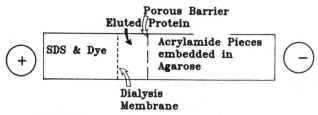

PROTOCOL:

1. Excise PROTEIN band from preparative gel
2. Mince & fuse into chamber with Agarose
   Use high pH buffer if native gel.
3. Electro–elute: Protein will remain trapped
   by dialysis membrane while SDS & Dye will be
   electrophoresed into final chamber.

Figure 2-10   Scheme of a preparative-scale electroelution.

performed using a flat bed of Sephadex as the ampholine carrier. Large quantities of protein can be handled, and after detection of activity, the desired protein can readily be eluted by scraping the appropriate region of the matrix from the bed, packing it into a column, and eluting as in any gel filtration experiment.

CHROMATOFOCUSING

Chromatofocusing combines the high resolution of isoelectric focusing separations with the high capacity of ion-exchange column chromatography. As in isoelectric focusing, the approach depends on the generation of a pH gradient. Since the charged group on an ion-exchange resin has a buffering action at a particular pH, it will, if eluted with a second buffer at a different pH, form a pH gradient. If proteins are bound to the ion-exchange resin, they elute as the generated gradient reaches the isoelectric point. Optimal resolution by the pH gradient is generally through linear gradients, which are achieved by ensuring that the eluting buffer and the ion-exchange resin have constant buffering capacity over the necessary pH range.

When a protein and eluting buffer first enter the column, the protein either adheres to the matrix of an anion exchange resin if the pH of the eluting buffer is initially higher than the pI of the protein, or it travels down the column if the pI is greater than the initial pH. As the eluting buffer travels through the column, its pH increases until it is greater than the pI of migrating proteins, at which point the formal charge on the protein reverses and it adheres to the matrix. As the pH gradient develops (with an anion-exchange resin an increasing hydrogen-ion gradient is used—that is, the pH decreases as the elution proceeds), the pH drops below the pI of the protein, which is therefore released from the resin and eventually elutes from the column at its isoelectric point. The initial migration of a protein through the matrix, followed by adsorption and release, results in a focusing effect for particular proteins.

SOME EXAMPLES OF PURIFICATION PROCEDURES

There is probably no such thing as a typical protein purification; proteins behave differently in each of the approaches we have discussed, and no individual protein purification of necessity uses all of them. As emphasized earlier, a variety of judgments must be made; sometimes yield will be sacrificed for purity, sometimes a step with a good yield or purification will be omitted for reason of speed required with an unstable protein. The figures and tables on the next few pages show several "typical" purifications, with some comments on the choices for the various steps used.

There are, however, some overall principles that can be followed as a guide to setting up a purification scheme. After considerations such as the rapid removal of proteases and the concentration of the sample, both of which are often achieved via a precipitation approach, it is usually advisable to employ a technique that is as selective as possible as early as possible. Since such a technique is early in the purification, it should have a high capacity. In general, succeeding stages should use different separating techniques and chromatographic steps should be linked to minimize handling. Ion exchange can precede hydrophobic chromatography since the high ionic strength used to elute proteins from ion-exchange resins gives optimal conditions for adsorption onto hydrophobic matrices. Steps involving dilution (such as gel filtration) should precede steps that increase concentration (ion-exchange chromatography), so that time and effect are not lost in concentrating the sample without benefit of purification.

High-resolution techniques should be used toward the end of a procedure since these tend to use small sample amounts and may be interfered with by contaminating proteins that can readily be removed during earlier stages.

Purification of a Metalloendoproteinase from Mouse Kidney

The endoproteinase activities in homogenates and at various stages of the purification were estimated using azocasein as substrate. Azocasein contains dye molecules covalently attached to amino acid side chains in the protein. When the protein is proteolytically degraded, dye-containing peptides are released. These are soluble in 4% trichloroacetic acid (TCA) while the parent molecule precipitates. The dye is quantitated by absorbance measurements on TCA-soluble material at various times of incubation.

1. Initial homogenates obtained using a Dounce homogenizer were centrifuged at $600g$ for 10 min at 4°C to give a supernatant (kept) and a residue that was rehomogenized and the second supernatant combined with the first prior to centrifugation at 100,000g. The sediment was resuspended.

2. Attempts to solubilize activity from 100,000g sediment with various salt concentrations or with mild detergents (e.g., 0.1% Triton X-100 or Brig-35) were unsuccessful and treatment with toluene–trypsin was necessary. This disrupts the membrane and proteolytically removes intrinsic membrane proteins, releasing membrane-bound proteases.

3. Stepwise ammonium sulfate fractionation was used. Twenty and forty percent caused protein precipitation, but the endoproteinase activity remained in the supernatant. Activity precipitated when the supernatant from the 40% step was raised to 80% saturation. Further increases in ammonium sulfate concentration produced no further activity precipitation.

4. Precipitated activity was redissolved and dialyzed against 20 mM Tris/HCl pH 7.5, which was the starting buffer for DEAE chromatography. After loading activity onto DEAE, the column was washed with 50 ml of starting buffer and activity eluted with a NaCl gradient as shown in Fig. 2-11.

5. Fractions containing more than 300 units of activity per milliliter were pooled, concentrated, and subjected to gel filtration on Sephadex G-200, which gave two peaks of activity (Fig. 2-12).

6. The first peak appeared pure when it was rechromatographed on Sepharose 6B: the activity eluted as a single symmetrical peak of constant specific activity. The

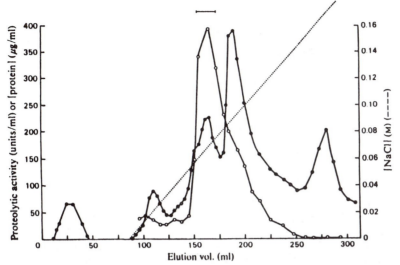

Figure 2-11   DEAE chromatography of toluene–trypsin-solubilized proteins from mouse kidney. After washing, elution was achieved with a linear gradient of 0 to 0.2 *M* NaCl. Protein concentration (●–●), proteolytic activity (○–○), and conductivity (---) were monitored. Fractions indicated as —— were pooled for further purification. (Reprinted with permission from: R. J. Beynon, J. D. Shannon, and J. S. Bond, *Biochem. J.*, *199*, 591–598. Copyright 1981 The Biochemical Society, London.)

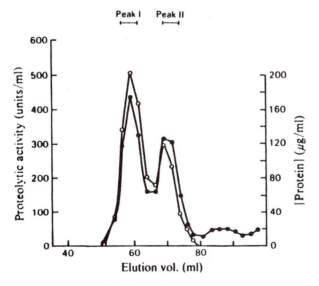

Figure 2-12   Gel filtration of DEAE purified material. Column was monitored for protein (●–●) and activity (○–○). Fractions indicated were pooled as peak I and peak II. (Reprinted with permission from: R. J. Beynon, J. D. Shannon, and J. S. Bond, *Biochem. J.*, *199*, 591–598. Copyright 1981 The Biochemical Society, London.)

TABLE 2-2   Summary of steps involved in the purification of metalloendoproteinase

Step	Protein (mg)	Activity[a] (units)	Specific activity (units/mg)	Purification (fold)	Yield (%)
Homogenate	2,936	27,028	9.2	1	100
100,000g sediment	1,041	22,846	21.9	2.4	85
Toluene–trypsin treatment and ammonium sulfate precipitation (40–80%)	32	18,314	572	62	68
DEAE	4.3	10,118	2,353	256	37
Sephadex G-200					
Peak I	1.3	4,185	3,219	350	15
Peak II	0.96	2,137	2,226	242	8

[a] Proteinase activity measured by azocasein hydrolysis at pH 9.5.

second peak had essentially identical enzymatic properties, but as shown in Table 2-2, has a significantly lower specific activity.

Reference: R. J. Beynon, J. D. Shannon, and J. S. Bond, *Biochem. J.*, *199*, 591–598 (1981).

Purification of an Endogluconase from Clostridium thermocellum

Often when proteins that are secreted by bacteria are being purified there is a very large volume of material to be handled. In this particular purification the cells were removed from the medium in late exponential phase by centrifugation at 10,000g for 30 minutes. An alternative approach employed in some cases is to culture the

TABLE 2-3   Summary of steps in endoglucanase purification

Step	Volume (ml)	Protein (mg)	Activity[a] (units)	Specific activity (units/mg)	Purification (fold)	Yield (%)
Cell-free spnt.	12,000	6,720	20,300	3.0	1	100
Ultrafiltration	1,000	4,750	17,800	3.7	1.2	87.6
Ammonium sulfate ppt.	400	3,370	13,800	4.2	1.4	68.3
DEAE						
Peak I	400	624	6,400	10.3	3.4	31.6
Peak II	60	119	2,750	23	7.6	13.6
SP-Sephadex	25	21	975	45.9	15.2	4.8
Prep. PAGE	10	7.9	514	65.1	21.5	2.5

[a] Units represent micromoles of reducing sugar formed per minute at 60°C.

bacteria *inside* dialysis tubing. A large amount of culture fluid is used *outside* the tubing to ensure a fresh supply of nutrients and the removal of small molecules that might slow growth. The secreted proteins are, however, retained (together with the cells) within the much smaller volume of the dialysis tubing. The following procedure is summarized in Table 2-3.

1. Although the initial cell-free supernatant was first concentrated by ultrafiltration, ammonium sulfate precipitation was also used to concentrate the large volumes, and was achieved by adding 80% ammonium sulfate and collecting the precipitate. After the precipitated material was redissolved the buffer was equilibrated with the start buffer for the DEAE step by continuous ultrafiltration. In this process fresh buffer was repeatedly added to the ultrafiltration cell until the conductivity of the dialysate matched that of the start buffer. The material was then adsorbed onto a DEAE column.

2. The first DEAE column was eluted with a step gradient using the indicated concentrations of ammonium acetate (Fig. 2-13). Two major areas of activity were obtained—indicated as pooled fractions I and III. Rechromatography of fraction I yielded further purification but still two main peaks of activity resulted, as shown in Fig. 2-14.

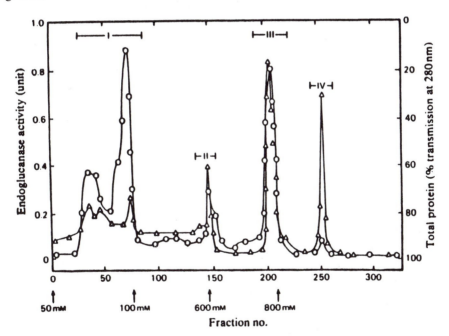

Figure 2-13 Separation of cellulase by DEAE chromatography using a step gradient of ammonium acetate at pH 5.0 with the indicated molarities. Fractions were monitored for protein (O–O) and for endoglucanase activity (△ △). (Reprinted with permission from: T. K. Ng and J. G. Zeikus, *Biochem. J.*, *199*, 341–350. Copyright 1981 The Biochemical Society, London.)

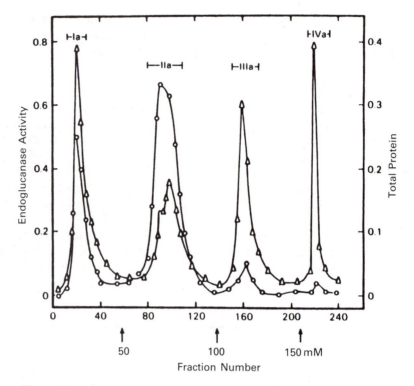

Figure 2-14   Re-chromatography of fraction II on DEAE. As in Fig. 2-13, a step-gradient elution protocol was used. Symbols as Fig. 2-13. (Reprinted with permission from: T. K. Ng and J. G. Zeikus, *Biochem. J.*, *199*, 341–350. Copyright 1981 The Biochemical Society, London.)

3. The major peak, IIa, was selected for further purification on SP-Sephadex, which was eluted with a sodium chloride gradient (Fig. 2-15).

4. The single major peak of activity was pooled and subjected to preparative gel electrophoresis on an 8% separating gel. This gave a major peak containing most of the activity and a minor peak with similar specific activity.

Reference: T. K. Ng and J. G. Zeikus, *Biochem. J.*, *199*, 341–350 (1981).

Purification of Uroporphyrinogen Decarboxylase from Human Erythrocytes

Uroporphyrinogen decarboxylase is a cystolic enzyme that decarboxylates a variety of porphyrinogens. In the purification described here cytosol was obtained from erythrocytes. The enzyme is present in tissues such as liver and in such a case, simple homogenization, in the appropriate buffer to disrupt the cells, releases the cytosol, which is obtained as the supernatant of an initial centrifugation step. The following procedure is summarized in Table 2-4.

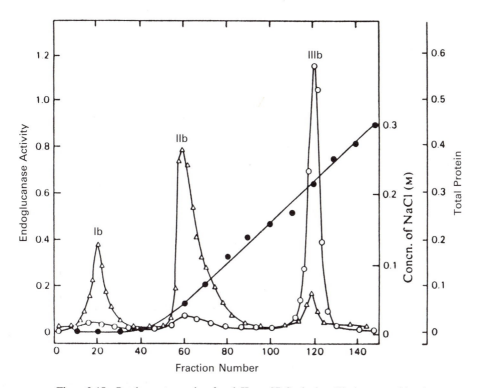

Figure 2-15   Re-chromatography of peak IIa on SP-Sephadex. Elution was achieved with a sodium chloride gradient up to 0.3 M (●–●). Fractions were monitored for protein (○–○) and for endoglucanase (△–△). (Reprinted with permission from: T. K. Ng and J. G. Zeikus, *Biochem. J.*, *199*, 341–350. Copyright 1981 The Biochemical Society, London.)

TABLE 2-4   Summary of the steps in the purification of uroporphyrinogen decarboxylase from human erythrocytes

Step	Protein (mg)	Activity (nmol/min)	Specific activity (nmol/min/mg)	Purification (fold)	Yield (%)
Hemolysate	111,212	1,468	0.0132	1	100
DEAE/ammonium sulfate	308	746	2.42	182	50.7
Gel filtration	49.7	218	4.39	333	24
Phenyl-Sepharose	1.64	30.5	18.6	1,409	4.3
Prep. PAGE	0.096	5.6	58.33	4,419	3.0

1. After washing the erythrocytes in saline, hemolysis was achieved by resuspending in ice-cold water for 2 hours at 4°C.

2. After hemolysis the pH was adjusted to pH 7.0 by addition of 2 volumes of 4 mM phosphate. Hemoglobin was removed by adsorbing activity to DEAE at pH 7.0 in a batchwise procedure. After the resin was obtained by centrifugation it was washed to remove hemoglobin, and activity was eluted by adding 0.5 M KCl.

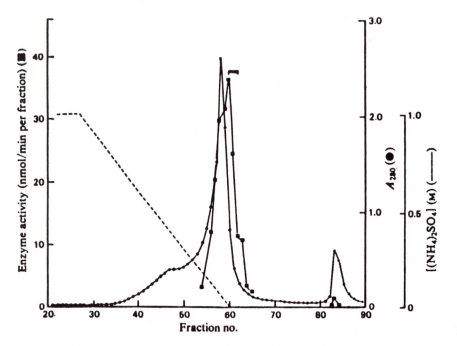

Figure 2-16 Hydrophobic chromatography of partially purified uroporphyrinogen decarboxylase using phenyl-Sepharose. (Reprinted with permission from: G. H. Elder, J. A. Tovey, and D. M. Sheppard, *Biochem. J.*, *199*, 45–55. Copyright 1981 The Biochemical Society, London.)

3. Ammonium sulfate precipitation with 25% (w/v) removed some protein, but not activity. Further ammonium sulfate addition (10%, w/v) precipitated the activity, which was then further fractionated using Sephacryl S-200. Active fractions were identified and pooled *if* they contained $> \frac{1}{3}$ the activity of the peak fraction. This protocol optimizes the purification achieved in this step. A greater yield could have been obtained by pooling *all* activity-containing fractions, but purity would have been considerably reduced.

4. Enzyme activity was concentrated by ammonium sulfate precipitation and resuspension in 50 mM phosphate buffer, diluted with an equal volume of 2 mM phosphate containing 2 M ammonium sulfate, and subjected to hydrophobic chromatography on a phenyl-Sepharose column. The column was eluted with a decreasing ammonium sulfate gradient, as shown in Fig. 2-16. The activity eluted just after the protein peak and the fractions with highest specific activity were pooled and subjected to preparative electrophoresis using a 6.2% separating gel, which produced a further three-fold increase in purification. Activity was recovered from the gel by slicing and eluting with phosphate buffer.

Reference: G. H. Elder, J. A. Tovey, and D. M. Sheppard, *Biochem. J.*, *215*, 45–55 (1983).

3

Protein Purification: Affinity Chromatography

INTRODUCTION

Many of the purification procedures discussed in Chap. 2 make use of one or more biophysical properties of proteins. The major drawback of these approaches is that proteins with similar properties can be separated only with difficulty by using methods based on minor differences. In general, affinity chromatography is based on the biological specificity a particular protein may have for a ligand. Proteins having very similar biophysical properties can have very divergent biological specificity, and if this can be used as a basis for separation, an effective purification procedure results. Although the term "affinity" chromatography has been popular for only about 20 years, the principle involved is much older, having first been used in 1951 in the form of "immunoadsorbents" for the separation of antibodies.

The general philosophy of affinity chromatography is that for a specific protein there are ligands that interact with that protein reversibly and which, if immobilized on a support material, specifically retard the chromatography of that protein and allow its purification. Depending on the strength of the immobilized ligand–protein interaction, the protein may effectively be immobilized onto the support material and measures to achieve its elution must be employed. There are many considerations that go into the selection and use of an immobilized ligand in affinity chromatography, and these are considered in this chapter. Although this method finds its widest application in the field of protein purification, it has also been useful in studying quantitative aspects of protein–ligand interaction, which is considered in detail in Chap. 18.

In its simplest conception affinity chromatography would appear to be the pan-acea for all problems in purifications of proteins. In a particular system, however, a variety of difficulties may be encountered. These include the inability to find a li-gand with an appropriate affinity for the protein, the lack of a suitable derivative of the ligand for immobilization, and the existence of a variety of proteins with similar specificity for the immobilized ligand. The work that has been done in the past 10 years or so has provided answers to many of these problems and has led to the development of new affinity chromatography approaches.

At the start of this section the term "reversible" was used to describe the interac-tion of a ligand with a protein in affinity chromatography. In recent years, as discussed at the end of this chapter, a variety of covalent affinity chromatography procedures have been developed which depend on the "reversibility" of various chemical reactions with specific residues in proteins or peptides.

REQUIREMENTS OF A LIGAND FOR EFFECTIVE USE

The optimal situation for an affinity chromatography system is one where non-specifically retarded proteins elute in the void volume of the matrix material, while specifically retarded proteins either remain associated with their specific ligand until elution is achieved by changing the chromatography conditions or are retarded sufficiently to achieve a complete resolution from void volume proteins or other proteins that may elute in the included volume of the column. Two parameters are important in determining the retardation of a protein on an affinity matrix: the dissociation constant, K_1, for the interaction of the protein and the ligand, L, given by

$$K_1 = \frac{[E][L]}{[EL]} \tag{3-1}$$

for the reaction $E + L \rightleftharpoons EL$, and second, the concentration of the immobilized ligand, L_0. If E_0 is the initial concentration of protein that can interact with the ligand, then

$$K_1 = \frac{[E_0 - EL][L_0 - EL]}{[EL]} \tag{3-2}$$

Where $L_0 \gg E_0$, this equation becomes

$$K_1 = \frac{[E_0 - EL]}{[EL]L_0} \tag{3-3}$$

We can now define the chromatographic distribution coefficient, K_d, as

$$K_d = \frac{[EL]}{[E_0 - EL]} = \frac{L_0}{K_1} \tag{3-4}$$

and the elution volume, V'_e, of the protein which interacts specifically with the ligand, is defined as

$$V'_e = V_0 + K_d V_0 \tag{3-5}$$

where V_0 is the void volume of the matrix. The retardation of the specifically interacting protein is therefore

$$\frac{V'_e}{V_0} = 1 + \frac{L_0}{K_1} \tag{3-6}$$

which is determined by the concentration of immobilized ligand and the dissociation constant of the protein–ligand complex. To achieve a retardation of 20 void volumes (i.e., $V'_e/V_0 = 20$), one would, of course, need a 19-fold excess of immobilized ligand concentration over the K_1 for the ligand–protein interaction. As discussed later, achievement of more than 1 to 2 mM ligand concentration on the matrix is usually impractical, giving useful upper limits for the dissociation constant K_1 of approximately 0.05 to 0.1 mM. If the immobilized ligand concentration can be increased, of course, larger values of K_1 can be accommodated.

If the matrix material has a fractionation range such that many of the proteins do not elute at the void volume, the parameter V'_e/V_e, where V_e is the elution volume of a protein of similar size but lacking specific interaction with the matrix, gives a better estimate of the efficacy of the immobilized ligand. From Eq. (2-4), V_e (the elution volume of the protein in the absence of specific interaction) is equal to K_{av} $(V_t - V_0) + V_0$, and hence the ratio V'_e/V_e is given by

$$\frac{V'_e}{V_e} = \frac{V_0 + K_d V_0}{V_0 + K_{av}(V_t - V_0)} \tag{3-7}$$

and knowledge of K_{av} is required to calculate the retardation. The retardation is still, of course, governed primarily by the ratio of L_0 to K_1.

As will be discussed shortly, these equations, and considerations of the affinity the desired protein has for the immobilized ligand, are often useful in designing effective conditions for the loading of protein onto, or the elution of protein from, an affinity column.

TYPES OF SUPPORT MATERIAL

Given that a suitable ligand can be found to be immobilized, a number of considerations go into the selection of the support material. The ideal support material for use in affinity chromatography has a variety of physical and chemical characteristics that give it optimal behavior. In terms of its physical properties it should have a high porosity, to allow maximum access of a wide range of macromolecules to the immobilized ligand. It should be of uniform size and rigidity to allow for good flow characteristics, and it must be mechanically and chemically stable to conditions used to immobilize the appropriate specific ligand. In terms of its chemical properties, it should have available a large number of groups that can be derivatized with the specific ligand, and it should not interact with proteins in general so that nonspecific adsorption effects are minimized.

A diverse variety of insoluble support materials have been used, including cellulose, polystyrene gels, cross-linked dextrans, polyacrylamide gels, and porous silicas, but by far the most popular support materials in use are beaded derivatives of agarose. The physical and chemical characteristics of agarose were described in Chap. 2 and are not repeated here, but in many ways agarose derivatives fulfill the characteristics required of the ideal matrix support.

Although agarose derivatives approach the ideal, a common problem with the support material (one that is not restricted to agarose) is steric hindrance. Once the ligand has been immobilized on the support, simple steric hindrance may prevent the interaction of proteins having binding sites for the ligand with the immobilized ligand. To circumvent this problem, a "spacer arm" is usually included in the support material to allow the ligand to be extended away from the physical surface of the support and permit unhindered interaction of protein with the ligand (although we are considering the spacer arm as part of the support, for practical purposes it is usually incorporated into the ligand prior to immobilization onto the support). In principle, such a spacer arm readily overcomes problems of steric effects. In practice, its inclusion *can* lead to a major problem in affinity chromatography—that of nonspecific interaction of proteins with the support material. The most commonly used spacer arms consist of 6 to 10 carbon atoms. However, these have quite a hydrophobic nature that they impart to the immobilized ligand, which can lead to nonspecific binding. Two approaches have been used to overcome this problem. In one, spacer arms with a hydrophobic nature are employed, but the buffer systems contain organic solvents such as N,N-dimethylformamide in low concentrations to help prevent these nonspecific interactions with the spacer-arm ligand system. In the other approach, spacer arms with hydrophilic natures are used. In particular 1,3-diaminopropanol, which can be coupled to the matrix by standard methods, has proven useful. Not only does this derivative (see Fig. 3-1) have a hydrophilic nature, but it is also easily lengthened or derivatized, as required.

A rather more pragmatic approach to this problem of nonspecific adsorption to the affinity matrix is to worry about the consequences during the elution phase of the purification rather than try to block the nonspecific adsorption during loading. This is dealt with later in the section on elution procedures.

Commonly Used Methods for the Immobilization of Ligands

The majority of matrices used in affinity chromatography are either agarose based, which have attendant hydroxyl groups that can be used in derivatization, or polyacrylamide based, which have reactive amide nitrogens available.

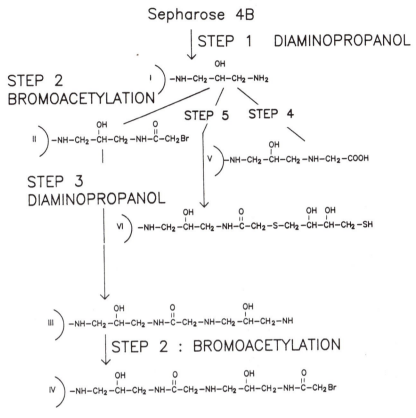

Figure 3-1   Utilization of 1,3-diaminopropanol in spacer-arm construction. In Step 1, Sepharose is activated by CNBr and 1,3-diaminopropanol is coupled. In Step 2, intermediate 1 is bromoacetylated by the N-hydroxysuccinimide ester of bromoacetic acid. In Step 3, the spacer arm is extended by addition of diaminopropanol. In Step 4, a carboxyl terminal is produced by incubation with bromoacetate in bicarbonate at pH 9.0, and in Step 5, a thiol terminal is produced by incubation with dithiothreitol in bicarbonate at pH 9.0.

With matrices having hydroxyl groups a variety of "activation" procedures are available that allow ligands with amino groups to be readily incorporated. In the case of protein ligands this represents no problem since usually, many amino groups are available. With small-molecule ligands, where spacer arms are usually incorporated into the ligand prior to immobilization, the spacer arm terminates in an amino group which is then used in the immobilization. The most commonly used activation procedures are (1) cyanogen bromide activation, (2) epoxide activation, (3) periodate oxidation, and (4) triazine activation. Each of these procedures is outlined briefly in Fig. 3-2.

Several methods are available for the activation and coupling of ligands to acrylamide, and these are outlined in Fig. 3-3. During coupling, especially with

Activation and Coupling of Amino-Containing Ligands to Agarose

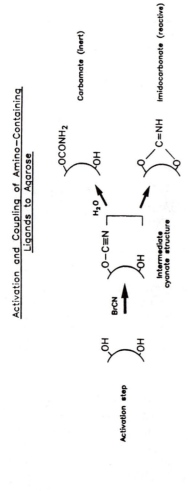

Activation step

BrCN

Intermediate cyanate structure

H_2O

OCONH$_2$

Carbamate (inert)

C=NH

Imidocarbonate (reactive)

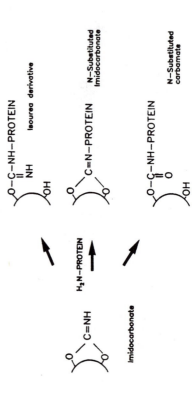

Coupling step

Imidocarbonate

H_2N-PROTEIN

O-C-NH-PROTEIN
‖
NH

Isourea derivative

C=N-PROTEIN

N-Substituted Imidocarbonate

O-C-NH-PROTEIN
‖
O

N-Substituted carbamate

a) by cyanogen bromide

50

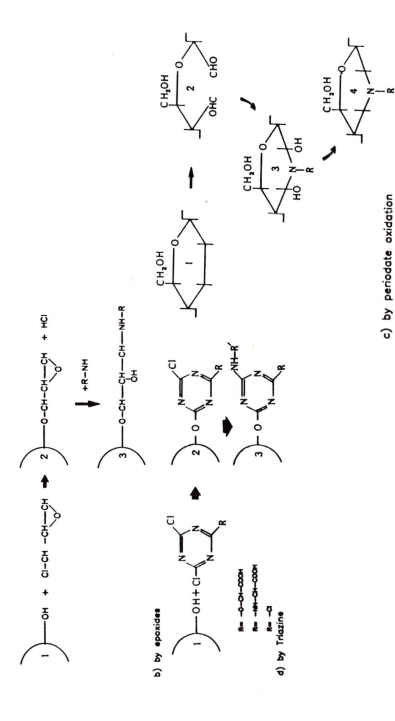

b) by epoxides

c) by periodate oxidation

d) by Triazine

Figure 3-2 Activation and coupling of amino-containing ligands by (a) cyanogen bromide, (b) epoxides, (c) periodate oxidation, and (d) triazine.

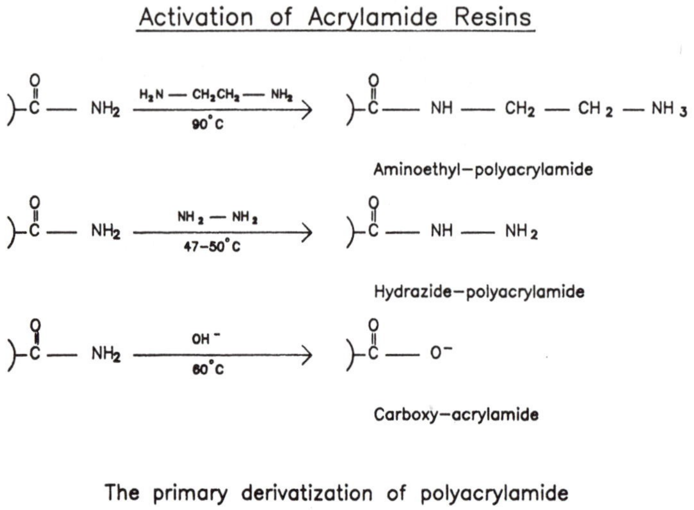

Activation of Acrylamide Resins

$$\text{)}\!-\!\overset{\overset{\displaystyle O}{\|}}{C}\!-\!NH_2 \xrightarrow[\text{90°C}]{H_2N-CH_2CH_2-NH_2} \text{)}\!-\!\overset{\overset{\displaystyle O}{\|}}{C}\!-\!NH\!-\!CH_2\!-\!CH_2\!-\!NH_3$$

Aminoethyl–polyacrylamide

$$\text{)}\!-\!\overset{\overset{\displaystyle O}{\|}}{C}\!-\!NH_2 \xrightarrow[\text{47–50°C}]{NH_2-NH_2} \text{)}\!-\!\overset{\overset{\displaystyle O}{\|}}{C}\!-\!NH\!-\!NH_2$$

Hydrazide–polyacrylamide

$$\text{)}\!-\!\overset{\overset{\displaystyle O}{\|}}{C}\!-\!NH_2 \xrightarrow[\text{60°C}]{OH^-} \text{)}\!-\!\overset{\overset{\displaystyle O}{\|}}{C}\!-\!O^-$$

Carboxy–acrylamide

The primary derivatization of polyacrylamide

Figure 3-3   Activation of acrylamide matrix material.

macromolecule ligands, it is often necessary to balance efficiency with the stability of the protein to be coupled. Efficiency is usually higher as the pH is raised; however, the protein ligand may not be stable at optimal coupling pH values, and efficiency must then be sacrificed to maintain the biological integrity of a protein ligand.

After coupling the ligand to the matrix, unreacted activated coupling groups are usually blocked by adding an excess of an "inert" blocking reagent. This step is sometimes unnecessary since the activated coupling groups are often hydrolyzed to inactivate derivatives as the coupled resin is washed at lower pH.

The amount of coupled ligand is established using one of several methods, the simplest being to quantitate the uncoupled ligand by thorough washing (and pooling the washes). The coupled amount is the difference between that added and the total amount recovered in the washes. The coupled ligand can also be estimated directly on the washed matrix by difference absorption spectral measurements, or if radioactive ligand is available, by radioactivity measurements. The final procedure that may be used is to hydrolyze the coupled ligand–matrix and determine the amount of ligand released.

All of these depend on the complete removal of noncovalently associated ligand prior to quantitation, and the washing protocols usually involve pH cycles and high salt concentrations.

In some cases of macromolecule ligands it is possible to determine the biological activity of the coupled macromolecule. Where, for example, the enzymatic function of the coupled macromolecule must be maintained, conditions may need to be adjusted to maximize its specific activity. This often involves coupling the enzyme in the presence of its substrates or ligands to "protect" the active-site region during the coupling process. It is usually possible to use an immobilized enzyme in assays to determine its specific activity.

SELECTION OF APPROPRIATE LIGANDS AND CONDITIONS

There are three categories into which "ligands" used in affinity chromatography can be placed. The most obvious is where the ligand is a small molecule with affinity for a specific protein. In such a case it may be either a substrate (or substrate analog) or a regulatory ligand. The second category is where the ligand is a macromolecule that has a particular affinity for a specific protein. It may be a molecule with very specific affinity for the desired protein (such as an antibody) or may have specificity toward a class of molecules (e.g., concanavalin A) into which the desired molecule falls. The final category is covalent affinity chromatography, where the ligand contains a moiety that reacts covalently but reversibly with the wanted protein. This category will be dealt with later.

Small-Molecule Ligands

Many proteins have small-molecule ligands which may make suitable affinity chromatography ligands. Apart from the consideration of the dissociation constant the protein has for the ligand, several others go into the selection of the appropriate ligand. An important corollary to the dissociation constant is the requirement that the linkage process not adversely affect the affinity of the ligand for its target protein. The simplest way to check this is to use the spacer-arm ligand derivative (either before or after immobilization) as a competitive inhibitor of the normal ligand. If a simple kinetic or ligand binding assay is available, this gives an easy way to screen not only the affinity the immobilized ligand has for the protein, but also conditions of buffer, pH, temperature, and so on, that give optimal binding. Often, inhibition studies using underivatized analogs can give much useful information regarding the contribution certain parts of the potential ligand play in binding, thus allowing an informed choice to be made regarding which part of the ligand to modify by incorporation of the spacer arm.

A common problem with small-molecule ligands is that the ligand, although it may bind tightly to the wanted protein, has affinity for other proteins that recognize the same ligand. Consider, for example, the use of immobilized NAD in the purification of a dehydrogenase. Many dehydrogenases, and a number of other proteins, bind NAD with similar affinity and are retarded by such an affinity matrix. The solution to this dilemma has two aspects: (1) one must establish conditions such as pH, ionic strength, and temperature where binding to the wanted protein is optimal compared to binding to other proteins with affinity for NAD, and (2) if possible, a coligand that specifically enhances the affinity of the wanted protein for the matrix should be used in the buffer; frequently, a substrate analog accomplishes this through the formation of a ternary complex, which is of course far more specific for the wanted protein and leads to increased affinity relative to other proteins that may bind the ligand.

From the earlier equations for V'_e/V_e it should be apparent that anything that can be done to increase the affinity of the wanted protein relative to the affinities of potential contaminating proteins increases the retardation of the wanted protein on

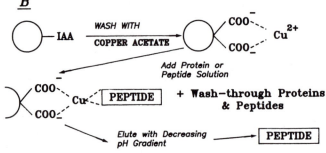

$$\Big)-CONH-(CH_2)_2-NH-CH_2-\underset{\underset{OH}{|}}{CH}-CH_2-N\Big\langle{}^{CH\ COOH}_{CH\ COOH}$$

Immobilized Iminodiacetic Acid (IAA)

Figure 3-4   Use of metal affinity ligands in peptide and protein purification: (A) structure of iminodiacetic acid (IAA); (B) scheme for use in purification procedures.

the matrix. In situations where a variety of proteins bind to the matrix ligand but do so with quite different affinities, specificity of purification can be achieved during the elution phase, as described in the next section.

A unique approach to immobilizing a small-molecule ligand has been introduced where the small molecule is a metal ion. By immobilization of a chelating agent such as iminodiacetic acid, a variety of different metal ions can be chelated and used as affinity ligands for metal binding proteins. Such affinity resins have also found use in the purification of peptides that contain certain amino acids whose side chains show affinity for particular metal ions. This application is outlined in Fig. 3-4.

Macromolecule Ligands

Macromolecular ligands that may be of use in affinity chromatography can be divided into three categories, all of which have found significant use in protein purifications. In the first category are proteins that have specific affinity (and often very high affinity) for moieties which may be covalently attached to the wanted protein. Most frequently, this is useful in the purification of glycoproteins, where a wide range of plant lectins are available with differing specificities in regard to the sugars that they recognize. Most plant lectins that have been characterized with regard to their sugar specificity have also been immobilized and used to purify glyco-protein derivatives. A number of these are listed in Table 3-1 together with their specificities. Of course, the major problem with this approach is that any protein with the appropriate sugar covalently attached binds to the immobilized lectin, al-though this can partially be overcome during the elution phase. Lectin affinity

TABLE 3-1 Lectin affinity chromatography

Lectin	Specificity (competing sugar)
Concanavalin A	Complex (α-methyl mannoside)
Dolichos biflorus	N-Acetylgalactosamine
Soybean	N-Acetylgalactosamine
Helix pomatia	Blood group A (NAGal)
Limulus	Galactose
Garden pea	Glucose, mannose
Winged pea	Fucose
Wheat germ	N-Acetylglucosamine

chromatography is widely used in both general glycoprotein isolation and fractionation, and in purification schemes for specific enzymes that have a glycoprotein nature.

Although many lectins do not have well-characterized specificities for ligands of related structure, much work has been done on the specificity of concanavalin A, where carbohydrate ligands of related structure show quite a range of affinity for this lectin. As a result, concanavalin A–Sepharose can be used to give information on the type of carbohydrate chain recognized. Into this category also fall proteins such as avidin, which is found covalently linked to many biotin-requiring enzymes.

Although they do not have affinity for a small molecule conjugated to the macromolecule as in the previous instances, affinity ligands such as immobilized staphylococcal protein A must also be included in this category. Protein A binds the Fc portion of the IgG molecule with high affinity and has been used as an affinity ligand for the purification of IgG molecules and is also useful in "sandwich" affinity chromatography—the isolation of IgG–protein complexes.

The second category involves the use of immobilized antibodies as the ligand. Because of the precise specificity of many antibodies for their particular antigen, an immobilized antibody makes the perfect affinity ligand—it recognizes only the wanted protein. There are, of course, instances where polyclonal antibodies to a particular protein cross-react with other proteins having similar or related structural features (antibody to, for example, lactate dehydrogenase often reacts weakly with other dehydrogenases). The advent and ease of preparation of monoclonal antibodies, however, have made it easier to find a suitable antibody for affinity chromatography purposes.

The availability of monoclonal antibodies allows for some quite interesting applications of affinity chromatography. For example, anti-AMP antibodies have been used in the separation of partially adenylylated forms of glutamine synthase. Antibody is immobilized, and adhered glutamine synthase (which is adenylylated) is eluted using AMP gradients. Different monoclonals have been used, allowing the separation of glutamine synthase with fewer than three adenylated subunits per dodecamer from those with higher amounts of adenylylation.

The third category includes any protein that has a unique affinity for another protein and can thus be used as an affinity ligand, provided that the affinity is high enough for that protein. A particularly interesting example of this type of mac-romolecule ligand is α-lactalbumin. This is a regulatory protein for galactosyl transferase, but interacts only weakly in the absence of coligands. However, when carbohydrate ligands such as N-acetylglucosamine are present, α-lactalbumin inter-acts strongly with the enzyme and has proved to be a most effective affinity ligand in its purification. Interestingly, galactosyl transferases from a wide variety of sources can be purified using immobilized α-lactalbumin, even though they do not physio-logically interact with this protein. Similarly, calmodulin has been used to purify calmodulin-binding proteins.

Finally, we consider an application of affinity chromatography recently developed that illustrates some of the potential for affinity resins using macromolecule ligands. Anhydrotrypsin is a catalytically inert derivative of trypsin containing dehydroalanine in place of the active-site serine residue. This derivative has been shown to have an approximately 20-fold higher affinity than substrate peptides for product peptides produced by tryptic cleavage, and immobilized anhydrotrypsin has been used as an affinity ligand for the purification of tryptic peptides. COOH-terminal arginine pep-tides adhere more tightly than COOH-terminal lysine peptides. Further specificity may be obtained during elution, which usually involves decreasing pH gradients.

Dye Molecule Ligands

Several commercially available "dye–ligand" affinity chromatography media have been used in a wide variety of protein purifications. Some of these, such as Cibacron-blue or Procion-blue, have been shown to have affinity for enzymes re-quiring adenylyl-containing cofactors. In general, with this type of "nonspecific" affinity ligand, the specificity is provided by the adsorption and elution conditions employed in a particular purification, and these must often be worked out by trial and error.

ELUTION PROCEDURES

As has been indicated in a number of places, the elution phase of affinity chromatog-raphy can be used to give enhanced specificity in a purification. First, however, we must consider the physical basis for elution. Except in the case of covalent affinity chromatography, the interaction of the protein with the matrix ligand is reversible, and it is this interaction that retards the passage of the specifically adsorbed protein. If enough of the buffer used to load the column were washed through, the protein would eventually elute from the matrix and its position of elution could be estimated from the expression for V'_e/V_e given earlier. Except for cases where the protein does not interact strongly with the matrix and V'_e/V_e is a fairly low number (hence elution occurs in a reasonable volume), it is usually necessary, once contaminating proteins have been washed through, to assist elution of the wanted protein. This course involves taking measures that *increase* the *dissociation constant* for the matrix–

ligand–protein interaction. In general, there are two ways in which this is achieved. Any change in conditions such as buffer pH, ionic strength, or temperature that weakens the affinity of the matrix for the protein assists the elution process. In some instances where the protein has a very high affinity for the matrix–ligand and can be reversibly denatured, mild denaturing conditions can be used. For the elution of carbohydrate-containing molecules from immobilized lectins such as concanavalin A, quite harsh denaturing conditions are sometimes necessary. This represents a real problem for the use of those affinity ligands in the purification of proteins that may not withstand such harsh elution conditions. These nonspecific approaches cannot be used where the purification may depend on specificity during elution.

The addition of a competing ligand to the elution buffer represents a far more controllable approach. In cases where a small-molecule ligand is used on the matrix, it is possible to elute by inclusion of the free ligand in the elution buffer. If the dissociation constant K_1 for the immobilized ligand and the dissociation constant K_f for the free, competing ligand are known, the amount of free, competing ligand that must be included in the elution buffer to achieve rapid elution can be calculated. Using a concentration of free competing ligand to give >90% of the protein bound to the free ligand gives rapid elution of the desired protein in a small volume. Lower amounts of competing ligand still increase the elution rate but give more dilute protein. It is, of course, not necessary that the competing ligand be an analog of the matrix ligand, although this is often the case. It is sufficient simply that the added ligand "compete" with the matrix ligand. For example, leucine effectively elutes glutamate dehydrogenase from a GTP-Sepharose affinity column by binding to a separate regulatory site on the protein and blocking, via a conformational change, the binding of GTP.

The use of a competing ligand to achieve elution provides a way of increasing the specificity of purification in cases where more than one protein may interact with the column via affinity for the matrix ligand. Provided that the various proteins have a different affinity for the matrix (either naturally or through the inclusion of a coligand), they can be eluted separately from the column by using a concentration gradient of the competing ligand. The example of glutamate dehydrogenase elution from GTP-Sepharose by leucine provides the basis of a second method of increasing specificity during elution. Provided that a competing ligand which is *not* a structural analog of the matrix–ligand is available, specific elution is possible even when more than one type of protein has specifically interacted with the matrix.

We have discussed using a competing ligand for cases where the matrix–ligand is a small molecule. However, provided that a small molecule competitor of a protein–protein interaction is available, these considerations also apply to affinity chromatography using macromolecular ligands. Two examples illustrate this approach. Galactose is an excellent competitor of the interaction between immobilized peanut agglutinin (PNA) and glycoprotein ligands containing terminal galactose. This is to be expected since the interaction is mediated by a "small" ligand—the carbohydrate chain of the glycoprotein. A more interesting example is provided by protein A–IgG interaction, where it has been found from the crystal structure of the

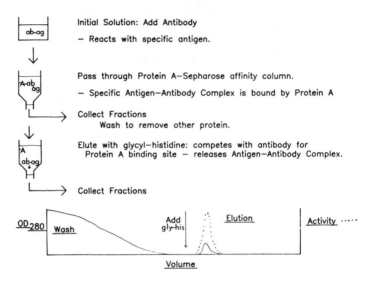

Affinity Chromatography of Antibody–Antigen Complexes

Figure 3-5   Use of proteinA–Sepharose in the isolation of antigen-antibody complexes.

complex that certain amino acid residues in each protein are involved in the interaction. As a result a variety of dipeptides, such as glycyltyrosine, glycylhistidine, and glycylphenylalanine, have been used as quite effective competing ligands for the protein–protein interaction that occurs with protein A–IgG. In this particular instance, and in other instances where such protein–protein interactions are mediated by hydrophobic bonds, reagents such as ethylene glycol, which disrupts hydrophobic interactions, have been used as nonspecific (but quite effective) elution procedures.

The use of protein A–Sepharose to isolate specific antigen–antibody complexes and the use of glycylhistidine to elute the complex is outlined in Fig. 3-5. In cases such as using α-lactalbumin–Sepharose to purify galactosyltransferase, where a coligand is required for affinity, elution can be achieved simply by omitting the coligand from the buffer.

SOME EXAMPLES OF AFFINITY
CHROMATOGRAPHY PURIFICATIONS

Glutamate Dehydrogenase

Although glutamate dehydrogenase can be purified to homogeneity by conventional means, the use of affinity chromatography on a GTP-Sepharose matrix increases the yield, cuts down on the number of steps, and results in a more rapid and efficient purification. After initial ammonium sulfate fractionation and gel filtra-

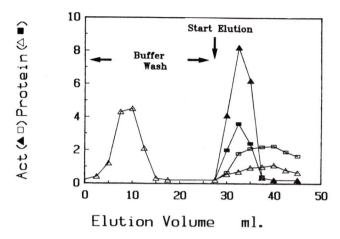

Figure 3-6 Typical elution profiles obtained during the purification of glutamate dehydrogenase with GTP-Sepharose: △, protein eluted during buffer wash or salt elution; ■, protein eluted by 5 mM leucine wash; □, activity eluted by salt wash; ▲, activity eluted by 5 mM leucine wash.

tion chromatography, active fractions are pooled, dialyzed against a 20 mM phosphate buffer at pH 7.0, and passed through a GTP-Sepharose column (5 ml bed volume). Nonspecifically retarded protein is washed through the column with starting buffer containing 50 mM sodium chloride (10 column volumes). Elution can be achieved in one of two ways: The pH of the buffer is raised to 8.0 and the NaCl concentration is raised to 100 mM, or leucine (5 mM) is included in the wash buffer. Figure 3-6 illustrates the different elution profiles obtained by these procedures and demonstrates the utility of the use of a competing ligand (in this case leucine) to achieve rapid elution.

Hexosaminidase

An affinity matrix suitable for the isolation of β-N-acetylhexosaminidase can be prepared by coupling asialofetuin to Sepharose 4B. This matrix has been used to enrich β-N-acetylglucosaminidase activity from rabbit sperm cytoplasmic droplets in a very simple procedure. The droplets, isolated by differential centrifugation, are freeze-dried, redissolved in buffer by sonication, dialyzed against 50 mM sodium acetate buffer (pH 4.5), and loaded onto the asialofetuin-Sepharose matrix. Figure 3-7 shows an elution profile of nonspecific protein washed through by the starting buffer, and a sharp elution of activity and its associated protein, using a gradient of NaCl from 0 to 500 mM.

Separation of Dehydrogenases

As discussed earlier, a problem often encountered in affinity chromatography is the resolution of a group of proteins all of which have affinity for the matrix ligand.

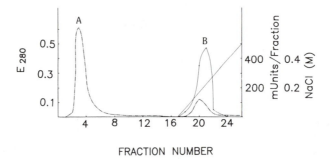

FRACTION NUMBER

Figure 3-7   Elution of nonspecific protein (peak A) with wash buffer, followed by elution of hexosaminidase activity (peak B) by salt gradient. (Reprinted with permission from: A. A. Farooqui and P. N. Srivastava, *Int. J. Biochem., 10*, 745–748. Copyright 1979 Pergamon Press, Elmsford, N. Y.)

Figure 3-8 illustrates the separation of a mixture of BSA, malate dehydrogenase, glucose-6-phosphate dehydrogenase, lactate dehydrogenase (H$_4$ isozyme), and yeast alcohol dehydrogenase using 5′-AMP–Sepharose (with an aminohexyl spacer arm). The mixture was applied in 10 mM phosphate, pH 6.0, and the column washed with this starting buffer. The BSA, which has no affinity for AMP, elutes immediately. The dehydrogenases are eluted from the matrix using a pH gradient as shown. Glucose-6-phosphate dehydrogenase elutes at low pH, yeast alcohol dehydrogenase at intermediate pH, and both malate dehydrogenase and lactate dehydrogenase at

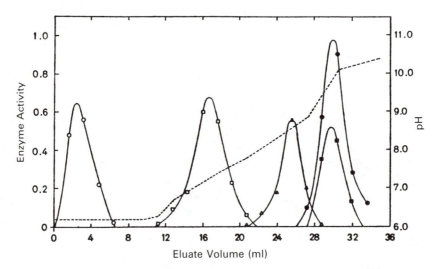

Figure 3-8   Differential elution of proteins adhering to AMP-Sepharose by use of a pH gradient: ○, BSA; ●, malate dehydrogenase; □, glucose-6-phosphate dehydrogenase; △, yeast alcohol dehydrogenase. [Reprinted with permission from: C. R. Lowe, M. J. Harvey, and P. D. G. Dean, *Eur. J. Biochem., 41*, 347–351 (1974).]

high pH. This example clearly shows the utility of gradient elution of affinity matrices to separate proteins with similar specificities but differing affinity characteristics.

COVALENT AFFINITY CHROMATOGRAPHY

Covalent affinity chromatography represents an aspect of affinity chromatography which has limited use in the purification of proteins but tremendous potential in the purification of peptides that might contain unique amino acids. The principle is simple: The matrix contains a reactive group that derivatizes reversibly a particular amino acid—any protein or peptide containing that particular amino acid reacts chemically with the matrix and is immobilized. Nonreactive peptides or proteins can be completely washed from the matrix without fear of eluting the immobilized material. Afterward, the immobilized material is eluted by inclusion in the buffer of a reagent that reverses the chemical modification. This idea was originally introduced using p-chloromercuribenzoate coupled to aminoethyl agarose, which reacts well with proteins having exposed, reactive sulfhydryl groups, and has been used in a variety of affinity separations, including the separation of calf thymus histone F_3 (which contains a cysteine) from other histones (lacking cysteine), and the fractionation of enzymatically active papain (intact sulfhydryl) from inactive papain (no intact sulfhydryl). In each case the reactive sulfhydryl containing material is eluted from the matrix by inclusion of β-mercaptoethanol or cysteine in the buffer.

Arylsulfenyl chlorides (specific chemical modification reagents for tryptophan) have been immobilized to cross-linked polyacrylamide and used to separate tryptophan containing peptides from peptides lacking tryptophan. The immobilized sulfenyl chloride reacts with the tryptophan side chain in the peptide as shown in Fig. 3-9,

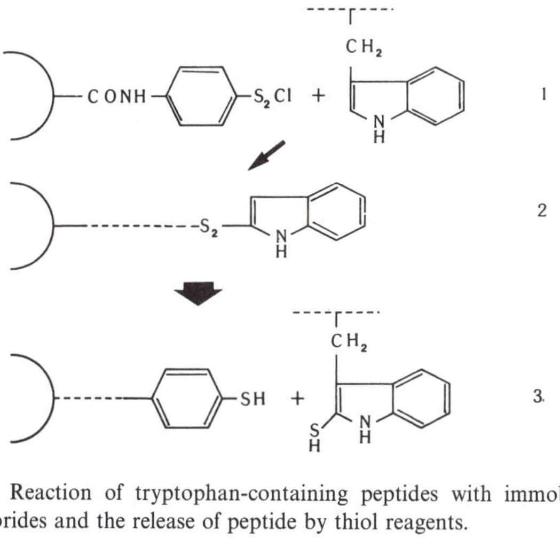

Figure 3-9 Reaction of tryptophan-containing peptides with immobilized arylsulfenyl chlorides and the release of peptide by thiol reagents.

and after elution of noncovalently immobilized peptides, the tryptophan-containing peptides are eluted by inclusion of a thiol in the buffer.

Although few such covalent affinity procedures have been detailed, the potential of immobilizing chemical modification reagents and using them for the purification of specific peptides containing unique amino acids remains.

AFFINOPHORESIS

The technique of affinophoresis is a recently introduced approach for the specific separation of macromolecules that makes use of some of the principles of affinity chromatography but uses as its separation method electrophoretic mobility. In this technique the specific affinity of a macromolecule is used to ligand to the macromolecule a polyelectrolyte containing an affinity ligand. The result of this specific interaction is that the target protein now carries increased charge and resultant increased electrophoretic mobility, allowing it to be readily identified and separated by electrophoretic methods. Cationic or anionic polyelectrolytes can be used and are linked to the affinity ligand in much the same way as a ligand would be immobilized on an insoluble matrix. Cationic polyelectrolytes are usually dextran derivatives with diethylaminoethyl groups as the cations, while polyacrylyl-β-alanyl carriers with sulfonate groups have been used as anions. As outlined in Fig. 3-10, affinophoresis has been successfully used to separate trypsin from "pronase" (which is a mixture of proteases, including trypsin) using a *p*-aminobenzamidine affinity ligand for trypsin on an anionic polyelectrolyte. In the absence of the affinophore (part A), trypsin, pronase (multicomponent), and the active-site-blocked TLCK-

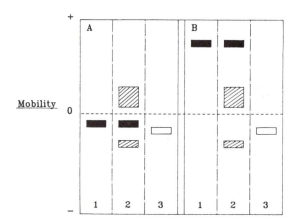

Figure 3-10   Schematic representation of the separation of trypsin from a pronase mixture by affinophoresis. (A) Track 1, trypsin; Track 2, pronase; Track 3, TLCK-trypsin. (B) Same as in part (A), but the electrophoresis is in the presence of an anionic affinophore. [Adapted from: K. Shimura and K.-I. Kasai, *Biochem. Biophys. Acta, 802*, 135–140 (1984).]

trypsin show similar electrophoretic mobilities. In the presence of the affinophore (part B), trypsin and the trypsin component of pronase show greatly increased mobility, while the remaining pronase components and TLCK-trypsin (which cannot bind the affinophore) show unaltered mobility.

The various affinity techniques that we have discussed in this chapter are usually adapted to column chromatographic approaches. Where an extremely specific ligand is available, they are ideally suited to batch procedures. Affinity techniques are almost always used with some form of prior fractionation, usually because the sheer amount of protein obtained in initial homogenization or extraction is too large to employ the generally small bed volumes of affinity resins. This can lead to problems with nonspecific protein interactions or to premature elution of the retarded protein as a result of the large volume of material initially loaded onto the resin.

Affinity techniques are, however, of major use in many protein purifications and are now adapted to HPLC techniques for both analytical and preparative purposes. Because the elution phase can be closely controlled, affinity resins also provide a convenient means of concentrating dilute protein solutions.

4

Molecular Weight Determination

INTRODUCTION

The determination of the molecular weight of a molecule would appear to be quite a straightforward procedure. For small molecules this is true: Elemental analysis and approaches using colligative properties yield easily interpretable answers. The situation with a protein is somewhat different: If the primary sequence is known, it is a quite simple matter to determine the molecular weight from the amino acid composition if a single chain is present. Many proteins, however, contain subunits that may or may not be chemically similar. The molecular weight is then either a multiple of the molecular weight of the monomeric polypeptide chain or a combination of different polypeptide chains. Thus, even if the sequence is known (or sequences if more than one type exist) for the polypeptide chain, one does not necessarily know the molecular weight of the native molecule. Distinctions must be drawn between the native molecular weight and the denatured molecular weight. The native molecular weight is that as the protein exists under normal in vivo conditions where it exhibits normal and full activity. The denatured molecular weight can be defined as that of the minimum covalently bonded structure that the molecule is broken down to under denaturing conditions such as the presence of SDS or guanidinium hydrochloride. This molecular weight must not be taken as the polypeptide chain molecular weight since two or more chains may be covalently linked by disulfide bridges and may contain covalently bound nonpolypeptide entities such as carbohydrates, which contribute to the molecular weight. As we shall see, carbohydrate substituents can lead to anomalous molecular weight estimates using some of the approaches described. Similarly, most of the methods for determining molecular weights make some assumption concerning either the shape or the behavior of the protein relative to some reference: deviations in behavior from this reference lead to misleading molecular weight estimates. These estimates are, in many instances, averages, and depending on how the averaging is done, different answers for mixtures of species of different molecular weights are obtained. The various types of molecular weight averages that can be experimentally determined are described in detail here. Different techniques give different molecular weight averages, and a comparison of those obtained by approaches giving different types of molecular weights average also yield an idea of purity.

The concept of protein purity is somewhat nebulous: As sensitivity of protein detection has been increased, especially in techniques such as polyacrylamide gel electrophoresis (PAGE), it has also been far easier to detect impurities. Many of the experimental approaches for determining molecular weights are also used to estimate purity. Careful consideration of these methods is thus helpful in the context of examining purity as well as determining molecular weights.

MOLECULAR WEIGHT AVERAGES

If a sample contains a distribution of species with different molecular weights (defined in the classical sense), the average molecular weight depends on how the averaging process is done.

If there are $n(M) \, dM$ moles of species with molecular weights between M and $M + dM$, the total number of species present in the sample, n^T, is

$$\int_0^\infty dM n(M) \tag{4-1}$$

where the function $n(M)$ is the molecular weight distribution function. Moments of this distribution function can be defined. The Kth moment is

$$m_k = \int_0^\infty dM n(M) M^K \tag{4-2}$$

The total number of species present, n^T, is the zeroth moment.

Average molecular weights are defined in terms of ratios of higher moments (Kth) to the ($K-1$)th moment. The *number-average molecular weight*, $M_n = m_1/m_0$:

$$M_n = \frac{\int_0^\infty dM n(M) M}{\int_0^\infty dM n(M)} \tag{4-3}$$

If we consider a discrete distribution of species containing n_i moles of components with molecular weight M_i, we can write the expression for M_n as

$$M_n = \frac{\sum_i n_i M_i}{\sum_i n_i} \tag{4-4}$$

Alternatively, to obtain an expression in terms of molar concentrations, divide by the volume V:

$$M_n = \frac{\sum_i n_i M_i / V}{\sum_i n_i / V} \tag{4-5}$$

Defining the weight concentration, $C_i = n_i M_i / V$, we get

$$M_n = \frac{\sum_i C_i}{\sum_i C_i / M_i} \tag{4-6}$$

The *weight-average molecular weight* is defined by the ratio m_2/m_1,

$$M_w = \frac{m_2}{m_1} = \frac{\sum_i n_i M_i^2}{\sum_i n_i M_i} \tag{4-7}$$

or in terms of weight concentrations,

$$M_i = \frac{\sum_i C_i M_i}{\sum_i C_i} \tag{4-8}$$

The third average that is commonly used is the so-called *Z-average molecular weight*, defined as m_3/m_2:

$$M_z = \frac{m_3}{m_2} = \frac{\sum_i n_i M_i^3}{\sum_i n_i M_i^2} \tag{4-9}$$

which in terms of concentration yields

$$M_z = \frac{\sum_i C_i M_i^2}{\sum_i C_i M_i} \tag{4-10}$$

As will be discussed, the various methods of determining molecular weights usually give either M_n or M_w. However, molecular weights estimated by viscosity measurements give a *viscosity-average molecular weight*, M_v, defined by

$$M_v = \left(\sum_i W_i M_i^a \right)^{1/a} \tag{4-11}$$

where W_i is the weight fraction of the ith species and a is an empirical constant that varies between 0.5 and 2.0, depending on the type of molecular weight average.

From the equation for M_v we get

$$M_v = \left(\frac{\sum_i n_i M_i^{(1+a)}}{\sum_i n_i M_i} \right)^{1/a} \tag{4-12}$$

and hence when $a = 1.0$ this expression is the same as that for M_w, and $M_v = M_w$. For $a < 1.0$, M_v is between M_n and M_w.

Effects of Purity

As indicated earlier, different experimental methods for determining molecular weights can give different types of molecular weight averages. This gives a basis for assessing the purity of a sample in a homogeneous sample $M_n = M_w = M_z$. However, with a mixture of different molecular weights the values for each of these averages *can* be quite disparate. Two examples illustrate this point.

Consider an equal mixture (by weight) of two molecules, one with a molecular weight of 10,000 and the other 100,000. In this instance,

$$M_n = \frac{1 + 1}{1/10,000 + 1/100,000} = 18,181$$

$$M_w = \frac{10,000 + 100,000}{1 + 1} = 55,000$$

and

$$M_z = \frac{10,000^2 + 100,000^2}{10,000 + 100,000} = 91,818$$

These values should, of course, easily be separated experimentally, allowing one to conclude that a mixture is indeed present.

However, if the mixture consists of equal weights of proteins with molecular weight of 80,000 and 100,000, the values for M_n, M_w, and M_z are 88,888, 90,000, and 91,111, respectively, which would be most difficult to resolve experimentally and could lead to the erroneous conclusion that one was dealing with a homogeneous solution of protein with a molecular weight of $90,000 \pm 1500$.

If, instead of equal weights of two species we have equal *numbers* of two types of polypeptide chains, then for our first example $M_n = 90,000$, $M_w = 91,111$, and $M_z = 92,195$, which, as before, would not easily be experimentally separated.

As we examine various methods for estimating molecular weights, these considerations should be remembered. In some of the techniques a clear indication of heterogeneity is obtained, but in others only the appropriate molecular weight average is obtained, and comparison of molecular weight averages can then be a valuable tool in estimating purity.

MOLECULAR WEIGHT DETERMINATION BY SITE TITRATIONS

One of the simplest and earliest methods employed to determine the molecular weight of a protein is based on the assumption that the protein contains a single specific site of some sort. This might be a binding site for a certain ligand, an N- or C-terminal amino acid residue, a cofactor such as a metal ion, or in an extreme case the amino acid present in the lowest amount in the amino acid composition. The molecular weight of the molecule is then calculated on the basis of 1 mol of this specific site per mole of protein. The determination of the number of specific sites present per unit weight of protein and the calculation of the average molecular weight by dividing the weight by the number of molecules present clearly indicates that these methods give a *number-average* molecular weight.

   Determination of molecular weight by these approaches has some pitfalls. Consider a situation where the molecular weight of a protein consisting of four chemically identical polypeptide chains is determined by end-group analysis or by titration of a single specific ligand binding site per polypeptide chain. The calculated molecular weight is one-fourth of the true molecular weight. It represents a minimum molecular weight defined in terms of the specific site. The types of measurements involved in this sort of determination are usually quite accurate, and with a rough estimate of the molecular weight by another approach, can be used to give an accurate estimate of the molecular weight. If the protein, instead of consisting of four chemically identical chains consists of two types of chains with different sizes, a quite erroneous molecular weight may be calculated depending on the exact situation. If an $(A - B)_2$ tetramer has a total of two binding sites for a ligand (one for each AB pair) and the molecular weight is determined by site titrations, the molecular weight is one-half of the true molecular weight no matter how disparate the sizes of the A and B chains are. If, however, the molecular weight is determined by the number of moles of dansyl chloride (an amino-terminal labeling reagent) incorporated per gram of protein, quite erroneous estimates are obtained.

METHODS BASED ON COLLIGATIVE PROPERTIES SUCH AS OSMOTIC PRESSURE

Although a number of colligative properties are used in the determination of low molecular weights, osmotic pressure measurements are the only type that have found much use with proteins. To determine the type of molecular weight average obtained by osmotic pressure measurements, let C_i be the concentration of the protein and OP the osmotic pressure. Then

$$OP = k \sum_i C_i \tag{4-13}$$

where k is the proportionality constant between osmotic pressure and concentration. Since the only quantity known is the total weight concentration of the protein in solution, C, which is equal to $\sum C_i$ (which in turn equals $\sum M_i C_i$), we get, from Eq. (4-13),

$$\frac{OP}{C} = \frac{k \sum_i C_i}{\sum_i C_i M_i} \qquad (4\text{-}14)$$

Since C_i is proportional to N_i one can write

$$\frac{OP}{C} = \frac{k \sum_i N_i}{\sum_i N_i M_i} \qquad (4\text{-}15)$$

Earlier we defined $M_n = \sum N_i M_i / \sum N_i$, and hence

$$\frac{OP}{C} = \frac{k}{M_n} \qquad (4\text{-}16)$$

and the molecular weight obtained from osmotic pressure (or any other colligative property) measurement is a *number-average* molecular weight.

Osmotic pressure, like all colligative properties, is a measure of the chemical potential, μ, of the solvent, and can thus be used to measure the molecular weight of the solute (at infinite dilution). In the experimental determination of the osmotic pressure of a solute molecule, the solute is separated from pure solvent by a semi-permeable membrane and the resultant "pressure" determined as outlined schematically in Fig. 4-1A. The osmotic pressure π results from the chemical potential of the solute and is defined as $\pi = P - P_o$.

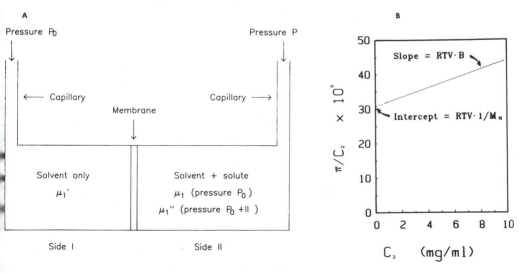

Figure 4-1   A: Schematic of apparatus for osmotic pressure determination. Osmotic pressure $\pi = P - P_0$. B: Plot of π/C_2 versus C_2 to show calculation of molecular weight and the virial coefficient, B.

At equilibrium in an osmotic pressure determination, molecules that can freely pass through the membrane do *not* affect the molecular weight estimates of those molecules retained by the membrane. From van't Hoff's limiting laws for osmotic pressure, extrapolation of the osmotic pressure to zero concentration allows the molecular weight, M_i, of the solute, to be obtained from Eq. (4-17)

$$\frac{\pi}{C_2} = \frac{RT}{M_i} \tag{4-17}$$

In reality, of course, the dependence of the osmotic pressure, π, on solute concentration is given by a virial equation of the form of Eq. (4-18). In this general equation the osmotic pressure $\pi = \mu_1, - F_1^0$.

$$\mu_1 - F_1^0 = -RTV_1^0 C_2 \left(\frac{1}{M_i} + B_{c_2} + C_{c_2}^2 + \cdots \right) \tag{4-18}$$

where F_1^0 is the molar free energy, C_2 the concentration of the solute, V the partial molar volume of the solvent, and B and C the virial coefficients. This equation shows that the effect of the solute on the chemical potential decreases as the molecular weight increases. From the limiting slope of the plot of π/C_2 versus solute concentration, an estimate of the virial coefficient, B, is obtained, together with the molecular weight, as shown in Fig. 4-1B.

Although, as indicated previously, the osmotic pressure at a given solute concentration decreases as the solute molecular weight increases, molecular weights up to 200,000–300,000 can be measured quite accurately.

MOLECULAR WEIGHT DETERMINATION BY LIGHT SCATTERING

The virial equation has an analogous equation for light-scattering measurements:

$$\frac{i}{I} = \frac{2\pi^2 \bar{n}^2 (d\bar{n}/dc)^2 (1 + \cos^2 \theta)C}{N\lambda^4 r^2 (1/M_i + 2B_c + 3C_{c^2} + \cdots)} \tag{4-19}$$

where i, I, and θ are defined by Fig. 4-2, r the distance between macromolecules, $\bar{n}$ the refractive index, and C the concentration.

The quantity determined is *Rayleigh's ratio*, R_θ,

$$R_\theta = \frac{r^2 i}{I(1 + \cos^2 \theta)} \tag{4-20}$$

which is independent of the scattering angle and is given by

$$R_\theta = \frac{K_c}{1/M_i + 2B_c + 3C_{c^2} + \cdots} \tag{4-21}$$

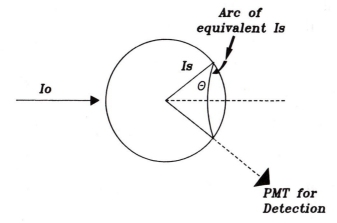

Figure 4-2 Scheme of light-scattering experiment. Scattered light is quantitated using a photomultiplier tube (PMT).

where

$$K = \frac{2\pi^2 \bar{n}^2 (d\bar{n}/dc)^2}{N\lambda^4}$$

The equation for R_θ can be rewritten as

$$\frac{K_c}{R_\theta} = \frac{1}{M_i} + 2B_c + 3C_{c^2} + \cdots \qquad (4\text{-}22)$$

and a plot of K_c/R_θ versus C gives an intercept of $1/M_i$.

When the scattering molecules are of comparable size to the wavelength of the scattered light, the equation becomes

$$\frac{K_c}{R_\theta} = \frac{1}{M_i}\left(1 + \frac{16\pi^2 R_G^2 \sin^2 \theta}{3\lambda^2}\right) \qquad (4\text{-}23)$$

where R_G is the radius of gyration of the scattering molecule. This equation is the basis of the Zimm plot (Fig. 4-3).

In the Zimm plot the molecular weight is obtained from the intersection point of the extrapolation at constant θ to $C = 0$ and at constant C to $\theta = 0$. The radius of gyration of the scattering molecule is also obtained from the initial slope of the curve at constant m extrapolated to $C = 0$. In Fig. 4-3 the x axis is plotted as $(\sin^2 \theta/2) + K'C$, where K' is an arbitrary constant added partly for convenience of presentation and partly to account for intermolecular effects.

It is quite apparent that light scattering is proportional to the weight concentration of the solute and its molecular weight, and one can write

$$LS = k \sum_i C_i M_i \qquad (4\text{-}24)$$

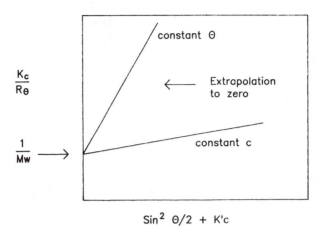

Figure 4-3 Outline of the Zimm plot for the extrapolation of light-scattering data.

Therefore,

$$\frac{\text{LS}}{C} = \frac{k \sum\limits_i C_i M_i}{\sum\limits_i C_i} = \frac{k \sum\limits_i g_i M_i}{\sum\limits_i g_i} \qquad (4\text{-}25)$$

and hence light-scattering measurements give a *weight-average* molecular weight.

SEDIMENTATION METHODS

There are two types of sedimentation methods commonly employed to study the molecular weight of a macromolecule: sedimentation velocity methods and sedimentation equilibrium methods. In each it is necessary to experimentally follow the solute protein during its sedimentation in an ultracentrifuge. In sedimentation velocity the rate of sedimentation is followed, while in sedimentation equilibrium the concentration gradient of the sedimenting macromolecule produced by a variety of opposing forces during centrifugation is established. As is inherently to be expected, sedimentation equilibrium uses lower centrifuge speeds than sedimentation velocity experiments. In most cases the sedimentation of the protein is experimentally followed by absorbance measurements. Although the rigorous derivation of the equations describing the behavior of a macromolecule during sedimentation is beyond the scope of this chapter, it is informative to examine some of them.

Sedimentation Velocity Experiments

In a two-component system containing the solvent (1) and the solute (2), a flux, J_2, of the macromolecule at a particular point X can be described by

$$J_2 = L_{22}\left(W^2 X - \frac{d\bar{\mu}_2}{dX}\right) \qquad (4\text{-}26)$$

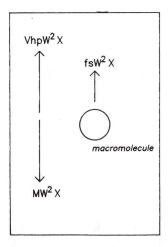

Figure 4-4   Forces on a macromolecule in an ultracentrifuge.

where L_{22} is a phenomenological constant, W the angular velocity, μ_2 the chemical potential of the solute, and X the radius to the point being considered (see Fig. 4-4). The forces due to radial acceleration and diffusion (as the result of a chemical potential gradient) are described respectively by the first and second terms within the parentheses. If there is no diffusion the applied force in the ultracentrifuge is W^2X and produces a constant velocity, V, which is a function of the size and shape of the sedimenting solute molecule. The proportionality constant is defined as

$$S = \frac{V}{W^2X} = \frac{1}{W^2X}\frac{dX}{dt} \tag{4-27}$$

and is known as the sedimentation coefficient. When diffusion occurs, the flux, J_2, becomes

$$J_2 = W^2XSC_2 - D\frac{dc_2}{dX} \tag{4-28}$$

where D is the diffusion constant, and we can now examine the relationship between S and the molecular properties of the solute. As illustrated in Fig. 4-4, there are three forces on the hydrated protein molecule during centrifugation.

The total acceleration force, F_a, on a single molecule is

$$F_a = \frac{M}{N_0}(1 + d_1)W^2X \tag{4-29}$$

where d_1 is the hydration of the molecule, M its molecular weight, and N_0 is Avogadro's number. The resultant acceleration is opposed by a frictional drag force, F_f, and a buoyant force, F_b, which is proportional to the mass of solvent, m_s, displaced by the protein molecule

$$F_f = -fSW^2X \tag{4-30a}$$

and

$$F_b = -V_h p W^2 X \tag{4-30b}$$

where V_h is the hydrated volume of the macromolecule and p is the solvent density.
In the steady state, the sum of these three forces is zero, and

$$\frac{M}{N_0}(1 + d_1 - V_2 p - d_1 V_1 p) = S'f \tag{4-31}$$

where f is the frictional coefficient of the macromolecule and V_h in the expression for F_b has been substituted for by

$$V_h = \frac{M}{N_0} V_2 + d_1 V_1 \tag{4-32}$$

where V_2 is the specific volume of the protein and V_1 is the partial specific volume of the solvent. Since in dilute solution $V_1 \cong 1/p$, we get, after rearranging,

$$S = \frac{M(1 - V_2 p)}{N_0 f} \tag{4-33}$$

Combining this equation with

$$D = \frac{kT}{f} \tag{4-33a}$$

for the diffusion, D, we get

$$M = \frac{SRT}{D(1 - V_2 p)} \tag{4-34}$$

which is the *Svedberg equation*.

So far we have assumed that the sedimentation coefficient (S) and the diffusion coefficient (D) are independent of concentration. In reality this is not true, and for accurate molecular weight determination it is necessary to extrapolate values of S and D to infinite dilution.

The Svedberg equation does not, however, allow the molecular weight of a protein to be calculated in the absence of knowledge of the diffusion and the partial specific volume. If the protein is spherical, it has been shown that S is proportional to $m_i^{2/3}$, which allows the molecular weight of an unknown protein to be estimated by determination of S if the sedimentation coefficient of a protein with known molecular weight is also determined. This approach depends on both the known and unknown proteins having spherical shapes and similar hydrations and partial specific volumes.

Equilibrium and Approach to Equilibrium Sedimentation

If centrifugation is performed using lower forces, a somewhat different situation holds. Rather than all of the material being transported to a packed band at the bottom of the cell, an equilibrium is set up.

At equilibrium, $J_2 = 0$, and using Eqs. (4-33) and (4-33a) for S and D, one can write

$$\frac{C_2 W^2 X M(1 - V_2 p)}{N_0 f} = \frac{kT}{f} \frac{1 + d(\ln \gamma_2)}{d(\ln C_2)} \frac{dC_2}{dX} \tag{4-35}$$

where C_2 is the concentration of the macromolecule and γ_2 is the velocity gradient. This rearranges to yield

$$M \left[\frac{1 + d(\ln \gamma_2)}{d(\ln C_2)} \right]^{-1} = \frac{2RT}{(1 - V_2 p)W^2} \frac{d(\ln C_2)}{dX^2} \tag{4-36}$$

At low solute concentrations (conditions where the activity coefficient term disappears) this equation can be integrated using as a boundary condition the solute concentration at a reference point X_0, to give

$$C_2(X) = C_2(X_0) \exp \left[\frac{M(1 - V_2 p)W^2}{2RT(X^2 - X_0^2)} \right] \tag{4-37}$$

suggesting that a plot of $\ln C_2$ versus X^2 gives as its slope the molecular weight. In practice, the absolute solute concentration gradient is hard to determine, and a form of this equation that employs concentration ratios is usually used:

$$\frac{C_b - C_m}{C_0} = \frac{M(1 - V_2 p)W^2(X_b^2 - X_m^2)}{2RT} \tag{4-38}$$

where C_0 is the uniform initial concentration and C_b and C_m are the equilibrium concentrations at the bottom and at the meniscus, respectively. Again, these ratios cannot be calculated unless the absolute concentration is known at some point for calibration. This is often overcome by using a rotor speed sufficiently high that at equilibrium the meniscus has zero solute concentration, allowing a reference point for detection of the concentrations. This approach is the *Yphanti's meniscus depletion method*.

In an alternative approach, Archibald noted that under an approach to equilibrium conditions, the flux, J_2, at the meniscus and at the bottom of the cell must be zero, hence

$$\frac{W^2 S}{D} = \frac{dC_2}{dX} \frac{1}{XC_2} \qquad \text{when } X = X_b \text{ or } X_m \tag{4-39a}$$

$$M_m = \frac{RT}{(1 - V_2 p)W^2} \left(\frac{dc_2}{dx} \right)_{X_m} \frac{1}{X_m C_m} \tag{4-39b}$$

$$M_b = \frac{RT}{(1 - V_2 p)W^2} \left(\frac{dc_2}{dx} \right)_{X_b} \frac{1}{X_b C_b} \tag{4-39c}$$

where C_m and C_b are the solute concentrations at the meniscus and at the bottom of the cell, respectively. The *Archibald method* gives two estimates of the molecular weight from a single experiment: these estimates for a homogeneous solution are equal, and this approach is useful for establishing homogeneity.

The weight concentration of solute is used in the derivations described previously and as a result, similar to the discussion on light scattering, it is evident that sedimentation approaches give a weight-average molecular weight.

Problem of Partial Specific Volume

From the previous sections it can be seen that many of these methods depend on a knowledge of the partial specific volume, V_2, of the protein. The equations developed indicate that a factor $(1 - V_2)$ is usually involved, meaning that an error in V_2 results in a *larger* error in the molecular weight estimate. Since V_2 is usually of the order of 0.7, the estimate of molecular weight is subject to about three times the error of the estimate of V_2.

The partial specific volume of a protein can be obtained in one of three ways:

1. *From Density Measurements*: The partial specific volume can be experimentally estimated by measuring the density of a series of solutions with different weight fractions of protein solute,

$$V_2 \cong \frac{1}{p_0}\left(1 - \frac{p - p_0}{wp}\right) \tag{4-40}$$

where w is the weight fraction of the solute, p_0 the solvent density, and p the measured density in the presence of solute.

2. *From Amino Acid Composition*: Estimates of V_2 from amino acid composition are based on the equation

$$V_2 = \sum W_i V_i \tag{4-41}$$

where W_i is the weight fraction of each type of amino acid present in the protein and V_i is the partial specific volume of the individual amino acids. Although this calculation takes no account of volume changes between free amino acids in aqueous solvent compared to that which might pertain to the environment of the residue in the protein, or to volume changes that might result from specific interactions of amino acid side chains in a protein, or to changes that might result from electrostatic interactions, the method appears to work quite well.

3. *From Alternate Solvent Measurements*: This approach is based on the use of H_2O and D_2O as solvents in sedimentation equilibrium experiments. From earlier,

$$M_i(1 - V_2 p_{H_2O}) = \frac{2RT}{w^2}\left(\frac{d \ln C}{d^2}\right) \tag{4-42}$$

when the experiment is run with H_2O as solvent. If the protein is dissolved in D_2O, two events take place: (a) the molecular weight of the protein is increased by deuterium exchange and the new molecular weight is related to the molecular weight in H_2O by the ratio

$$k = \frac{M_{i_{D_2O}}}{M_{i_{H_2O}}} \tag{4-43}$$

and (b) the partial specific volume of the protein is decreased by the same relative amount. As a result,

$$k_M = \left(1 - \frac{V_2}{k} p_{D_2O}\right) = \frac{2RT}{w^2}\left(\frac{d \ln C}{d^2}\right) \tag{4-44}$$

These equations can be solved simultaneously to yield

$$V_2 = \frac{k - [(d \ln C/d^2)_{D_2O}/(d \ln C/d^2)_{H_2O}]}{p_{D_2O} - p_{H_2O}[(d \ln C/d^2)_{D_2O}/(d \ln C/d^2)_{H_2O}]} \tag{4-45}$$

To estimate V_2 all that is required is sedimentation equilibrium data with H_2O and with D_2O as solvent, and a value for k: k can be quite reliably estimated from a knowledge of the exchangeable protons in the protein, which are the one amide hydrogen per residue and to a minor extent some side-chain protons. In pure D_2O, $k = 1.0155$ for proteins in general. If the solvent contains D_2O at lower percentages, the value of k is reduced proportionally.

As discussed previously, the accuracy of molecular weight estimates obtained from techniques that require a knowledge of the partial specific volume of the protein depends on the accuracy of the partial specific volume. Most proteins that contain only amino acids (i.e., excluding glycoproteins, etc.) have partial specific volumes between 0.69 and 0.75 cm^3 g^{-1}. If we assume an average value of 0.72 cm^3 g^{-1}, the maximum error for V_2 can be 4.2%, which gives a potential error in the molecular weight estimate of approximately $\pm 12\%$.

GEL FILTRATION METHODS

In Chap. 2 we discussed the basic principles of gel filtration and some of the fundamental equations used to describe the elution behavior of proteins in a gel filtration experiment. The total volume, V_t, of a gel filtration column is expressed by

$$V_t = V_0 + V_{\text{gel matrix}} + V_s \tag{4-46}$$

When a solute is introduced to the column it partitions between the internal and external solvent regions and the distribution can be described by a partition coefficient (PC): The solute mass, SM, found in the internal regions is

$$SM = PC \cdot V_s \cdot C \tag{4-47}$$

where C is the solute concentration in the external regions. In terms of the elution volume of the solute, V_e,

$$V_e = V_0 + PC \cdot V_s \qquad (4\text{-}48)$$

when a solute is *totally* excluded from the gel phase, $PC = 0$ and $V_e = V_0$. When the solute can diffuse freely with no size restrictions in the gel phase, $PC = 1$ and $V_e = V_0 + V_s$.

The elution position, V_e, can also be written in terms of K_{av} by rearranging Eq. (2-4):

$$K_{av} = \frac{V_e - V_0}{V_t - V_0} \qquad (4\text{-}49)$$

which yields

$$V_e = V_0 + K_{av}(V_t - V_0) \qquad (4\text{-}50)$$

For a particular gel filtration column,

$$PC = -A \log M + B \qquad (4\text{-}51)$$

where A and B are constants according to the nature of the gel filtration material and column size and are established for a particular column by measuring PC for a series of proteins with known molecular weights. As discussed in Chap. 2, K_{av} is usually used in place of PC in such experiments. Figure 4-5 is a plot of K_{av} versus $\log M$ for a series of standard proteins. The molecular weight of an unknown protein

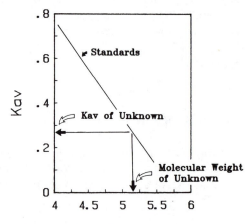

Figure 4-5 Relationship between K_{av} and log molecular weight for molecular weight standards in a gel filtration experiment. Standard proteins used were: ribonuclease (13.7 kD), chymotrypsinogen (25 kD), ovalbumin (43 kD), albumin (67 kD), aldolase (158 kD), catalase (232 kD), ferritin (440 kD), and thyroglobulin (669 kD).

is estimated by determining its V_e, and hence K_{av}, and using a calibration plot from the same column.

Although these equations fit experimental data for a reasonable number of proteins, a better fit to experimental data is obtained if the molecular weight in these equations is replaced by the effective hydrated radius, R_h.

$$PC = -A' \log R_h + B' \tag{4-52}$$

From this it is easy to understand why very large or small proteins deviate from the earlier relationship—they have hydrated radii that are not strictly related to their molecular weights. Similarly, glycoproteins or very asymmetric molecules also deviate from the earlier relationship because of "anomalous" effective hydrated radii.

The molecular weight of proteins can be expressed in terms of the molecular radius, a, as

$$M = Ka^p \tag{4-53}$$

where p varies with the shape of the protein. These relationships form the basis of using gel filtration data to determine shape parameters for proteins, as discussed in more detail in Chap. 13.

POLYACRYLAMIDE GEL ELECTROPHORETIC METHODS

Polyacrylamide gel electrophoresis (PAGE), because of its experimental ease, has become one of the commonest ways of determining the molecular weight of a protein. Because in general one can directly visualize protein after staining of a gel, it is also an excellent method for establishing a level of purity for a sample. Two types of PAGE are most often used: native and SDS-PAGE, which are based on different parameters and assumptions.

Native PAGE

Three parameters influence the movement of a protein in a native PAGE experiment: the charge on the protein (Q), the electric field (E), and the frictional coefficient (f). The charge on the protein depends on its pH and the pI, the electric field is experimentally set for a particular experiment, and the frictional coefficient depends on the pore size of the gel and the size and shape of the protein. The limiting velocity for movement of a particular protein in an experiment is $Q(E/f)$. The total gel concentration, T (which is the sum of the acrylamide and cross-linker concentrations), affects the porosity (and hence frictional effect) of the gel. Using a *relative* mobility, R_f, the mobility of the protein relative to some small molecule that encounters no sieving effect, we can use the *Ferguson equation*,

$$\log R_f = -K_r T + \log R_0 \tag{4-54}$$

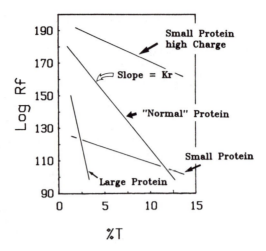

Figure 4-6   Ferguson plot. Typical lines are shown for the various types of protein referred to in the text.

where R_0 is the relative mobility of the protein in the absence of sieving (frictional) effects and K_r is the retardation coefficient. These two parameters are determined experimentally by determining R_f at a series of different gel concentrations and plotting a *Ferguson plot* (Fig. 4-6).

Different overall ranges of slopes of the Ferguson plot are expected for proteins of various types. In Fig. 4-6, line 1 typifies a large protein subject to considerable molecular sieving. Line 2 would be expected for the opposite extreme, a protein with low molecular size. Line 3 would be obtained for a protein of similar size to line 2 but with a much higher charge. Line 4 represents a typical medium-sized protein of normal charge.

The molecular weight of an unknown protein is obtained from the empirical linear relationship between K_r and molecular weight, M_i (see Fig. 4-7):

$$K_r = A + BM_i \qquad (4\text{-}55)$$

As with gel filtration, A and B are system-dependent parameters determined by using proteins of known molecular weight. As with any method that depends on the properties of standards, the validity of a molecular weight determined by this method is based on the unknown protein having molecular characteristics (i.e., partial specific volume, shape) similar to the standard proteins.

SDS-PAGE

The same fundamental parameters (Q, E, and F) influence polyacrylamide gel electrophoresis in the presence of the anionic detergent sodium dodecylsulfate (SDS) as in its absence. The difference between the two approaches is that several variables

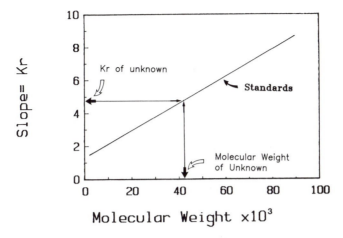

Figure 4-7 Determination of molecular weight from plots of the Ferguson plot versus molecular weight of the standard proteins. The molecular weight of an unknown protein is obtained by extrapolation of its Ferguson plot slope to intersect with the standard plot.

that can affect these parameters are more defined when SDS is present. It has been observed experimentally that most proteins bind a uniform amount of SDS per gram of protein (1.4 g) and that in the presence of a reducing agent (to break disulfide bonds), proteins tend to assume a uniform, "extended-rod" shape when SDS is present. As a result, proteins in the presence of SDS have the same charge per unit mass (and approximately the same charge per unit length). The result is that during electrophoresis only sieving effects are important in determining the mobility of a particular protein (as illustrated in Fig. 4-8). As a result, a plot of R_f versus log M_i can be used directly to give an estimate of the molecular weight of an unknown protein. The determinations based on SDS-PAGE depend on the validity of the assumptions made concerning the amount of SDS bound and the shape of the resultant denatured protein, and factors that affect these (such as glyco or lipo conjugates or a preponderance of basic or acidic side chains in the protein) lead to anomalous molecular weight estimates.

Native versus Denatured Molecular Weights. With several of the techniques that have been discussed here, a choice can be made between determining the molecular weight under native conditions or under denaturing conditions. A comparison of molecular weights determined under native or denaturing conditions can give significant information regarding any subunit structure the protein may have. In particular, sedimentation and gel filtration methods lend themselves to a direct comparison of native and denatured molecular weights since the appropriate solvent systems can readily be used. SDS-PAGE gives a denatured molecular weight of course, but since for the most reproducible results a reducing agent such as β-mercaptoethanol

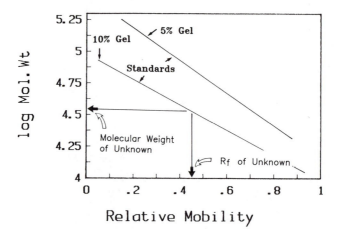

Figure 4-8   Calibration plots of molecular weight standards in SDS-PAGE: the effects of acrylamide concentration. In line 1 a 10% gel is used with molecular weight standards ranging from 14 kD (lysozyme) to 68 kD (bovine albumin). In line 2 a 5% gel is used, with molecular weight standards ranging from 29 kD (carbonic anhydrase) to 210 kD (myosin).

is used, this method does not give information concerning subunit molecular weights where the subunit contains two polypeptide chains covalently linked by one or more disulfides. Although it is possible to run SDS-PAGE in the absence of a reducing agent, it should be kept in mind that polypeptide chains that contain intra-peptide disulfide bonds run anomalously faster than if the intra-peptide disulfide is reduced. This is because such a molecule does not form an extended rod of the same size in the presence of SDS as the comparable reduced molecule and has a resultant smaller radius and value for f. The result is that such a protein has a high R_f value in a particular system. It is this that causes a protein with intra-peptide disulfide bonds to run above the diagonal in a two-dimensional diagonal map used to detect proteins with inter-polypeptide chains (see Fig. 4-9).

Determination of the Molecular Weight of Active Species. A frequently encountered problem in determining molecular weights is the question of the minimum molecular weight of a protein's active species. This is particularly difficult with a subunit containing protein, where the oligomer may itself undergo an association reaction that could possibly affect its activity. The problem is best illustrated by an example: Glutamate dehydrogenase is a hexamer that undergoes a concentration and regulatory ligand-dependent polymerization. The enzyme can utilize either glutamate or norvaline, and it was suggested that the hexamer preferentially uses glutamate, whereas norvaline is utilized only by higher polymers. The problem was resolved by determining the molecular weight of the active enzyme while it was catalyzing the oxidative deamination of either glutamate or norvaline, and showing that the molecular weights of the active species are in fact the same.

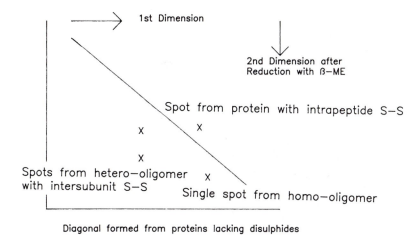

1st Dimension

2nd Dimension after
Reduction with β—ME

Spot from protein with intrapeptide S—S

X   X

X

X

Spots from hetero—oligomer   X
with intersubunit S—S   Single spot from homo—oligomer

Diagonal formed from proteins lacking disulphides

Figure 4-9   Schematic representation of a two-dimensional SDS-PAGE of disulfide-containing proteins.

Several of the methods for determining molecular weights lend themselves to the question: What is the molecular weight of the catalytically active species of an enzyme? Sedimentation methods have been used in what is known as *reactive enzyme centrifugation*. The protein is in effect sedimented through a solution containing all the necessary substrates for reaction to take place and the rate of sedimentation followed by following a reaction product rather than the protein. Thus the rate of sedimentation of the active enzyme is followed. Gel filtration methods have also been employed to allow determination of V_e for the reactive species of the enzyme and hence its molecular weight. The gel filtration matrix is equilibrated with buffer containing all the necessary substrates prior to the protein sample being loaded. V_e is determined by measuring the point at which reaction product is first detected in the eluent.

If one considers the basis of each of these techniques, it should be readily apparent that in circumstances where more than one molecular form of the protein is active, a molecular weight of the *smallest* active species is obtained from sedimentation methods, while the molecular weight of the *largest* active species is obtained using the gel filtration approach.

5

Determination of the Primary Structure of Proteins and Peptides

INTRODUCTION

The determination of the primary structure (sequencing) of a polypeptide chain by chemical means represents a major investment of time for all except relatively short polypeptide chains. With small proteins it is possible that the complete sequence can be established at least semiautomatically using a "sequenator": With larger proteins, proteins containing disulfides (either internal or between subunits), or proteins with covalent substituents such as glycoproteins, the straightforward approach is not always possible. In such instances it is necessary to fragment the polypeptide chain in some defined and controllable way, purify the fragments, and characterize them in terms of their amino acid sequences. Once the primary structure of each fragment is obtained the fragments must be aligned in the correct order to give the linear sequence of the protein. The cysteine residues involved in disulfide bonds must be matched and a number of other assignments, such as the position of carbohydrate or lipid substituents, must be made, as well as the positions of amides determined. Each one of these aspects represents separate challenges to the protein chemist.

Although many proteins have been sequenced using the approaches described in this chapter, the main aim is not to prepare the protein chemist to sequence a polypeptide chain completely since, as outlined briefly at the end of the chapter, there are faster approaches available using the tools of molecular biology. The approaches described are aimed at another problem, that of the isolation and identification of a particular peptide in a protein. A frequently encountered problem is identifying a peptide containing a residue that may have been specifically modified by a chemical reagent or which might contain a disulfide bond. If the primary sequence of the protein is known (which increasingly *is* the case), the peptide can be identified by establishing the sequence of a relatively small number of amino acids in the fragment (usually four or five residues are sufficient). Of course, prior to sequencing the fragment, it must be (1) generated and (2) purified, hence the need for an understanding of the topics covered in this chapter. Some of the techniques discussed here involving the chemical modification of particular amino acid side chains are discussed in detail in Chap. 7.

Before proceeding, we must digress briefly to discuss the problem of the purity of fragments, which must be achieved prior to attempting sequence work. If we consider a hypothetical situation where the yields of peptides obtained in purification may vary between 5 and 90%, we can discuss the problem of contaminating proteins. During peptide purification it is a natural assumption that the peptides present in the highest amounts are those that arise from the intended protein rather than from impurities. If, however, we have a 10% contaminant in the original protein preparation (which could easily escape detection), and after fragmentation we isolate a peptide from this contaminant with a 90% yield, we may overlook a peptide from the real protein which was isolated in only 5% yield and consider it as a minor contaminant, or isolate both peptides, sequence them, and assume that both arise from the real protein. Obviously, under such conditions small amounts of contaminating protein can be a real problem. If we are attempting to isolate a specific peptide that has been labeled with a site-specific modification reagent, it is less likely that such a problem will be encountered unless there is a significant amount of nonspecific labeling. Even in such favorable circumstances, however, contaminating proteins can hinder the purification of the desired peptide: Simply by being there they will increase the work (and probably decrease the yield) of purifying the wanted peptide.

FRAGMENTATION METHODS

There are two different approaches that can be used to cleave a polypeptide chain at fairly defined positions. These involve chemical and biochemical methods using specific proteolytic enzymes. The phrase "fairly defined" implies that the cleavage specificities of both approaches may vary with the particular protein. In this section we discuss the generalities of the specificity. In the context of using proteolytic enzymes, the question of whether to perform the cleavage under conditions where the protein is in its native state or has first been denatured must be addressed. In some cases limited proteolysis of a native protein can give considerable useful information, as emphasized in Chap. 10.

Prior to attempting cleavage with any of these various reagents, another question must be addressed which pertains to disulfide bonds. There are two problems encountered: (1) Do you want to leave them intact during fragmentation, and (2) can you prevent spurious disulfide bonds forming as a result of oxidation events occurring either during fragmentation or subsequent peptide purification? Usually, sulfide bonds that exist in the protein are reduced and alkylated prior to fragmentation—overcoming problem 2. Of course, when one is attempting to establish the location of disulfide bonds, cleavage is performed without first reducing the disulfides. Any free sulfhydryls are usually blocked from reaction by prior alkylation of the denatured protein.

Chemical Methods

Although a number of quite esoteric chemical methods of cleaving polypeptide chains at specific sites have been described, the ones mentioned here are those most commonly encountered.

Cyanogen Bromide Cleavage. Cyanogen bromide cleaves polypeptide chains at methionyl-X peptide bonds via the mechanism shown in Fig. 5-1. The reaction is usually performed in a 70% formic acid or trifluoroacetic acid to water (v/v) mix with a 50- to 100-fold molar excess of cyanogen bromide to methionine. The total reaction time is usually 24 hours and in many cases the cyanogen bromide is added in two batches during the incubation.

While cyanogen bromide cleaves at methionine-X peptides, the nature of X can influence the rate of cleavage. In particular, the hydroxyl-containing side chains of serine and threonine interfere by attacking the iminolactone intermediate of Fig. 5-1, leading to the conversion of methionine to homoserine lactone without the cleavage of the polypeptide chain (see Fig. 5-2). However, the use of high CNBr-to-methionine ratios (up to 500:1) overcomes this effect and cleavage of Met-Ser or Met-Thr up to 80% is usually achieved under the experimental conditions described.

The acid conditions employed in CNBr cleavage can cause cleavage of other peptide bonds in the protein, especially Asp-Pro bonds, and if a particular sequence contains this bond, more fragments than would be predicted based on the number of methionine residues may be obtained.

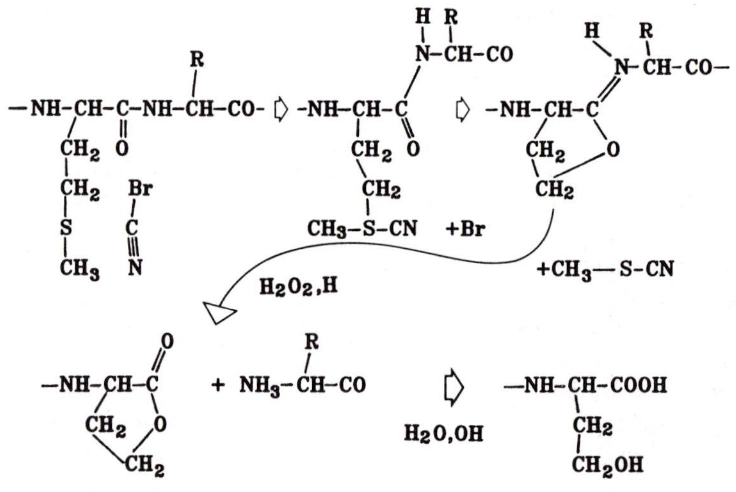

Figure 5-1 Mechanism of cleavage of methionyl-X bonds by cyanogen bromide.

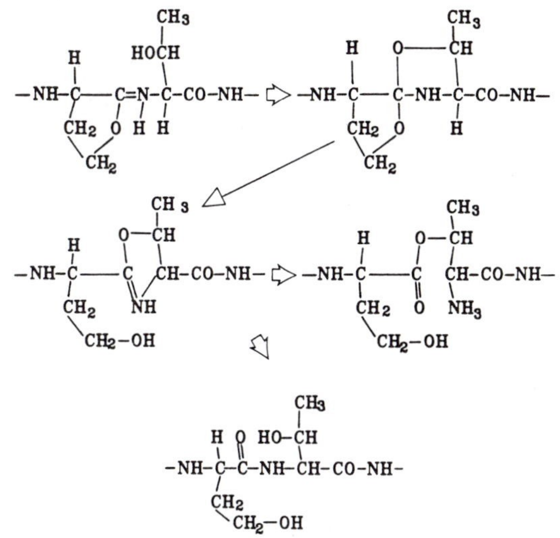

Figure 5-2   Interference of cleavage of methionyl-serine or methionyl-threonine bonds by hydroxyl attack on the iminolactone intermediate.

BNPS-Skatole. For a number of years brominating reagents in acidic media have been used to cleave polypeptide chains. Reagents such as *N*-bromosuccinimide will cleave polypeptides at a variety of sites, including tryptophan, tyrosine, and histidine, but often give side reactions which lead to insoluble products. BNPS-skatole [2-(2-nitrophenylsulfenyl)-3-methylindole] is a mild oxidant and brominating reagent that leads to polypeptide cleavage on the C-terminal side of tryptophan residues, as shown in Fig. 5-3.

Although reaction with tyrosine and histidine can occur, these side reactions can be considerably reduced by including tyrosine in the reaction mix. Typically, protein at about 10 mg/ml is dissolved in 75% acetic acid and a mixture of BNPS-skatole and tyrosine (to give 100-fold excess over tryptophan and protein tyrosine, respectively) is added and incubated for 18 hours. The peptide-containing supernatant is obtained by centrifugation.

Apart from the problem of mild acid cleavage of Asp-Pro bonds, which is also encountered under the conditions of BNPS-skatole treatment, the only other potential problem is the fact that any methionine residues are converted to methionine-sulfoxide, which cannot then be cleaved by cyanogen bromide. If CNBr cleavage of peptides obtained from BNPS-skatole cleavage is necessary, the methionine residues can be regenerated by incubation with 15% mercaptoethanol at 30°C for 72 hours.

Cleavage with o-Iodosobenzoic Acid. *o*-Iodosobenzoic acid cleaves tryptophan-X bonds under quite mild conditions. Protein, in 80% acetic acid containing 4 *M* guanidine hydrochloride, is incubated with iodobenzoic acid (approximately 2 mg/ml of protein) that has been preincubated with *p*-cresol for 24 hours in the dark at room temperature. The reaction (which can be terminated by the addition of dithio-erythritol) proceeds via the scheme shown in Fig. 5-4.

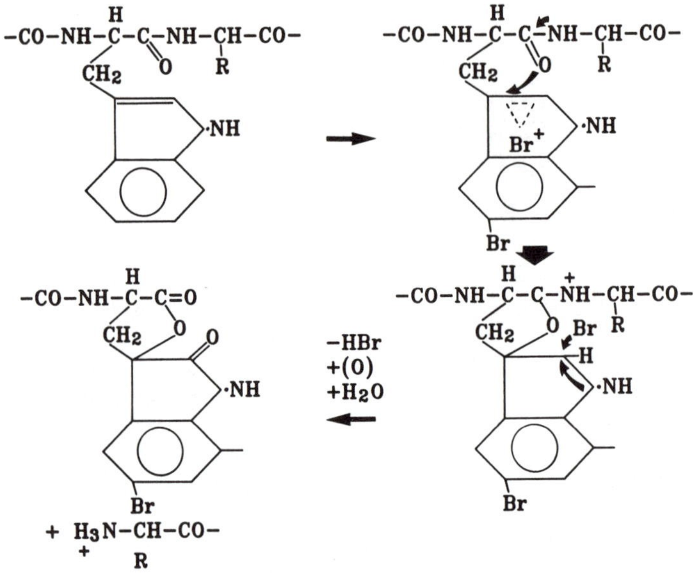

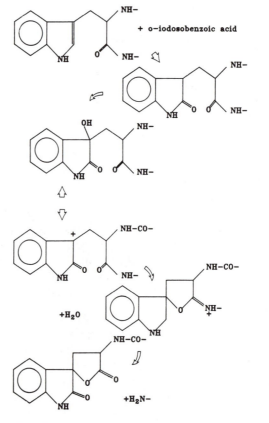

Figure 5-3 Cleavage of tryptophan-X peptide bond by brominating reagents such as BNPS-Skatole or N-bromosuccinimide.

Figure 5-4 Cleavage of tryptophan-X peptide bonds by O-iodosobenzoic acid.

Care must be taken to use purified *o*-iodosobenzoic acid since a contaminant, *o*-iodoxybenzoic acid, will cause cleavage at tyrosine-X bonds and possibly histidine-X bonds. The function of *p*-cresol in the reaction mix is to act as a scavenging agent for residual *o*-iodoxybenzoic acid and to improve the selectivity of cleavage.

Cleavage of X-Cysteinyl Bonds. Two reagents are available that produce cleavage of peptides containing cysteine residues. In both cases cleavage occurs on the amino-terminal side of the cysteine. The mechanisms of these reagents, (2-methyl)-*N*-1-benzenesulfonyl-*N*-4-(bromoacetyl)quinone diimide (otherwise known as Cyssor, for "cysteine-specific scission by organic reagent") and 2-nitro-5-thiocyanobenzoic acid (NTCB), are shown in Fig. 5-5.

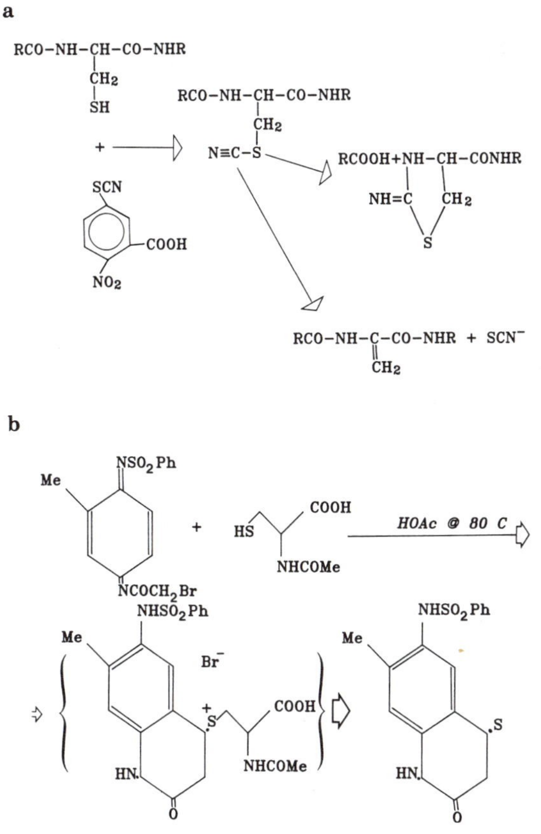

Figure 5-5. Cleavage of X-cysteinyl peptide bonds by reagents such as NTCB (a) or "Cyssor" reagents such as Cyssol-1 (b).

Hydroxylamine Cleavage. Hydroxylaminolysis slowly leads to cleavage of a number of peptide bonds; however, it has been found that the asparaginyl–glycine bond is particularly susceptible and cleavage can be achieved by the reaction outlined in Fig. 5-6. The reaction occurs by incubating protein, at a concentration of about 4 to 5 mg/ml, in 6 M guanidine hydrochloride, 20 mM sodium acetate + 1%

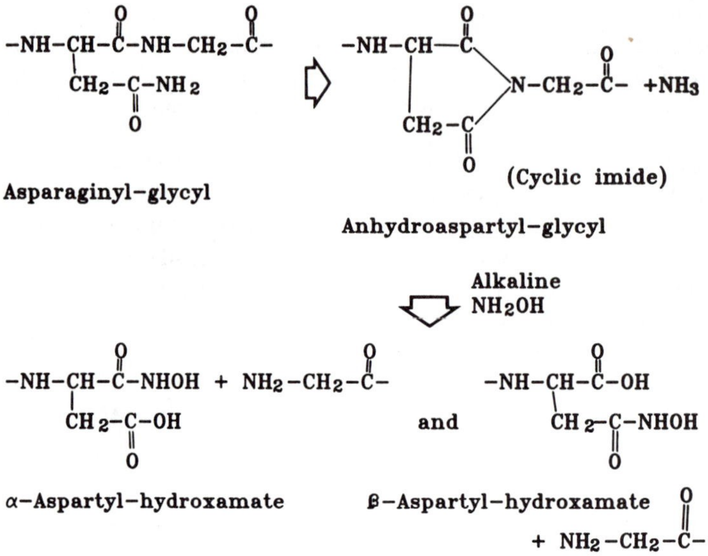

$$-NH-CH-\overset{\overset{O}{\|}}{C}-NH-CH_2-\overset{\overset{O}{\|}}{C}- \qquad\Rightarrow\qquad -NH-CH-\overset{O}{C}\diagdown$$

$$\underset{\overset{|}{CH_2-\underset{\overset{\|}{O}}{C}-NH_2}}{} \qquad\qquad\qquad \underset{CH_2-C\diagup}{\overset{|}{}}\overset{\diagup}{N-CH_2-\overset{\overset{O}{\|}}{C}-}\ +NH_3$$

Asparaginyl–glycyl $\qquad\qquad\qquad\qquad$ (Cyclic imide)

Anhydroaspartyl–glycyl

Alkaline
NH$_2$OH

$$-NH-CH-\overset{\overset{O}{\|}}{C}-NHOH\ +\ NH_2-CH_2-\overset{\overset{O}{\|}}{C}- \qquad -NH-CH-\overset{\overset{O}{\|}}{C}-OH$$

$$\underset{\overset{\|}{O}}{\overset{|}{CH_2-C-OH}} \qquad\qquad\qquad and \qquad\qquad \underset{\overset{\|}{O}}{\overset{|}{CH_2-C-NHOH}}$$

α–Aspartyl–hydroxamate $\qquad$ β–Aspartyl–hydroxamate

$$+\ NH_2-CH_2-\overset{\overset{O}{\|}}{C}-$$

Figure 5-6 Hydroxylamine cleavage of asparaginyl–glycine peptide bonds.

mercaptoethanol at pH 5.4, and adding an equal volume of 2 *M* hydroxylamine in 6 *M* guanidine hydrochloride at pH 9.0. The pH of the resultant reaction mixture is kept at 9.0 by the addition of 0.1 *N* NaOH and the reaction allowed to proceed at 45°C for various time intervals; it can be terminated by the addition of 0.1 volume of acetic acid. In the absence of hydroxylamine, a base-catalyzed rearrangement of the cyclic imide intermediate can take place, giving a mixture of α-aspartylglycine and β-aspartylglycine *without* peptide cleavage.

Cleavage of Asp-Pro Bonds. Peptide bonds containing aspartate residues are particularly susceptible to acid cleavage on either side of the aspartate residue, although usually quite harsh conditions are needed (Fig. 5-7). Asp-Pro bonds have been found to be susceptible under conditions where other Asp-containing bonds are quite stable. Suitable conditions are the incubation of protein (at about 5 mg/ml) in 10% acetic acid, adjusted to pH 2.5 with pyridine, for 2 to 5 days at 40°C. The enhanced reactivity of Asp-Pro bonds relative to other Asp-containing peptides under these conditions is presumably due to the greater basicity of the imino ring in the proline.

Enzymatic Methods

A wide variety of different endopeptides have been described and purified, and many have found use in the enzymatic fragmentation of polypeptide chains. Some of these proteases are specific for peptide bonds containing certain defined side chains on adjacent amino acids, and the more common are given in Table 5-1 together with their specificities.

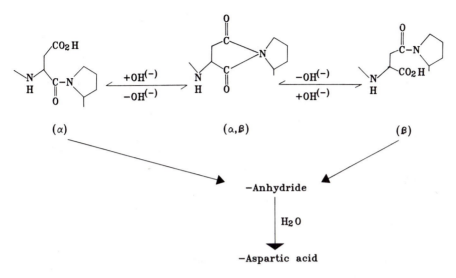

Figure 5-7 Hydrolysis of aspartyl–proline peptide bonds at acidic pH.

TABLE 5-1 Specificity of endopeptides used in fragmentation

Enzyme	B Site–C site specificity[a]	Other requirements
Trypsin	B = Lys or Arg	No hydrolysis if C = Pro; hydrolysis slowed if A or C are acidic
Chymotrypsin	Preferentially, B = Trp, Tyr, Phe; some hydrolysis if B = Met, Leu, or His	As with trypsin
Pancreatic elastase	B = Ala, Val, Gly, or Ser	
Thermolysin	Preferentially, C = Phe, Leu, Val, Tyr, Ile, Met, or Trp; some hydrolysis if C = Ala, Asn, Thr, or His	No hydrolysis with C = Gly or with D = Pro
Pepsin	B or C = Phe, Tyr, Trp, Leu (cleaves others slowly)	No cleavage if B = Pro
Subtilisin	B or C = hydrophobic	
Papain	Broad specificity	
Streptococcal proteinase	A, B = bulky (e.g., Phe, Tyr, Leu, His)	No cleavage if B = Gly
Staphylococcal protease	B = Glu or Asp	Hydrolysis slowed if C = hydrophobic; C = Pro does *not* prevent cleavage

[a] Specificities given in terms of a tetra-peptide A-B-C-D with *site* of cleavage between B and C.

The conditions used for enzymatic fragmentation are usually governed by the optimal activity of the protease. Many are quite stable to either somewhat elevated temperatures or the presence of mild denaturants. It is common practice when using proteolytic enzymes to fragment a polypeptide chain to include low concentrations of SDS. This breaks the native tertiary structure of the target protein, exposes susceptible bonds in a uniform manner, and leads to more rapid and uniform cleavage.

In many instances, the utility of a particular protease can be enhanced by using specific chemical modifications to alter the cleavage sites: With trypsin, for example, which has specificity for lysine and arginine side chains, the cleavage pattern can be altered by chemical modification of lysine side chains to block those sites. This process is outlined schematically in Fig. 5-8.

USE OF CHEMICAL MODIFICATIONS TO ALTER TRYPTIC CLEAVAGE PATTERNS

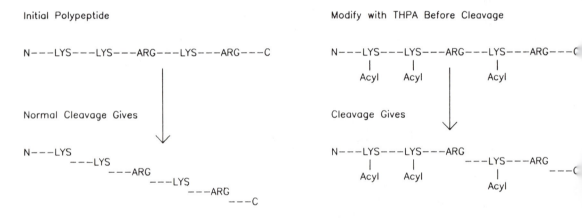

Figure 5-8 Modification of tryptic cleavage patterns by chemical modification of lysine side chains with tetrahydrophthalic anhydride (THPA).

In addition, reagents can be used that modify arginine and not lysine residues. If reversible modification reagents are selected, the blocked sites can be restored, thus allowing subsequent cleavage. With trypsin it is also possible to create additional cleavage sites if cysteine residues are present.

As shown in Fig. 5-9, the cysteine side chain can be chemically modified with aminoethylating reagents such as *N*-(β-iodoethyl)trifluoroacetamide or ethyleneimine, which produces *s*-(2-aminoethyl)cysteine, which is homologous to the side chain of lysine and is recognized by trypsin. Such manipulations can be quite useful in generating different types of fragments for use in establishing overlaps.

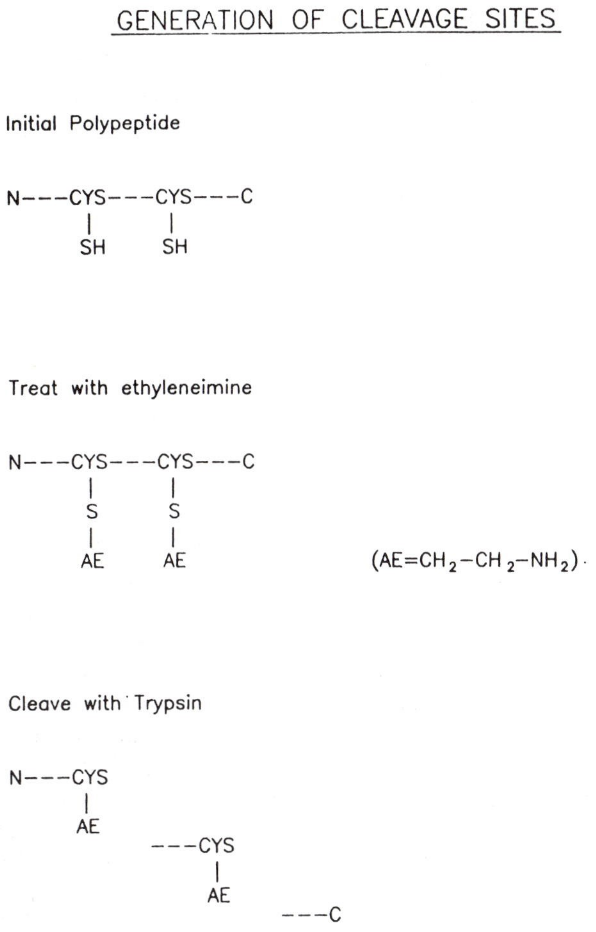

GENERATION OF CLEAVAGE SITES

Initial Polypeptide

```
N---CYS---CYS---C
      |      |
      SH     SH
```


Treat with ethyleneimine

```
N---CYS---CYS---C
      |      |
      S      S
      |      |
      AE     AE          (AE=CH₂−CH₂−NH₂).
```

$(AE = CH_2-CH_2-NH_2)$.

Cleave with Trypsin

```
N---CYS
      |
      AE
            ---CYS
              |
              AE
                  ---C
```

Figure 5-9 Creation of tryptic cleavage sites by modification of cysteine side chains with ethyleneimine.

ISOLATION AND CHARACTERIZATION OF FRAGMENTS

In many ways the problems encountered in the isolation of the polypeptide chain fragments are the same as those discussed in earlier chapters on protein purification: gel filtration, ion-exchange chromatography, HPLC (especially RP-HPLC), and even affinity chromatography are all employed to purify fragments. Polyacrylamide gel electrophoresis and paper electrophoresis or chromatography are also widely used. Although all of these approaches have been discussed in Chaps. 1 to 4, several general comments must be made. Preparative work with proteins usually involves buffered aqueous systems: When working with peptides, harsher conditions are often used. With peptides it is not necessary to maintain "native" structure and they (especially those from hydrophobic regions of a protein) may be quite insoluble in aqueous buffers. It is not unusual to use formic acid or acetic acid solvents for column chromatography of peptides or to use solvents with low dielectric constants for hydrophobic peptides. The solubility problem is also overcome in some cases by chemical modification of the peptide. With disulfide-containing peptides, sulfitolysis [Eq. (5-1)] is often used to enhance solubility by introducing a group with high polarity without leading to the oxidation of residues such as tryptophan or methionine.

$$R_1-S-S-R_2 + SO_3^{2-} \longrightarrow R_1-S^- + R_2-S-SO_3^-$$

then   (5-1)

$$R_1-S^- + SO_3^{2-} + 2Cu^{2+} \qquad R_1-S-SO_3^- + 2Cu^+$$

Another problem, which is somewhat different from those encountered in the protein purification, involves detection and quantitation of the fragments. Absorbance measurements at 280 nm are frequently used to follow elution of proteins from columns. However, with peptide fragments many do not contain tyrosine or tryptophan (the primary contributors to absorbance at 280 nm) and thus are not detected. Measurements taken at wavelengths where the peptide bond absorbs can be used, but quantitation is still a problem. Absorbance measurements also present a problem in terms of sensitivity, and for accurate work have largely been replaced by methods using fluorescence detection or radioactivity. Ninhydrin, long the standard chemical reagent of choice for peptide detection, has been replaced by reagents such as fluorescamine and o-phthalaldehyde. Both give fluorescent adducts of amino groups: fluorescamine has an excitation at 390 nm with emission at 475 nm, while o-phthalaldehyde gives a derivative with an excitation at 340 nm and emission at 455 nm.

Fluorescamine reacts with primary amino groups in peptides at pH 7.5 to 9.0 in aqueous solution to give fluorescent products, and the reaction can be carried out to monitor a column effluent. The method is readily adaptable to the detection of peptides on paper or thin-layer chromatograms or on electrophoretograms that are first treated with a triethylamine solution and then sprayed with a fluorescamine in acetone solution. After drying, peptides are detected by their fluorescence under a long-wavelength UV lamp. Peptides with proline as their amino terminus become observable only after heating at 110°C for 3 hr.

o-Phthalaldehyde is used in essentially the same way, with the advantage that it is more soluble and more stable in aqueous solutions. The most popular radioactive compound used for peptide detection is phenylisothiocyanate, which is discussed in more detail later. There are several preparative methods which find use in peptide purification that are somewhat unique to peptides.

Countercurrent Methods

Countercurrent methods are based on phase partition of peptides: A mixture in a particular solvent is partitioned using a nonmiscible second solvent. Separation is achieved if one or more peptides has a higher solubility in one of the phases than it has in the other: If the two phases are separated after mixing, peptides with higher solubility in, for example, the top phase are extracted into that phase, while other peptides with higher solubility in the bottom phase remain preferentially in that phase. If the process is now repeated—fresh top phase is added to the bottom phase, and vice versa—further extraction and enrichment takes place. This cycle can be repeated until, for all practical purposes, the peptides have been completely separated. The process is shown diagrammatically in Fig. 5-10 for an example where two

Outline of Counter Current Purification

Starting Mixture: 1:1 Mix   Purity of A=50%

Start with 100mg of each
Partition Coefficient(I/II) A = 4
                             B = 0.5

Dissolve in solvent I
Add immiscible solvent II

1st Cycle: at Equilibrium

80mg A	33mg B	I
20mg A	66mg B	II

Purity of A in top phase (I) = 71%
TAKE TOP PHASE: REPEAT

2nd Cycle

64mg A	11mg B
16mg A	22mg B

Purity of A =85%

3rd Cycle

51.2mg A	7.3mg B
12.8mg A	3.6mg B

Purity of A =88%

Similar procedure with bottom phase gives increased
purity of B.

Figure 5-10 Schematic outline of the separation of peptides by countercurrent purification.

peptides have quite different distribution coefficients between the two solvents. Although this example might be considered to represent an optimal situation, all that is necessary for separation to be eventually achieved is that the peptides to be separated have different distribution coefficients.

Diagonal Methods

The basis of diagonal methods was introduced in Chap. 4. The method depends on a two-dimensional electrophoresis or chromatography with both dimensions carried out under identical conditions except that they are rotated through 90° between the dimensions. If nothing is done to the mixture of peptides separated in the first dimension prior to the second, the peptides behave just as they did during the first dimension, giving a diagonal pattern. However, if between the two dimensions some of the peptides are altered in such a way that their mobility is different in the second dimension than in the first dimension, they do not fall on the diagonal and can thus be identified.

Previously, we discussed the use of diagonal methods for identifying proteins containing inter-subunit disulfide bonds, where reduction was employed between the dimensions. This approach is very useful with peptides for identifying disulfide-bonded fragments. This is not, however, the limit of the usefulness of diagonal methods in peptide purification and characterization. A number of treatments between dimensions have been developed that allow identification of a number of different peptides types. Most depend on specific chemical modification of certain amino acid side chains and detect peptides containing these amino acids.

Sulfhydryl-Containing Peptides. A mixture of peptides, some of which contain free sulfhydryl groups, is alkylated with *N*-ethylmaleimide, giving *N*-ethylsuccinimide cysteine derivatives that are quite stable at acid pH but readily hydrolyzed at alkaline pH. After electrophoresis or chromatography in the first dimension the paper is exposed to ammonia, which leads to hydrolysis and the generation of an additional negative charge at the succinimide derivative:

$$(5\text{-}2)$$

The generated *N*-ethylsuccinamic acid derivative moves differently in the second dimension than in the first and moves off the diagonal. If two-dimensional electrophoresis is used, the sulfhydryl-containing peptides move more toward the positive in the second dimension. In another approach involving cysteine residues, performic acid oxidation after the first dimension produces a cysteic acid derivative which electrophoreses more to the anodic side of the diagonal in the second dimension.

Methionine-Containing Peptides. If after separation in the first dimension the peptides are alkylated with iodoacetamide, sulfonium derivatives are formed from any methionine-containing peptides that migrate to the cathodic side of the diagonal during electrophoresis in the second dimension.

Lysine-Containing Peptides. If a protein *prior to fragmentation* is acylated with a reagent such as tetrahydrophthalic anhydride, negative charges are introduced onto lysine side chains and the amino-terminal residue (unless the protein has a blocked amino terminal). The protein is then fragmented and electrophoresed in the first dimension. Prior to the second dimension the pH is lowered to 4–5, which produces deacylation of the modified amino residues and removes negative charge from those peptides, leaving derivatives that electrophorese on the cathodic side of the diagonal in the second dimension. This approach identifies lysine-containing peptides *and* the amino-terminal peptide.

Identification of the C-Terminal Peptide after Tryptic Digestion. Unless the protein in question has a C-terminal lysine or arginine residue, all but one of the tryptic peptides of the protein will have a C-terminal lysine or arginine residue. As a result, treatment of these peptides with pancreatic carboxypeptidase B removes a positively charged residue from the C terminal, altering the electrophoretic mobility. If tryptic peptides, after electrophoresis in the first dimension, are subjected to carboxypeptidase B digestion prior to electrophoresis in the second dimension, all except the C-terminal peptide will have altered mobility. As a result, the only peptide remaining on the diagonal will be the C-terminal peptide.

Although each of the techniques described were designed for use in paper electrophoresis in two dimensions, they can be used with paper chromatography or with native PAGE. With paper chromatography the movement away from the diagonal of the appropriate peptides is not in a consistent direction; the direction away from the diagonal is determined by the solvent system used and the effects produced by the altered charge of the various peptides.

Amino Acid Analysis

The basic principles of amino acid analysis were described earlier and are not reiterated here. Several points need to be made, however, concerning the effects produced by various procedures that may be used during the generation or isolation of peptide fragments.

The sulfur-containing amino acids are oxidized during acid hydrolysis, and performic acid oxidation prior to acid hydrolysis gives stable derivatives of methionine, cysteine, or cystine. Methionine is converted to methionine sulfone, and cysteine or cystine are converted to cysteic acid.

A number of proteins contain derivatives of the normal amino acids, and two problems can arise from acid hydrolysis used to break down the peptide bonds in such fragments. If a fragment contains phosphoserine or γ-carboxyglutamic acid, the derivative is converted to the parent amino acid during acid hydrolysis. The second problem arises from the fact that some amino acid derivatives, such as 3-methyl histidine and ε-N-methyl lysine, although not broken down during acid hydrolysis, are poorly resolved from their parent amino acids. These derivatives behave very similarly to histidine and lysine, respectively, in most systems used to separate amino acids during analysis.

These problems have been overcome in a variety of ways. Analysis of γ-carboxyglutamate involves alkaline hydrolysis of the protein or peptide (2 M KOH, 110°C for 24 hours) prior to ion-exchange chromatography of the hydrolysate. Although γ-carboxyglutamic acid elutes quite differently from glutamate, it can coelute with other ninhydrin-positive material, making quantitation a problem. This is overcome by independently quantitating the amount of γ-carboxyglutamic acid: The γ proton readily exchanges in titrated water at *slightly* acid pH, allowing specific titration of the γ-carboxyglutamic acid in a protein. The proton is stable at pH 8 and above, and provided that subsequent steps are performed above pH 8, the specific activity of the titrated water allows quantitation. An alternative method is based on the altered electrophoretic mobility that a γ-carboxyglutamic acid–containing peptide has after exposure to acid conditions (50 mM HCl) and heat, which converts the acid to glutamate.

Resolution of methylated amino acids from their parent amino acids has been achieved by altering the temperature of the ion-exchange columns used in analysis. Figure 5-11 shows the separation of various standards, together with the separations achieved during analysis of hydrolysates of myosin containing mono- and trimethyl lysine and methyl histidine in small amounts.

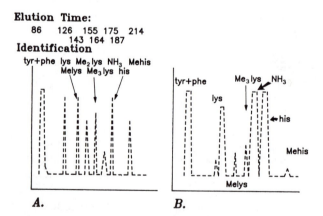

Figure 5-11   (A) Outline of the results expected for an HPLC separation of various standard amino acids and methylated derivatives of lysine and histidine by ion-exchange chromatography; (B) separation of hydrolyzates of myosin, indicating the presence of methylated lysine and histidine residues.

End-Group Analysis

The final step in the characterization of a peptide prior to sequencing is the determination of the end groups (C terminal and N terminal). In many cases one of the two terminal groups will have been determined by the cleavage method; for example, all tryptic fragments from a protein will have C-terminal lysine or arginine (with the exception of the C-terminal protein fragment). Terminal residue determination serves two functions: (1) it can establish purity, and (2) it can in some instances be sufficient to indicate which fragment has been isolated if the sequence of the protein is shown. Before discussing these two points in more detail, we will first examine some of the procedures commonly used to determine terminal residues.

Amino Terminus. There are four commonly used methods; all involve a chemical modification specific for free amino groups followed by hydrolysis and identification of the labeled amino acid. In all four cases some of the derivatives are not particularly stable to acid hydrolysis, and accurate quantitation requires time-course corrections where hydrolysis is carried out for three or four different time periods (usually ranging from 12 to 48 hours) and the analysis extrapolated to $t = 0$. The methods involve labeling with 2,4-dinitrofluorobenzene (DNFB), dansyl chloride, phenylisothiocyanate (PITC), or cyanate, and are shown schematically in Fig. 5-12. With DNFB, dansyl chloride, and PITC, chromophoric derivatives are obtained that are identified by TLC. The sensitivity of these methods can be increased by using radioactive DNFB, dansyl chloride, or PITC. With cyanate labeling, isotopically labeled cyanate is usually used.

With DNFB most amino acid derivatives are quite stable to acid hydrolysis, with the exception of serine and proline, which require time-course corrections. Dansyl proline is somewhat acid labile; however, a number of dansyl derivatives are difficult to resolve by TLC. Dansyl arginine, histidine, and cysteine can present problems. A typical TLC separation of dansyl amino acids is shown in Fig. 5-13. Resolution of some of the closer derivatives can be achieved by using different solvent systems.

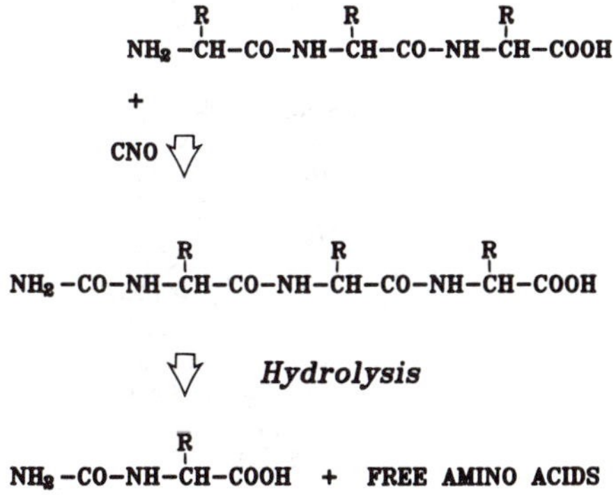

$$NH_2-\overset{\overset{\textstyle R}{|}}{CH}-CO-NH-\overset{\overset{\textstyle R}{|}}{CH}-CO-NH-\overset{\overset{\textstyle R}{|}}{CH}-COOH$$

$$+$$

$$CNO \;\triangledown$$

$$NH_2-CO-NH-\overset{\overset{\textstyle R}{|}}{CH}-CO-NH-\overset{\overset{\textstyle R}{|}}{CH}-CO-NH-\overset{\overset{\textstyle R}{|}}{CH}-COOH$$

$$\triangledown \quad \textit{Hydrolysis}$$

$$NH_2-CO-NH-\overset{\overset{\textstyle R}{|}}{CH}-COOH \quad + \quad \text{FREE AMINO ACIDS}$$

Figure 5-12A

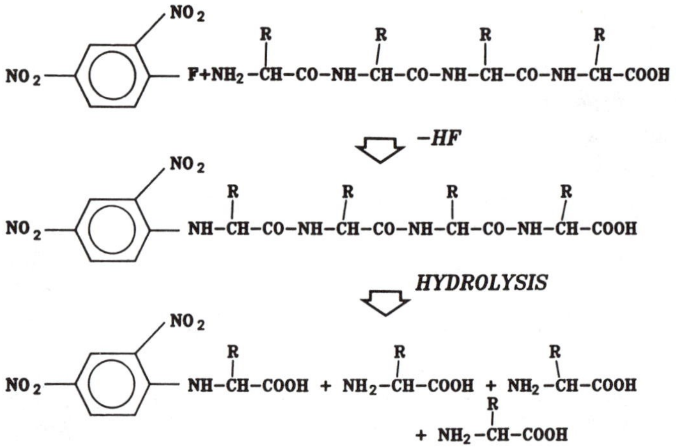

NO$_2$—⬡(NO$_2$)—F + NH$_2$—CH(R)—CO—NH—CH(R)—CO—NH—CH(R)—CO—NH—CH(R)—COOH

⬇ *−HF*

NO$_2$—⬡(NO$_2$)—NH—CH(R)—CO—NH—CH(R)—CO—NH—CH(R)—CO—NH—CH(R)—COOH

⬇ *HYDROLYSIS*

NO$_2$—⬡(NO$_2$)—NH—CH(R)—COOH + NH$_2$—CH(R)—COOH + NH$_2$—CH(R)—COOH

+ NH$_2$—CH(R)—COOH

Amino Terminal Residue Gives 2,4—Dinitrophenylamino acid
Other Residues Give Free Amino Acids

Figure 5-12B

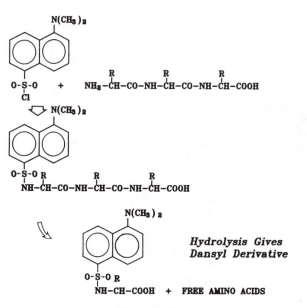

Hydrolysis Gives
Dansyl Derivative

Figure 5-12C

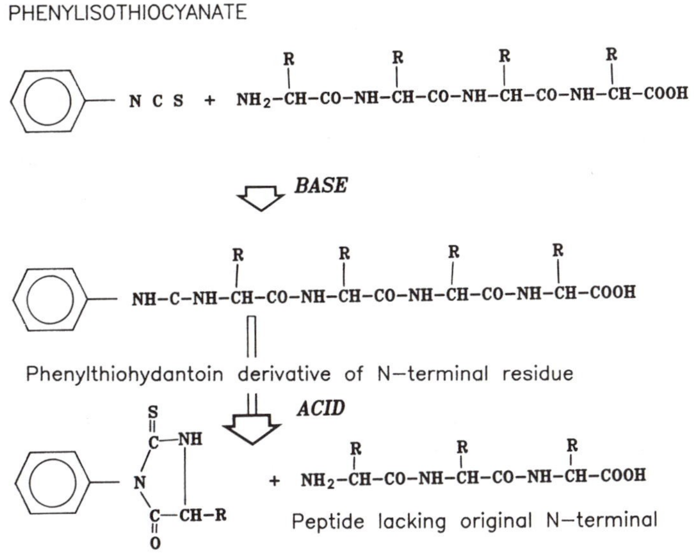

PHENYLISOTHIOCYANATE

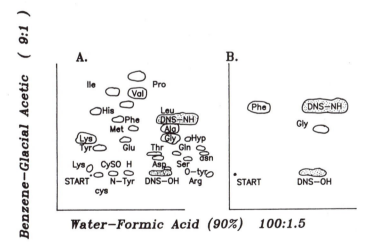

$$\text{NCS} + \text{NH}_2\text{-CH-CO-NH-CH-CO-NH-CH-CO-NH-CH-COOH}$$

BASE

$$\text{NH-C-NH-CH-CO-NH-CH-CO-NH-CH-CO-NH-CH-COOH}$$

Phenylthiohydantoin derivative of N−terminal residue

ACID

$$+ \quad \text{NH}_2\text{-CH-CO-NH-CH-CO-NH-CH-COOH}$$

Peptide lacking original N−terminal

Figure 5-12D

Figure 5-12 Chemistry of amino-terminal labeling by (A) cyanate, (B) 2,4-dinitro-fluorobenzene, (C) dansyl chloride, and (D) phenylisothiocyanate.

Figure 5-13 (A) Schematic separation of dansyl amino acids by two-dimensional thin-layer chromatography; (B) identification of the amino-terminal residues of insulin by hydrolysis after dansylation. In each case the dotted areas of the chromatograms indicate the by-products of the dansylation.

The cyanate method gives better accuracy than those using chromophore quantitation but needs large time-course corrections for serine and threonine.

Blocked Amino Terminals. In a number of proteins an attempt to use one of these methods to detect the amino-terminal residue of a *protein* will be met by failure. A number of proteins (e.g., cytochrome C, ovalbumin, bovine superoxide dismutase, various immunoglobulins) have blocked amino-terminal residues, where the group is blocked by some type of derivatization.

In many instances the blocking results from acylation of the α-amino group to give either acetyl or formyl derivatives. There are two basic ways to identify the blocking group in these cases: (1) acid hydrolysis releases the carboxylic acid of the blocking group, which can then be identified by gas chromatography, and (2) hydrazinolysis (see Fig. 5-11 for the chemistry of this process) gives the hydrazide derivative of the acyl group, which can then be identified chromatographically.

In some cases (e.g., certain immunoglobulins) intramolecular acylation occurs,

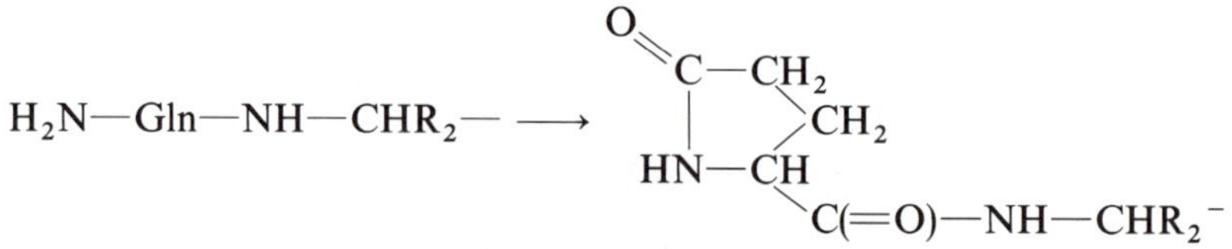

$$H_2N-Gln-NH-CHR_2- \longrightarrow \begin{array}{c} O \\ \diagdown \\ C-CH_2 \\ | \quad \diagup CH_2 \\ HN-CH \\ \diagdown \\ C(=O)-NH-CHR_2^- \end{array} \qquad (5\text{-}3)$$

giving a 2-pyrrolidone-5-carboxylic acid (PCA) derivative that can be demonstrated by alkaline cleavage of the PCA ring to give glutamic acid.

Carboxyl Terminus. Two chemical approaches to identifying the C-terminal residue of a peptide are often used: hydrazinolysis and tritium labeling. Hydrazinolysis, outlined in Fig. 5-14, leads to the hydrazide derivative of all the carboxyl functions originally in amide linkage: only the C-terminal residue remains as the free amino acid, which is thus identified. Unfortunately, if the C-terminal residue contains an amide in its side chain (i.e., asparagine or glutamine), it is not converted to its free

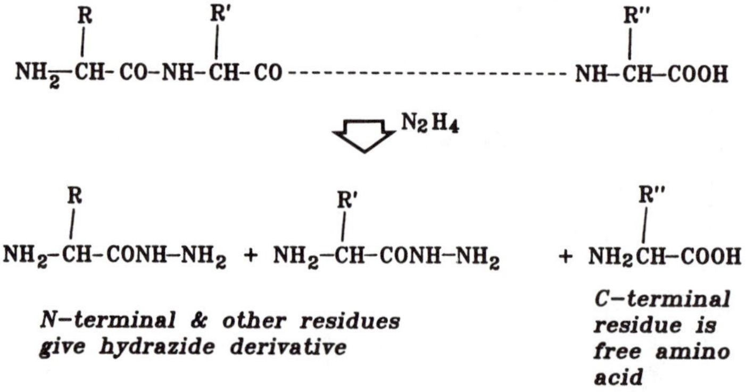

$$\begin{array}{ccccc} R & & R' & & R'' \\ | & & | & & | \\ NH_2-CH- CO-NH-CH- CO & \text{-----------} & NH-CH-COOH \end{array}$$

$$\Downarrow N_2H_4$$

$$\begin{array}{ccccc} R & & R' & & R'' \\ | & & | & & | \\ NH_2-CH-CONH-NH_2 & + & NH_2-CH-CONH-NH_2 & + & NH_2CH-COOH \end{array}$$

N−terminal & other residues give hydrazide derivative   *C−terminal residue is free amino acid*

Figure 5-14   Determination of carbonyl-terminal residue by hydrazinolysis.

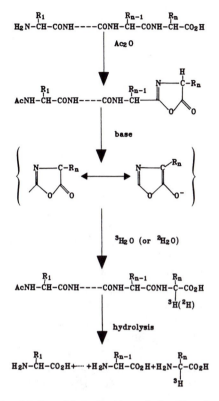

Figure 5-15   Tritium labeling of the carboxyl-terminal residue of a peptide or protein.

amino acid during hydrazinolysis. In addition, C-terminal arginine or cysteine are fairly unstable under hydrazinolysis conditions.

Tritium labeling of the C-terminal residue is a result of the lability of the hydrogen on the C-terminal residue α-carbon. This hydrogen is made particularly labile by the formation, via intramolecular cyclization of the free COOH with the C$=$O of the adjacent peptide bond, of an oxazolone (see Fig. 5-15). As a result, a base-catalyzed exchange of this proton occurs in the presence of tritiated water, and after acid hydrolysis the C-terminal residue is identified by its tritium content. Because of its lack of a proton on the α-carbon, C-terminal proline is not labeled in this manner.

In addition to these chemical methods, the C-terminal residue can also be identified enzymatically using carboxypeptidase. There are several types of carboxypeptidase available (see Table 5-2), and the ideal enzyme for use in C-terminal identification would hydrolyze *all* terminal residues with equal facility. Unfortunately, as shown in Table 5-2, this is not the case. Several steps can be taken to help remedy this: The enzymes can be used in combination or denaturants can be employed. The carboxypeptidases are active in denaturants such as dilute SDS or 6 *M* urea, and their presence makes the rate of release of different carboxy terminal residues more

TABLE 5-2   Carboxypeptidase specificities

Enzyme	Specificity	Comment
A	Preference for aromatics and aliphatic hydrophobes	Pro and Arg not released
B	Best with Lys and Arg	
C	Less specific than A or B	Fairly uniform rate of release
Y	Will hydrolyze Pro	Gly and Asp released slowly

even by breaking secondary and tertiary structures. The pH of the incubation can also affect the rate of release; when lowered sufficiently to suppress the charge on the carboxyl, the rate of release is enhanced. Carboxypeptidase Y, although it releases glycine and aspartate slowly, hydrolyzes all C-terminal residues, including proline. In addition, it is possible to immobilize carboxypeptidase Y in a stable, active form that considerably enhances its utility in C-terminus determination.

Finally, let us return to the somewhat less obvious uses of end-group determination referred to earlier. If, upon end-group determination of a peptide more than one end group is detected, two possibilities must be considered: (1) there is impurity present, and (2) a disulfide-bonded entity has been isolated which in fact contains two peptides joined by a disulfide. In this case reduction and alkylation, followed by repurification, should give two peptides, each with a single terminal residue. Because of the yield variability in many of these detection methods, it is not always easy to distinguish an end group arising from a minor contaminant from that of the "real" peptide.

Where the primary sequence of a protein is known and peptides are being characterized solely for the purpose of identifying which peptide has been isolated, an amino acid analysis and end-group determination are often sufficient to allow identification of a particular peptide. For example, when the peptides have been obtained by a specific fragmentation procedure of a known sequence, the amino acid composition and N-terminal and C-terminal residues of all possible fragments are known. With the smaller fragments it is likely that this information will be unique to a particular peptide. Even with larger peptides, knowledge of the terminal residues and which amino acids may be *missing* from a peptide's composition may allow identification of the peptide.

DETERMINATION OF THE AMINO ACID SEQUENCE
OF FRAGMENTS

The phenylisothiocyanate method of determining the amino terminal of a peptide is the basis of the most popular method of sequencing a peptide. Reexamination of Fig. 5-12 will show that once the amino acid has been released by mild acid hydrolysis, the remaining peptide is left intact. It is thus a simple procedure to rederivatize the remaining peptide with PITC and repeat the process used for identifying the original N-terminal residue. With each cycle a new N terminal is generated that can subsequently be identified; this is the basis of sequencing a peptide by *Edman degradation*. Because of the yields at each step, the cycle cannot be carried on indefinitely. Using manual procedures with reasonable amounts of starting material, it is possible to go through up to about 15 cycles. With the automated methods available, using immobilized peptides, it is not unusual to sequence 50 to 100 amino acids.

A particularly useful alternative to the Edman system utilizes the same general approach to PITC but involves dimethylaminoazobenzene isothiocyanate (DABITC), which reacts with amino groups to give a highly fluorescent reagent that is readily identified by TLC after release from the peptide. The release involves anhydrous trifluoracetic acid, which does not cleave remaining peptide bonds. After the extraction, the residual peptide (in the aqueous phase) is dried and can be subjected to a further cycle of derivatization. The whole process requires no specialized equipment and is sensitive enough to permit several rounds of identification on nanomole amounts of material.

When the amino acid sequence of prospective peptides is known from the primary structure of the protein, the DABITC method allows for rapid sequence determination of short regions of a peptide and thus identification of the peptide.

An alternative approach involves the use of dipeptidyl peptidases. A variety of these peptidases are available, which split dipeptides either from the amino terminus (DAPs) or from the carboxyl terminus (DCPs). The principle of the method is quite straightforward:

Step 1. Digestion of the peptide with the dipeptidyl peptidase is performed (DAP or DCP).

Step 2. The dipeptides produced are identified: usually by derivatization to their trimethylsilyl derivatives and identification by gas chromatography or mass spectrometry.

Step 3. The original peptide is modified, either by the addition of one amino acid (usually with DAP) or by the removal of an amino acid (usually with DCP).

Step 4. Steps 1 and 2 are repeated.

Step 5. The original C-terminal and N-terminal residues of the peptide are determined.

Step 6. The dipeptides obtained from the native peptide and the modified peptide are listed separately, and the dipeptides are aligned by alternately picking from one set and then the other.

The N-terminal dipeptide is known and is used as the starting point. First let us consider the sequence

Leu-Lys-Cys-Met-Arg-Glu-Thr-Leu-Phe-Val-Ala-Leu

Aminodipeptidase cleavage of this peptide gives the following dipeptides:

Group A:

> Leu-Lys
> Cys-Met
> Arg-Glu
> Thr-Leu
> Phe-Val
> Ala-Leu

After modification of the N-terminal by addition of radioactive glycine and subsequent aminodipeptidase cleavage, we get:

Group B:

> 14C-Gly-Leu   (therefore, N terminal)
> Lys-Cys
> Met-Arg
> Glu-Thr
> Leu-Phe
> Val-Ala
> Leu   (therefore, C terminal)

Starting with the N-terminal dipeptide from group B,

> Gly-Leu

we alternate picks between groups A and B. The next dipeptide must be Leu-Lys, giving (Gly-Lys) (Leu-Lys). With this particular peptide the picks from each table are unequivocal, and the sequence of the starting peptide is easily obtained.

However, consider the related sequence

> Leu-Lys-Cys-Met-Arg-Glu-Leu-Ala-Val-Ala-Leu

The original dipeptidase (DAP) gives:

Group A:

> Leu-Lys
> Cys-Met
> Arg-Glu
> Leu-Ala
> Val-Ala
> Leu   (must be C terminal)

Modification by addition of an amino acid to N-terminal end (e.g., glycine) gives

> Gly-Leu-Lys-Cys-Met-Arg-Glu-Leu-Ala-Val-Ala-Leu

Subsequent DAP treatment gives:

Group B:

Gly-Leu
Lys-Cys
Met-Arg
Glu-Leu
Ala-Val
Ala-Leu

Knowing that the Gly-Leu dipeptide from group B is the N-terminal dipeptide, we start from there, as shown in Fig. 5-16.

After all the dipeptides have been used, we find that two sequences satisfy the pattern and a unique sequence cannot be assigned. If, however, we use a C-terminal

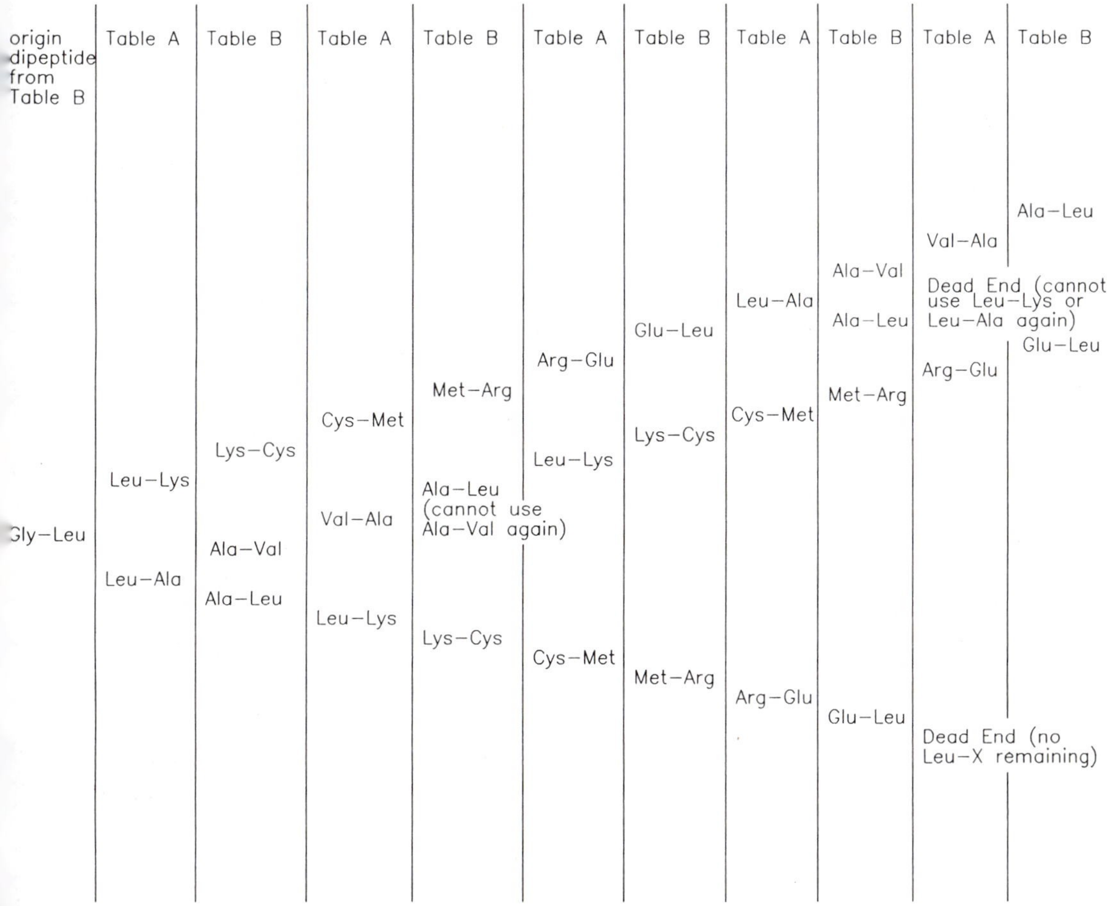

origin dipeptide from Table B	Table A	Table B	Table A	Table B	Table A	Table B	Table A	Table B	Table A	Table B
										Ala–Leu
									Val–Ala	
								Ala–Val		Dead End (cannot use Leu–Lys or Leu–Ala again)
							Leu–Ala	Ala–Leu		Glu–Leu
						Glu–Leu			Arg–Glu	
					Arg–Glu			Met–Arg		
				Met–Arg			Cys–Met			
			Cys–Met			Lys–Cys				
		Lys–Cys			Leu–Lys					
	Leu–Lys		Val–Ala	Ala–Leu (cannot use Ala–Val again)						
Gly–Leu		Ala–Val						Arg–Glu		
	Leu–Ala	Ala–Leu	Leu–Lys	Lys–Cys	Cys–Met	Met–Arg	Arg–Glu	Glu–Leu		Dead End (no Leu–X remaining)

Figure 5-16 Alignment of dipeptides obtained from diaminopeptidase treatment.

dipeptidase before and after addition of radioactive glycine to the C-terminal end, we get the following dipeptides:

Group A:

Ala-Leu
Ala-Val
Glu-Leu
Met-Arg
Lys-Cys
Leu   (therefore, must be N terminal)

Group B:

Leu-Gly   (therefore, C terminal)
Val-Ala
Leu-Ala
Arg-Glu
Cys-Met
Leu-Lys

This time (Fig. 5-17) we can start from the C-terminal dipeptide, Leu-Gly, and add dipeptides as before. From this scheme it is readily apparent that the sequence of the peptide is

Leu-Lys-Cys-Met-Arg-Glu-Leu-Ala-Val-Ala-Leu

The aligning of the dipeptides is readily computerized and can be done in similar fashion with tables of dipeptides obtained from either DAP or DCP digestion.

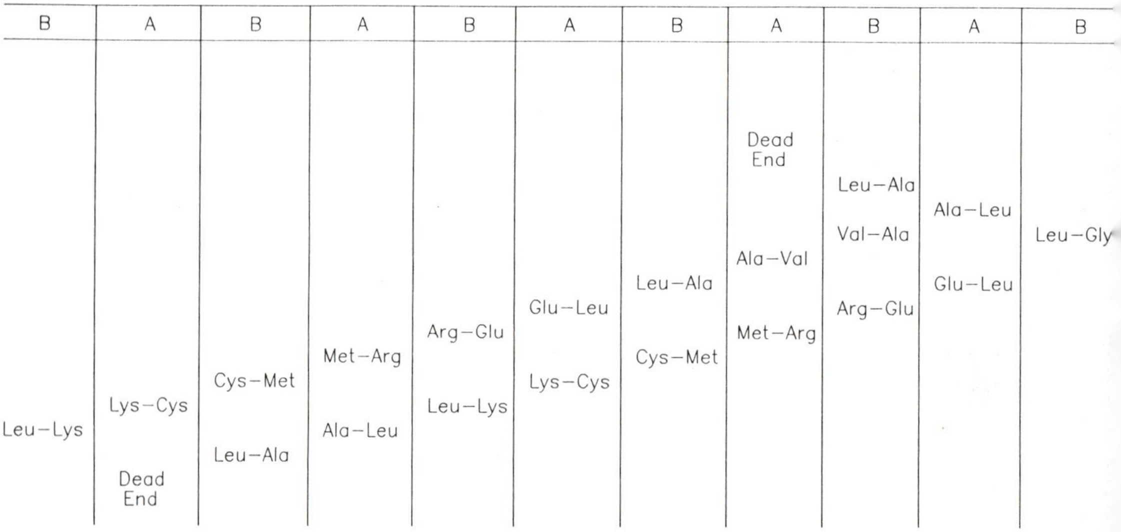

B	A	B	A	B	A	B	A	B	A	B
							Dead End			
								Leu–Ala		
									Ala–Leu	
								Val–Ala		Leu–Gly
							Ala–Val			
						Leu–Ala			Glu–Leu	
					Glu–Leu					
				Arg–Glu			Met–Arg	Arg–Glu		
			Met–Arg			Cys–Met				
		Cys–Met			Lys–Cys					
				Leu–Lys						
	Lys–Cys		Ala–Leu							
Leu–Lys		Leu–Ala								
	Dead End									

Figure 5-17   Alignment of dipeptides obtained from dicarboxypeptidase treatment.

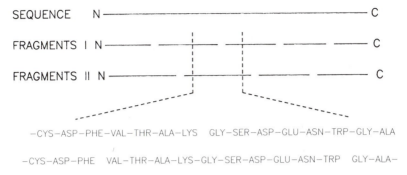

Figure 5-18 Alignment of sequenced fragments by use of overlapping fragment. Fragments I are obtained by tryptic cleavage of the original peptide, while fragments II are obtained by chymotryptic cleavage.

ALIGNMENT OF THE FRAGMENTS

Once all possible peptides from a particular fragmentation procedure have been isolated and sequenced, the question remains as to the order in which the fragments appear in the primary sequence of the protein. Usually, the original N-terminal and C-terminal peptides can be determined by labeling the particular terminal residues *prior* to fragmentation so that these two fragments are readily assigned. It is not possible, however, to align the remaining fragments (assuming that there is more than one) without further work. The process (shown in Fig. 5-18) requires a second fragmentation method to be employed and the resultant fragments purified and sequenced. The sequences of this second set are then used to align the fragments obtained in the first fragmentation.

ASSIGNMENT OF DISULFIDE BONDS

Consider a protein with the structure shown schematically in Fig. 5-19. Isolation of tryptic fragments from this protein results in two rather interesting fragments in addition to a number of "normal" fragments. A fragment with a single amino and carboxyl terminal is obtained (fragment I in the diagram), and a second fragment (II) with two amino and two carboxyl terminals per mole of fragment is also found. Consider first the characteristics of fragment I. This fragment, prior to reduction, has a single —SH group; however, amino acid analysis of the reduced and carboxymethylated fragment indicates the presence of three cysteine residues and thus the existence of a disulfide bond. Sequencing shows the positions of the three cysteines but not which two are involved in the disulfide. This ambiguity is overcome by chemically labeling free sulfhydryl groups, for example, by carboxymethylating the peptide *prior* to reduction with C-14 iodoacetamide. After reduction, C-12 iodoacetamide is used prior to sequencing to prevent reformation of disulfide bonds.

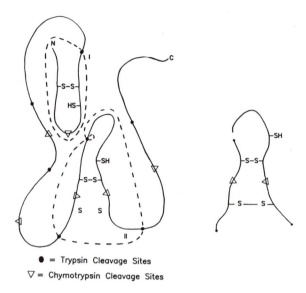

● = Trypsin Cleavage Sites
▽ = Chymotrypsin Cleavage Sites

Figure 5-19   Cleavage and isolation required in the identification of disulfide-bonded residues. Inset at right shows detail of tryptic fragment II.

With fragment II other problems are encountered. The fragment as isolated has two amino and two carboxyl terminal residues, indicating the presence of two poly-peptides covalently linked by disulfides. Determination of the free and total cysteine content shows that two disulfides are present. As before, the free cysteine can be radioactively labeled prior to the two chains being separated. Each chain on se-quencing has two unlabeled cysteines, each of which is involved in a disulfide bond; the question that remains is, which is bonded to which? In the absence of additional information, this cannot be known. However, if tryptic fragment II is subsequently digested with chymotrypsin *prior* to reduction of the disulfides, two further frag-ments are obtained (see Fig. 5-19), each of which contains a disulfide bond that can be analyzed as already outlined and allow the unique assignment of the disulfides.

It is entirely possible, of course, that a second cleavage site that allows separation of the disulfides in this manner may not exist. In such a case ambiguity remains concerning the exact partners in each disulfide bond.

OTHER ASSIGNMENTS

Four other types of assignment with regard to the covalent structure of a polypeptide chain may be necessary. These involve the position of amides and the location of covalently attached groups such as carbohydrate chains, lipids, or phosphate groups.

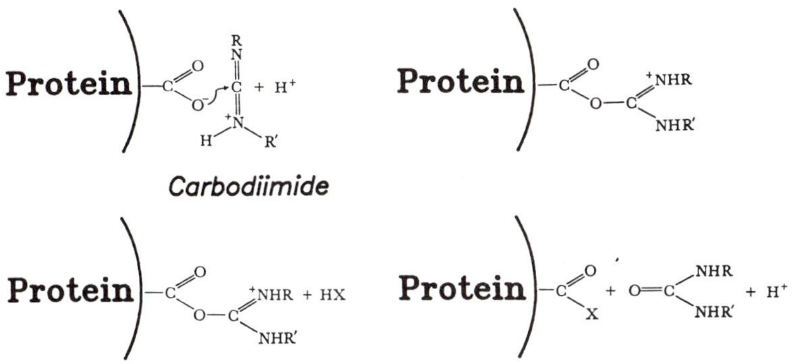

Figure 5-20 Carbodiimide activation of free carboxyl groups to allow labeling with radioactive nucleophile.

Amides

There are three means used to establish whether an amide (Gln or Asn) is at a particular position or whether the free acid (Glu or Asp) occurs. Where ambiguity remains, the residue is often listed as Asx or Glx. Where the peptide has not been exposed to conditions that might convert the amide to the free acid, it is possible to separate the phenylthiohydantoin derivatives of the amide from the free acid by TLC or by gas chromatography during Edman sequencing. The second approach involves the electrophoretic mobility of isolated peptides that contain a single Asx or Glx residue, which usually indicates whether the amide or the free acid is involved. If the Asx or Glx is close to either the amino or carboxyl terminus, the use of the appropriate exopeptidase removes the ambiguous residue and a comparison of the electrophoretic mobility before and after exopeptidase treatment distinguishes between Asn or Gln and Asp or Glu. In cases where a peptide contains multiple Asx or Glx, it is necessary to use further fragmentation methods to produce peptides, each of which contain a single ambiguous residue, before this method can be used.

The final method depends on the covalent derivatization of free carboxyl groups via the carbodiimide–glycine methyl ester coupling method, which is outlined in Fig. 5-20. This approach gives radioactive derivatives of the free carboxyls, which are then readily distinguished from the amides even if the amides are subsequently converted to the free carboxyls.

Carbohydrates, Lipids, and Phosphate

These covalent derivatives are usually linked to quite unique amino acid side chains: carbohydrates through asparagine or serine (N-linked and O-linked can be distinguished on the basis of their acid lability: O-linked is acid labile), fatty acids via Ser, Thr, or Cys residues via ester linkages, and phosphate groups via serine (though tyrosine and histidine derivatives can occur). Identification of the exact residue involved requires extensive degradation and isolation of the derivatized

TABLE 5-3 Specific sequences for covalent substitution

Substitution	Specific sequences
N-linked CHO	Asn(CHO)-X-Ser
O-linked CHO	Not known
Fatty acids and Lipids	Not known
Phosphate	Lys(Arg)-Ser-Asn-Ser(PO_4),
	Lys(Arg)-Arg-Ala-Ser(PO_4),
	Arg-Thr-Leu-Ser(PO_4)

peptide such that only a single copy of the appropriate linkage amino acid remains in the peptide. Assignment is then done by inference. In certain cases these covalent substituents are associated with a specific sequence that may be identified in the primary structure, and thus represent potential sites for the appropriate derivatization. Some of these specific sequences are summarized in Table 5-3.

ALTERNATIVE STRATEGY TO SEQUENCING A PROTEIN

As indicated at the outset of this chapter, the protein chemist can resort to the tools of the molecular biologist in order to obtain a complete linear amino acid sequence. Sequencing of a specific DNA fragment is extremely rapid and relatively inexpensive. Consequently, once cloned in a suitable vector and isolated, the sequence of a gene is readily determined, and from this the primary sequence can be derived. However, this approach is complicated by the fact that within a given nucleotide sequence, there can exist more then one potential reading frame. A partial protein sequence of only five to six amino acid residues, perhaps from a chymotrypic peptide, can quickly serve to orient the direction and reading frame for transcription and translation of the protein of interest. Uncertainty regarding the site of translational intitation is readily overcome with an N-terminal analysis of the protein by standard techniques. Problems exists if a pre-protein is proteolytically processed to yield a mature protein, and in this case N-terminal analysis of the precursor protein is necessary to determine the actual translational start site.

The first stage in obtaining a nucleotide sequence requires that the gene of interest be cloned, identified, and isolated using recombinant DNA techniques. There are three basic approaches to obtaining the desired gene and many variations on them have been successfully utilized. They are: (1) selection by complementation, (2) detection with antibody, and (3) the oligonucleotide approach.

Selection by complementation involves a direct screening for the activity of the desired gene product via expression in a host lacking the activity (complementation). This approach has been very useful for obtaining genes from prokaryotic sources and some lower eukaryotes, but in general is not applicable to higher eukaryotes. It usually requires efficient expression of the gene in order to work, and low-level expression is often difficult to detect, thereby limiting the sensitivity of this approach.

The second method involves the use of antibody either to detect clones producing the protein of interest or to obtain (or enrich for) nucleic acid that is actively being translated. The detection of clones producing the desired protein requires that part or all of the gene be accurately transcribed and translated. Clones producing the protein of interest are detected after lysis and immobilization of macromolecules on a filter membrane, usually nitrocellulose. Antibody is added under conditions that allow specific binding, and positive reactions are disclosed by addition of either radio-labeled staphylococcal A protein or a secondary labeled or enzyme-conjugated antibody directed against the primary antibody. Positive clones are then isolated and propagated from a master plate. This technique is extremely sensitive, and depending on the affinity of the antisera can easily detect nanogram quantities of a gene protein.

Specific antibody is also useful in significantly enriching for the mRNA encoding the desired protein. In vitro, one can obtain translation of mRNA from either eukaryotic or prokaryotic sources. Selective precipitation of mRNA–ribosome–protein complexes (polysomes) can be done with antibody and the mRNA deproteinized. In this way, nucleic acid encoding the gene of interest can be significantly purified or enriched.

The oligonucleotide approach for detecting or identifying specific cloned sequences is perhaps the most sensitive and most powerful. It does not require transcription or translation of the cloned gene and is readily adaptable to screen either eukaryotic cDNA or genomic libraries as well as gene libraries from prokaryotic sources. In this method a partial protein sequence is needed. From this, one can derive a number of possible nucleotide sequences. Since the triplet code is degenerate, it is advantageous if the protein sequence used contains amino acids that are coded for by the least number of triplets. Particularly suitable are methionine and tryptophan, since each has a single triplet codon. Phenylalanine, tyrosine, histidine, glutamine, asparagine, lysine, aspartate, glutamate, and cysteine are also desirable since each has just two triplet codons. Serine should be avoided, if possible, since it shows degeneracy at both the 2 and 3 positions in the triplet. The goal of this approach is to synthesize a probe oligonucleotide which will hybridize strongly to the appropriate segment of DNA, and maximum complementarity will be obtained with the least degenerate probe possible. Mixed oligonucleotide probes are frequently used in instances where ambiguity exists so that a complementary sequence will be represented in the probe mixture. These probes are generally end-labeled with ^{32}P and used to screen clones after lysis and immobilization on a filter membrane.

Cloning Hosts and Vectors

The gram-negative bacterium *Escherichia coli* is the most readily employed host for cloned genes from all sources for a variety of reasons: It is easily grown and maintained, and foreign DNA is easily introduced into this organism by transformation, transfection, or infection with bacteriophage. DNA is stably maintained and readily recovered, and a plethora of "genetic tricks" exist that allow the investigator to manipulate cloned DNA efficiently and inexpensively. Other prokaryotes, yeast, and mammalian cells are also used as hosts for cloned DNA, but it is desirable to do most DNA manipulation using *E. coli*, for the reasons already given.

There are a huge variety of cloning vectors, which vary in their host range, in the amount of DNA that can be accommodated, and in the regulatory functions contained on the vector. Vectors exist which are designed for the generation of deletions in DNA, for DNA sequencing, or for the expression of foreign DNA. In general, there are four classes of vectors: plasmids, viral (prokaryotic and eukaryotic), cosmids, and integration vectors.

Plasmid cloning vectors are small, autonomously replicating elements that usually contain a marker for selection, such as an antibiotic-resistance determinant, as well as a mechanism to detect if foreign DNA is inserted. The pUC plasmids developed in Joachim Messing's laboratories are perfect examples. The ampicillin-resistance gene allows positive selection of transformants, and cloning into a number of unique restriction sites destroys β-galactosidase activity, so that clones containing recombinant plasmids can be differentiated on appropriate medium. Plasmids can accommodate a wide range of sizes of DNA fragments, from a few base pairs to several kilo-base pairs (kbp).

E. coli bacteriophage vectors are extremely useful cloning vectors and have played a central role in molecular biology. Bacteriophage lambda has been widely used for cloning cDNA or genomic DNA from all sources. Varying amounts of a nonessential region of the phage DNA can be exchanged for foreign DNA, and the recombinant DNA molecule can be packaged into viable phage particles in vitro. The recombinant phage can be propagated in an appropriate *E. coli* host. Up to 23 kbp of DNA can be cloned in some vectors.

The single-stranded filamentous *E. coli* phage, m_{13}, has been developed by Messing for use in the Sanger dideoxy chain-termination DNA-sequencing system. Foreign DNA is inserted into the double-stranded replicative form of the phage at a polylinker region containing a number of unique restriction endonuclease sites. High yields of single-stranded DNA necessary for the Sanger sequencing method are easily obtained after transfection with the recombinant replicative form and propagation.

Eukaryotic viral vectors have been very useful in studies on gene regulation and development. They allow the study of gene function and DNA sequences and have already been useful in examining sequence–function relationships among some of the oncogenes.

Cosmids are hybrids of plasmids and bacteriophages capable of autonomous replication and can be recognized and packaged into phage particles in vitro. They are useful for cloning very large DNA fragments (approximately 38 kbp).

Insertion vectors have been developed for both prokaryotes and eukaryotes. They have no replication origin for a particular host or a replication origin with a conditional mutation that allows inactivation of replication functions, generally by a temperature shift. These vectors facilitate the stable integration of foreign DNA into the host's chromosome. These types of vectors are very powerful tools in regulation studies.

Of all the cloning vectors, plasmids and single-stranded bacteriophage can be used directly in experiments where one wants to create specific sequence alterations in a nucleic acid coding region.

Cloning Strategies

Genes from most prokaryotes and some eukaryotes do not contain introns and can therefore be cloned directly from partial or complete digestion of chromosomal DNA into any of the vectors described previously. If the desired gene is expressed, clones synthesizing the desired gene product can be detected by screening for activity, or by using antibody against the gene product, as already described. In the absence of sufficiently strong expression, the oligonucleotide probe approach is a feasible alternative.

Genes containing introns can be cloned directly, but more often than not they are not transcribed; and even with transcription, prokaryotes cannot properly process mRNA for translation. It is therefore desirable to isolate a copy of the gene representative of the mature mRNA. This can be done by isolating mature, spliced, polyadenylated mRNA from cells that are actively producing the gene product of interest. Because most mature eukaryotic mRNAs are polyadenylated at their 3' end, oligo-dT can be employed to prime the message for reverse transcription. In the presence of dNTPs, reverse transcriptase will synthesize a copy of DNA from the mRNA. DNA polymerase can then be used to complete the population of double-stranded copy DNA. Synthetic linkers can be inserted into a compatible restriction site in a plasmid or phage vector to construct a cDNA "bank" or "library".

In instances in which the mRNA of interest is present as a significant portion of total mRNA, it is sufficient to prepare a cDNA from whole mRNA. In this case, the desired clone should represent a significant amount of the cDNA library and detection should be possible. For very low abundance message, the mRNA encoding the desired clone can be significantly enriched by polysome precipitation using specific antisera, as described previously. Then a cDNA library can be prepared from the precipitated mRNA, thereby increasing the representation of the clone of interest in the total population. Another ingenious technique has been used to clone low-abundance mRNAs which are cell-type specific. In this "subtraction cloning" protocol, mRNA from a population of mRNAs containing the desired message is allowed to hybridize on filters with a cDNA library from cells that do not make the desired mRNA. Message that does not hybridize is washed off the filter and collected. A DNA library can then be constructed from this RNA. This method "subtracts" the mRNAs common to both populations and facilitates recovery of cell-type-specific message. It has been successfully employed in obtaining genes expressed only in T cells by subtracting T-cell mRNA with a cDNA bank from B cells.

The detection of a successful cloning event from among the many clones generated when a cDNA or genomic library is constructed is often a difficult and labor-intensive task. Two methods are available and widely employed. A clone *can* be detected in a cDNA library by the use of antibody as described previously, but because eukaryotic promotor and translational initiation sites are not usually recognized by *E. coli*, a prokaryotic version of these regulatory sequences must be provided. For example, the popular bacteriophage vector λGT11 allows the expression of in-frame fusions of the *E. coli* β-galactosidase gene and foreign proteins. Plasmid vectors containing λ or *E. coli* regulatory sequences are also available.

The most sensitive method for identifying recombinants from cDNA libraries cloned in either phage or plasmid vector is the use of mixed oligonucleotide probes, as described previously. Expression of the gene is not required and therefore this method is also useful for identifying genomic clones. Specific oligonucleotides which are complementary to the desired gene can also be employed to prime synthesis of a cDNA library, so that the majority, if not all, of the cDNA thus generated originated from the sequences related to the desired genes. Also, oligonucleotides can be used to prime synthesis from a specific mRNA in the presence of radiolabeled precursors, and this transcript can be used directly to probe a cDNA library.

Another very useful strategy for detecting cloned genes is to use clones of genes from other species as radiolabeled probes. For example, porcine factor VIII was used to detect the human gene, and mammalian *ras* oncogene has been used to identify and clone a related protein from yeast.

Sequencing the Gene

The success of sequencing a gene depends on the experimental separation of single-stranded DNAs having different lengths by polyacrylamide gel electrophoresis in the presence of urea. Two basic techniques have been developed for the generation of the required sets of fragments, one dependent on random termination of the labeled oligonucleotide during synthesis in the presence of $[^{32}P]$dNTP, and the other dependent on random chemical cleavage after labeling the 5' end with ^{32}P. The Sanger method is an enzymatic approach relying on the ability of dideoxynucleoside triphosphates to function as efficient nascent DNA chain terminators. The Maxam–Gilbert method involves the random chemical cleavage of end-labeled DNA fragments. Polyacrylamide gels with between 8 and 12% acrylamide give good separation of oligonucleotide fragments between 10 and 300 nucleotides in length generated in the aforementioned sequencing methods. Since oligonucleotides differing in length by a single nucleotide can readily be separated in this manner, it is possible to sequence an oligonucleotide if some way of differentially identifying the terminal residue of each fragment is available. If a separate series of fragments, each terminating in A, C, G, or T, can be generated and separately electrophoresed, it is a simple matter (Fig. 5-21) to read off the nucleotide sequence. The urea prevents the formation of base pairing, which leads to double-stranded DNA or secondary structure that could result in anomalous migration.

Sanger Method. This method requires a single-stranded DNA template of the gene of interest. In general, this is obtained by cloning into the replicative form of

Fragments Terminating
in

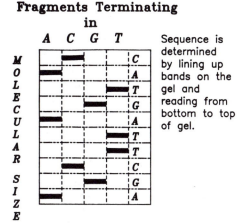

Figure 5-21   Outline of PAGE method of sequencing an oligonucleotide.

the single-stranded ϕm_{13} bacteriophage by transfecting an appropriate host, and allowing the phage machinery to produce particles carrying the single-stranded DNA. A short oligonucleotide (15 to 22 kbp) complementary to a region upstream of the DNA to be sequenced is used to prime DNA synthesis. The 79-kDa fragment of *E. coli* polymerase I, the Klenow fragment, containing the DNA-polymerizing activity is used to synthesize a complementary DNA strand. Reaction mixes containing the single-stranded DNA primer, all four dNTPs $[\alpha^{32}P]$dNTP labeled, and one of four dideoxynucleotides (ddNTP) at appropriate concentrations and the Klenow fragment are constructed. The ddNTPs are inserted randomly by the polymerase, resulting in termination of DNA synthesis. In this way a random pool of different-size labeled fragments which all terminate at a specific nucleotide are generated. The double-stranded DNA can be melted and applied to a urea gel and the sequences read off as previously. This method is rapid and efficient. About 300 to 400 bases can be read from a single gel. By cloning and sequencing different portions of a gene in different orientations, one can quickly obtain an entire gene sequence. Alternatively, a single bacteriophage carrying the entire gene can be sequenced by synthesizing primers for different regions of the DNA.

Maxam–Gilbert Method. In an alternative approach, the piece of DNA to be sequenced is first labeled at the 5′ end by using polynucleotide kinase and ^{32}P. The radioactive DNA is then subjected to various chemical cleavages with differing base specificity to generate randomly cleaved fragments, which are then analyzed similar to the dideoxynucleoside triphosphate approach. Because of the chemical similarity of the purines it is not possible to obtain unique cleavage at either A or G; however, after methylation of the DNA with dimethylsulfate, heating at neutral pH preferentially cleaves at G, while incubation with acid leads to preferential cleavage at A. As a result, alternating light and dark bands are seen on the autoradiograph, the dark bands corresponding to sites of preferential cleavage and the light bands corresponding to the dark bands obtained with the other purine cleavage pattern. Treatment with hydrazine (followed by piperidine) leads to cleavage (with equal facility) at cytosine and thymine. If 2 M NaCl is included in the hydrazine reaction, cleavage at the thymine is suppressed.

These four chemical reactions give four sets of fragments: with cleavage at G > A, with cleavage at A > G, with cleavage at C + T, and with cleavage at C > T. As before, the nucleotide sequence is easily deduced from the gel patterns.

Although these tools from molecular biology do provide an easy approach for obtaining the linear amino acid sequence of a protein, there is much essential description of the covalent structure of a protein that demands the classical approaches of the protein chemist. The DNA sequence of a protein gives no information regarding covalent modifications of particular residues, or the location of disulfide bonds, and a complete description of proteins containing such features depends on the isolation and characterization of peptide regions containing such modifications. The aid to these tasks comes from the fact that if the complete linear sequence of a protein is known, it is easier to identify particular fragments once they have been purified. It is even possible in some instances to uniquely identify, for example, tryptic fragments of a protein on the basis of their amino acid composition. In most cases, however, when the linear sequence is known, several cycles of Edman degradation are sufficient to completely assign a particular fragment to its location in the sequence. It is fair to say that the development of gene cloning and sequencing techniques have aided but not replaced, the classical aims of protein chemistry.

6

Peptide Synthesis

INTRODUCTION

The chemical synthesis of a peptide presents a variety of challenges to the organic chemist. Many side-chain groups in the constituent amino acids have to be protected and subsequently unblocked, peptide bonds synthesized without racemization of the optically active constituents, side reactions prevented, and at the end of the synthesis a homogeneous product must be obtained from a mixture that may well contain highly analogous side products. The classical criteria of success has to be applied to peptide synthesis just as in any other organic synthesis; that is, success is judged by the observation that the physical, chemical, and biological properties of the product match those of the natural parent compound.

Peptide synthesis has evolved to the stage where small peptides (up to 20 to 30 residues) have become a cornerstone in examining structure–function relationships in peptide hormones (almost all of which have been chemically synthesized) by the systematic variation of one or more residues. In another application, the putative epitope of a viral antigen has been explored by systematic synthesis of a series of analogs with different amino acid substitutions in each position of a heptapeptide epitope, first identified by synthesis of all 208 possible overlapping peptides covering the 213-residue viral coat protein antigen.

Synthesis of long peptides has not been successfully achieved by chemical means and with the advent of cloning and site-specific mutagenesis as an approach to manipulating protein structures, much of the impetus to long peptide synthesis has been removed. Among the larger peptides successfully synthesized is the 57-residue apolipoprotein C-I, the protein constituent of the very low density human plasma lipoproteins. The synthesized protein activated lecithin:cholesterol acyltransferase to the same extent as the native protein and bound similar amounts of dimyristoyl phosphatidylcholine. In an ambitious, yet to be completed project, the synthesis of a hypothetical protein that would have a predicted β-barrel secondary–tertiary structure has been undertaken. The synthesis of this approximately 80-residue protein has been reported, but the determination of the crystal structure is still in progress.

In this chapter we discuss some of the problems encountered in peptide synthesis, the basic protocols that have been developed to accomplish it, and some of the alternative approaches that are being used to provide the answers to protein structure–function questions that the original workers in the field of peptide synthesis envisaged being able to answer about proteins in general.

CHEMICAL APPROACHES

Two different approaches have been developed which share many of the experimental challenges of synthesis, and they are schematically shown in Fig. 6-1: the solution method (part A) and the solid-phase support method (part B). In both methods the carboxyl group of the amino acid on the soon-to-be N-terminal side of the peptide must be activated to allow peptide-bond formation during the coupling phase of the reaction. In the solution method the carboxyl group of the C-terminal side and the amino group of the N-terminal side must both be protected to prevent unwanted side reactions. In the solid-phase support method the carboxyl group of the C-terminal side is involved in immobilization to the support and does not need protection. With both approaches any reactive groups in the amino acid side chains have to be protected. After coupling, the blocking groups (and in the case of the solid-phase method the immobilization linkage) must be removed to give the final dipeptide product. To achieve larger peptides, these processes can be repeated sequentially, although purification of intermediates to remove unwanted-side-product polypeptides is advisable.

It is quite a simple process to synthesize a peptide up to five or six residues by such procedures. If the final end product is a much larger peptide, the fragment condensation approach becomes attractive. In this approach, shown schematically in Fig. 6-2, two half-fragments are synthesized, purified, and subsequently ligated.

The advantage is that purification of the final product, peptide C in the scheme, is quite easy since the reaction mixture contains only peptides A and B (or derivatives thereof produced during the ligation reaction) and the desired product peptide C, which is much larger than the reactants. Although this process sounds simple, it has some potential problems associated with the ligation procedures.

Reactive Group Protection

We now review briefly some of the procedures used in carboxyl or amino-terminal protection, in coupling and ligation, and in side-chain protection, as well as immobilization processes employed in the solid-phase method.

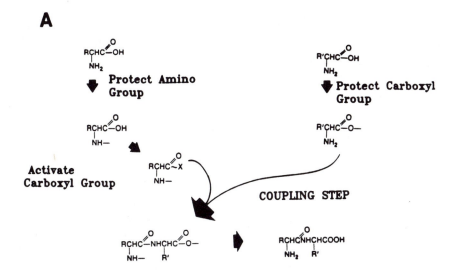

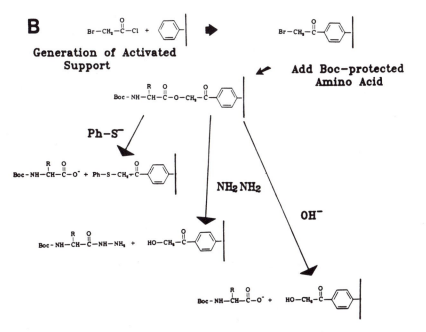

Figure 6-1   Outline of alternative methods for peptide synthesis: (A) solution method; (B) solid-phase support method.

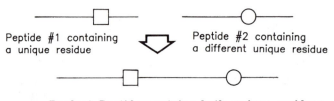

Product Peptide contains both unique residues

Figure 6-2 Fragment condensation used to synthesize long peptides from precursors.

Carboxyl Protection. Three commonly used carboxyl protecting groups are shown in Fig. 6-3. Ethyl esters are removed at the termination of synthesis by saponification, or if the product is to be used in a fragment condensation scheme, they can be converted to the hydrazide by hydrazinolysis. If complex peptides are being synthesized, however, exposure to alkali or hydrazine can lead to unwanted side reactions. *tert*-Butyl esters, on the other hand, are readily removed acidolytically. Nitrobenzyl or benzyl esters are frequently employed and are removed by hydrogenolysis. As a result, such protection is not suited to the synthesis of peptides containing methionine or cysteine.

Amino Protection. Figure 6-4 shows some of the amino-protecting groups used in peptide synthesis. The major problem encountered with them is that during a multistage synthesis it is necessary to selectively remove the α-amino protecting group from the N-terminal amino acid in the growing chain to allow elongation.

This problem is compounded by the fact that as is discussed in the next section, many of these groups are used to protect reactive side-chain moieties. The general protocol is to derivatize the side chains with one type of protecting group, for example, the carbobenzoxy protection group, and the α-amino group with a different protecting

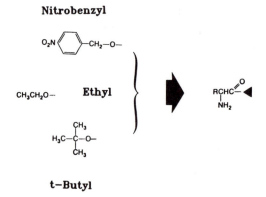

Figure 6-3 Carboxyl protecting groups used in peptide synthesis.

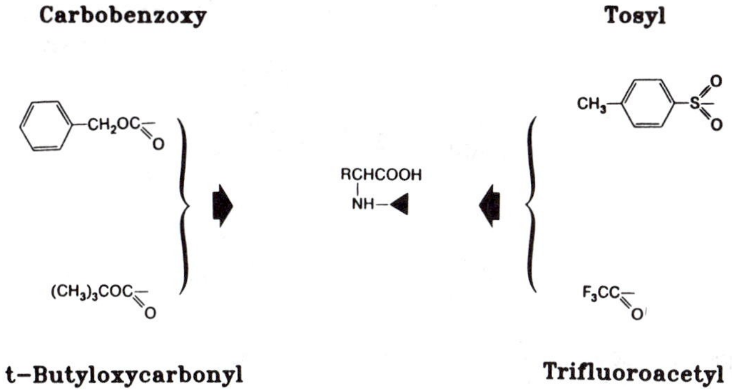

Carbobenzoxy   **Tosyl**

RCHCOOH
|
NH—◄

t–Butyloxycarbonyl   **Trifluoroacetyl**

Figure 6-4 Amino protecting groups used in peptide synthesis.

group, for example, the t-butyloxycarbonyl protection group (or vice versa), and selectively remove the α-amino protecting group.

In peptides lacking cysteine or methionine the carbobenzoxy group can be removed in the presence of t-butyloxycarbonyl groups by hydrogenolysis. The converse situation, selective removal of t-butyloxycarbonyl in the presence of carbobenzoxy, is more difficult, but has been achieved by treatment with 98% formic acid or β-mercaptoethanesulfonic acid.

Side-Chain Protection. Table 6-1 shows a variety of approaches for the protection of the reactive side chains in trifunctional amino acids. As can be seen, different protection groups have been used in different cases for the same amino acid side chain. There are no general rules governing protection; in some cases protection is not even used, but this can lead to increased risk of unwanted side reactions.

TABLE 6-1 Protection and deprotection of trifunctional amino acids[a]

Amino acid	Protection[b]	Deprotection
Lys	Boc(1) Z(4,5)	TFA(1) HF(4,5)
Arg	NO_2(2,3,4)	H_2/Zn/HCl(2) H_2/Pd(3) HF(4)
His	Trt(2) Z(2,3)	H_2/HOAc(2) H_2/Pd(3)
Asp	Bu^t(1,3) Bzl(3,4)	TFA(1,3) H_2/Pd(3) HF(4)
Glu	Bu^t(2) Bzl(3,4)	HCl(2) H_2/Pd(3) HF(4)
Ser	Bu^t(1,2) Bzl(3,4)	TFA(1) HCl(2) HF(3,4)
Thr	Bu^t(1,2) Bzl(4)	TFA(1) HCl(2) HF(4)
Tyr	Bu^t(1,2) Bzl(4)	TFA(1) HCl(2) HF(4)
Cys	Trt(2) Bzl(2,4) Acm(5)	Hg(OAc)$_2$(2,5) Na/NH_3(2) HF(4)

[a] 1, glucagon; 2, calcitonin; 3, secretin; 4, RNAse-A; 5, RNAse-S.
[b] Bu^t, t-butyl; Z, carbobenzoxy; Trt, trityl; Bzl, benzyl; Acm, acetamidomethyl.

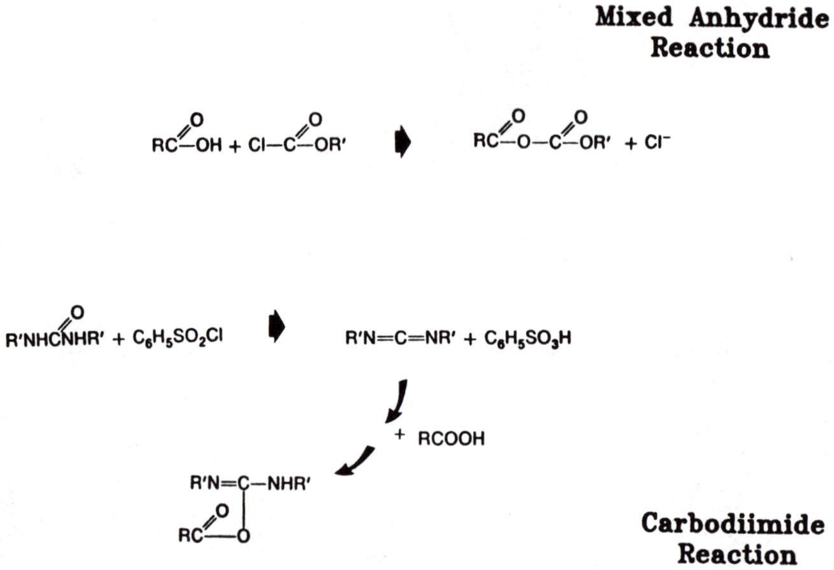

Mixed Anhydride Reaction

$$RC\overset{O}{\diagup}OH + Cl-C\overset{O}{\diagup}OR' \quad \blacktriangleright \quad RC\overset{O}{\diagup}O-C\overset{O}{\diagup}OR' + Cl^-$$

$$R'NHC\overset{O}{\diagup}NHR' + C_6H_5SO_2Cl \quad \blacktriangleright \quad R'N=C=NR' + C_6H_5SO_3H$$

$$+ \ RCOOH$$

$$\begin{array}{c} R'N=C-NHR' \\ RC\overset{O}{\diagup}O \end{array}$$

Carbodiimide Reaction

Figure 6-5 Activation of carboxyl groups for coupling in peptide synthesis.

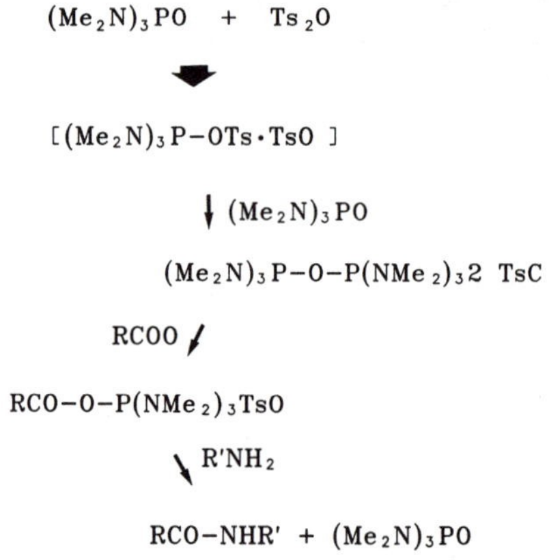

$$(Me_2N)_3PO \quad + \quad Ts_2O$$

$$\blacktriangleright$$

$$[(Me_2N)_3P-OTs \cdot TsO]$$

$$\downarrow (Me_2N)_3PO$$

$$(Me_2N)_3P-O-P(NMe_2)_3 2 \ TsC$$

$$RCOO \diagup$$

$$RCO-O-P(NMe_2)_3TsO$$

$$\diagdown R'NH_2$$

$$RCO-NHR' \quad + \quad (Me_2N)_3PO$$

Figure 6-6 Hexamethylphosphoramide approach to coupling.

Coupling and Ligation Procedures

Several methods have been used for the sequential coupling of activated amino acids to the growing chain of a synthetic peptide. The two most common are the mixed-anhydride method and the carbodiimide method, both illustrated in Fig. 6-5. After the activation step the coupling proceeds as in Fig. 6-1. These methods, especially the mixed-anhydride method, are quite adequate for the stepwise addition of amino acids without the danger of racemization. However, anhydrides of peptides (as compared to anhydrides of single amino acids) frequently undergo racemization, making this method of coupling not very useful in fragment condensation reactions (i.e., ligation). When fragments are ligated using the mixed-anhydride approach, the peptide to be added to the existing amino terminal usually has a C-terminal glycine or proline which cannot undergo racemization on anhydride formation.

An alternative approach involves activated derivatives of hexamethylphosphoramide and is outlined in Fig. 6-6. This method produces little or no racemization and the products are readily separable. The reaction has been used successfully to couple asparagine, glutamine, methionine, tyrosine, or tryptophan containing peptides without the need for side-chain protection. When serine, threonine, or histidine is present, protection is necessary.

Attachment to Solid-Phase Support

In the solid-phase support method of peptide synthesis, several types of linkage of the growing chain to the immobile support have been used.

Ester Linkage. The phenacyl ester linkage procedure, illustrated in Fig. 6-7, is typical of this approach. The matrix copoly(styrene-divinylbenzene) is activated by bromo acetylation and then esterified with amino-protected amino acid. The resultant ester linkage is stable to acidolytic removal of the amino protecting group, but after completion of the synthesis is readily cleaved by sodium hydroxide, ammonia, or hydrazine, as indicated.

Amide Linkage. A variety of linkages have been used, including the carboxamide and sulfonamide linkages illustrated in Figs. 6-8 and 6-9, respectively. In these procedures it should be noted that the carboxamide linkage is cleaved by hydrofluoric acid (HF) to give an amide derivative, while the sulfonamide linkage gives the sodium salt of the C-terminal amino acid.

Other Linkages. Although the examples discussed use the carboxyl group of the first (i.e., C terminal) amino acid in the linkage, it is possible to immobilize via the amino group as indicated in Fig. 6-10, although synthesis must then proceed in an N-to-C direction as opposed to the more usual C-to-N manner.

Some syntheses have been reported which even use a functional side chain of an amino acid as the point of attachment to the solid support. Such procedures are somewhat esoteric and are not recommended as general approaches to peptide synthesis since they lead to increased problems in protecting the backbone amino and

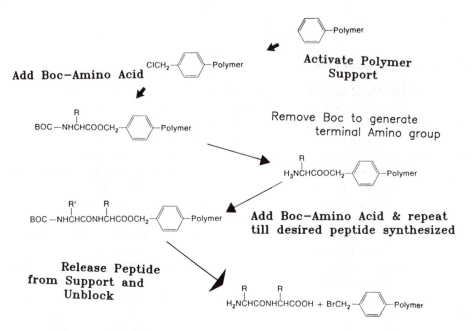

Figure 6-7 Preparation and use of bromoacetyl resin.

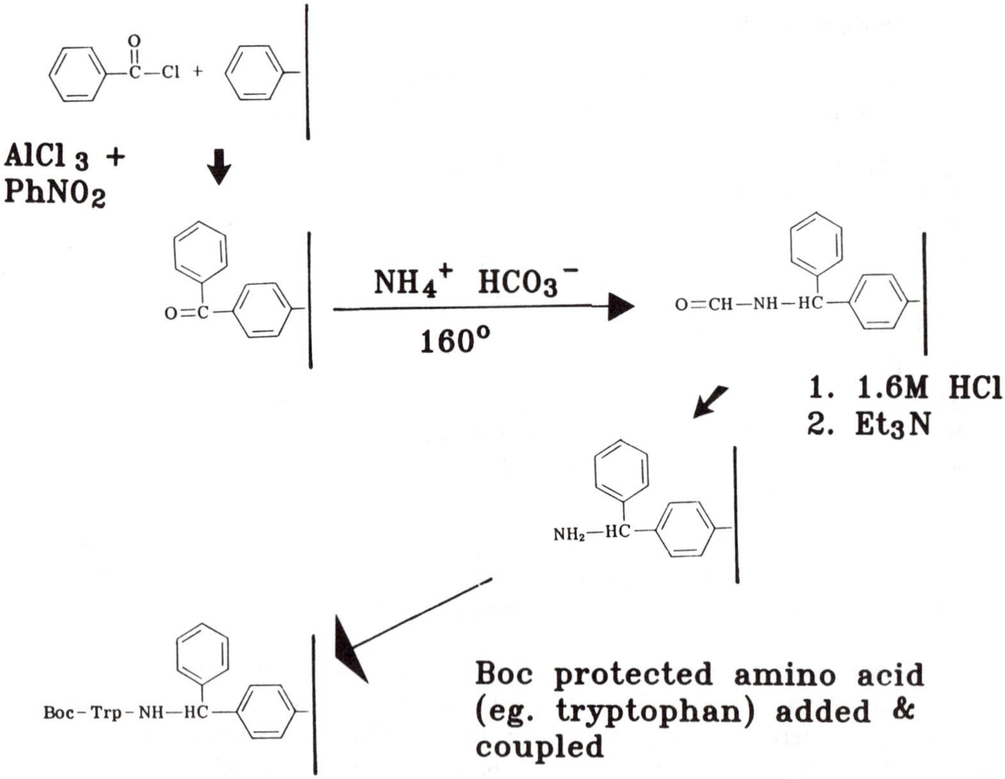

Figure 6-8 Carboxamide linkage used in benzhydrylamine support.

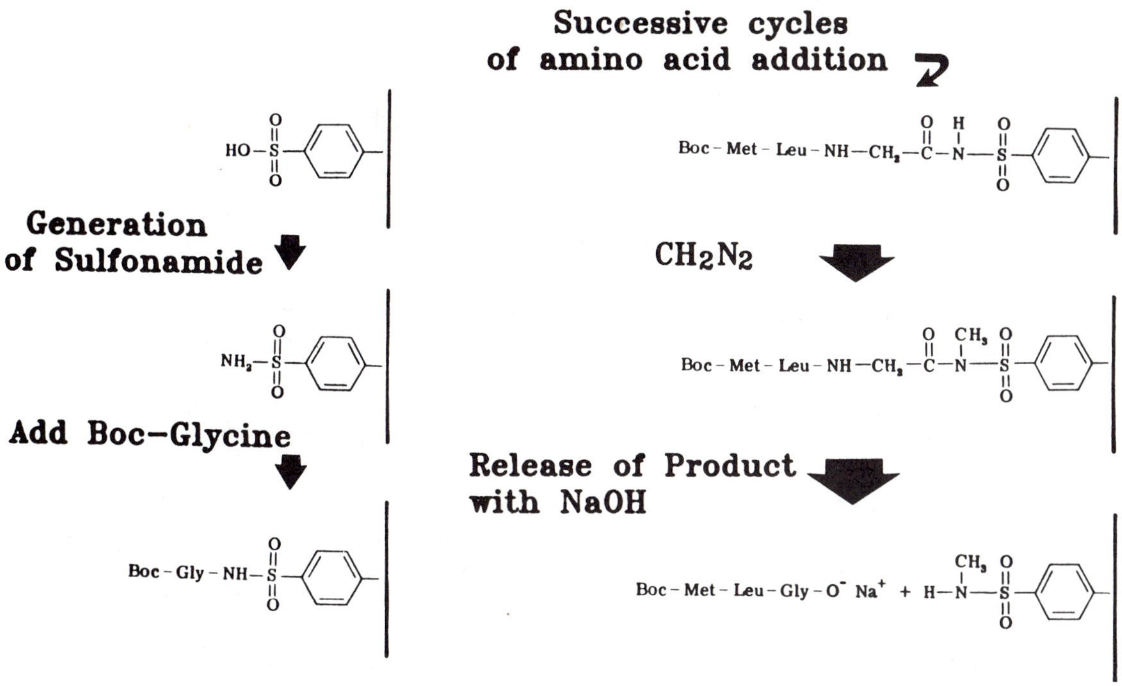

Successive cycles of amino acid addition ⟲

Generation of Sulfonamide

Add Boc–Glycine

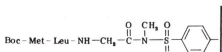

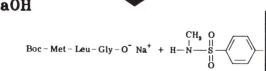

$HO-S(=O)_2-$ ⎯⎯ (aromatic ring, support)

Boc – Met – Leu – NH—CH₂—C(=O)—N(H)—S(=O)₂— ⎯⎯

CH₂N₂

$NH_2-S(=O)_2-$ ⎯⎯

Boc – Met – Leu – NH—CH₂—C(=O)—N(CH₃)—S(=O)₂— ⎯⎯

Release of Product with NaOH

Boc – Gly – NH–S(=O)₂– ⎯⎯

Boc – Met – Leu – Gly – O⁻ Na⁺ + H–N(CH₃)–S(=O)₂– ⎯⎯

Figure 6-9 Sulfonamide linkage used in tripeptide synthesis.

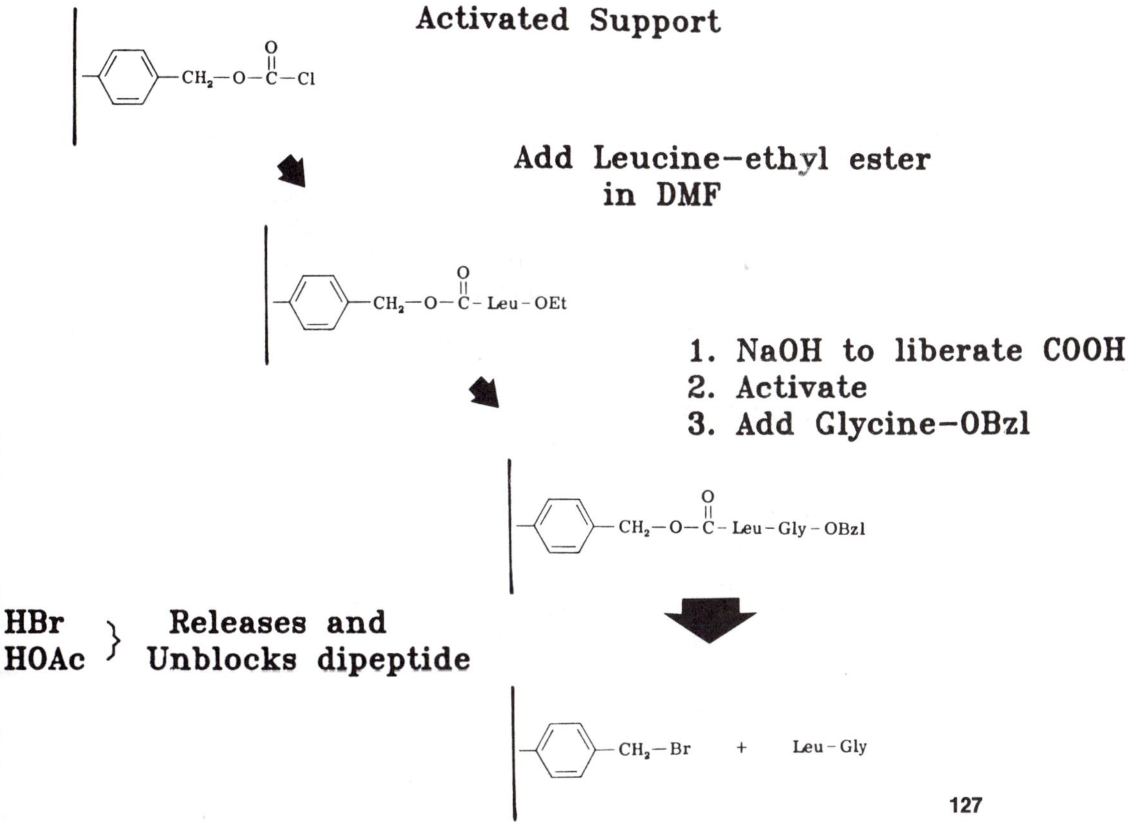

Activated Support

⎯⎯ CH₂—O—C(=O)—Cl

Add Leucine–ethyl ester in DMF

⎯⎯ CH₂—O—C(=O)—Leu – OEt

1. NaOH to liberate COOH
2. Activate
3. Add Glycine–OBzl

⎯⎯ CH₂—O—C(=O)—Leu – Gly – OBzl

HBr ⎫
HOAc ⎬ **Releases and Unblocks dipeptide**

⎯⎯ CH₂—Br + Leu – Gly

127

Figure 6-10 N linkage used in dipeptide synthesis.

carboxyl groups. In addition, specifically deprotecting the appropriate group to allow synthesis in the correct direction represents a problem.

ENZYMATIC APPROACHES

A number of proteolytic enzymes exist which can hydrolyze polypeptide chains at specific points, and it is attractive to consider that such enzymes might be of use in peptide synthesis since enzyme-catalyzed reactions are reversible. Although it has not proved feasible to consider the synthesis of a peptide from individual amino acids via enzymatic means, the ligation of peptide fragments by such processes has been developed.

There are two barriers to enzymatic fragment ligation, one thermodynamic and the other kinetic. The equilibrium position of proteolytic enzymes tends to favor (for obvious reasons) the cleavage of a polypeptide chain. However, the equilibrium constant can be shifted toward ligation, thus overcoming the thermodynamic barrier. This shift is achieved by the inclusion of an organic cosolvent. In the presence of the cosolvent, the concentration of one of the products (of the ligation reaction), water, is reduced, and at the same time the pK values of the terminal carboxyl groups are raised, which at a particular pH acts to raise the concentration of the protonated form of these carboxyls; it is the protonated form that participates in the ligation reaction. The presence of an organic cosolvent thus simultaneously reduces a product concentration and increases a substrate concentration, which leads to a shift in the equilibrium position toward ligation. The organic solvent of choice is glycerol since it does not act as a protein denaturant at high concentrations.

The kinetic barrier results from the limited solubility of the reactant molecules in solvent systems suitable for enzymatic ligation. This problem can be partially overcome if the two fragments to be ligated interact in an appropriate manner to increase the effective local concentration of the reactants. It is possible to resynthesize ribonuclease from the S-protein and the S-peptide quite simply with the proteolytic enzyme acrolein because of the complex formed from the two fragments.

GENETIC APPROACHES TO THE MANIPULATION
OF PRIMARY SEQUENCES

In theory, a tremendous amount of information regarding structure–function relationships in proteins could be obtained if particular amino acid side chains in proteins could be altered and the effects on structure and function observed. As indicated earlier, chemical synthesis of peptides has allowed such studies with relatively small peptides, but the complexity of peptide synthesis precludes such studies with all but the smallest proteins.

An early alternative approach employed the chemical conversion of a residue in situ into a different residue. The hydroxyl of the serine at the active site of sub-

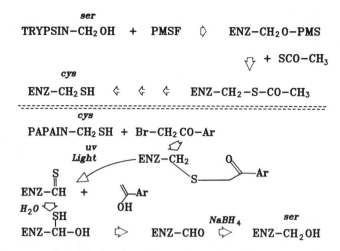

Figure 6-11   Chemical alteration of the serine or cysteine residue at the active site of trypsin and papain, respectively.

tilisin or trypsin could be converted to a sulfhydryl, and the sulfhydryl at the active site of papain converted to a serine residue. The major problem with such interconversions is specificity (actually lacking) and the limited range of alterations that can be attempted.

Although, as shown in Fig. 6-11, it has become possible to chemically alter serine or cysteine residues in active sites, other serine or cysteine residues may also be modified, and the harsh conditions employed for such alterations may lead to secondary reactions of other residues.

In an approach designed to radically alter the activity of the parent protein, some recent chemical modification work has been directed toward derivatizing, for example, the active-site cysteine in papain with flavin coenzymes. Such methods are designed to create new enzyme activities rather than provide insight into the functioning of the parent protein. In the particular case of papain it has become possible to derivatize the cysteine using 8-bromoacetyl-10-methylisoalloxazine with the resultant loss of proteolytic activity, but also with the generation of an effective oxidoreductase activity.

Genetic engineering techniques have allowed a wide range of such interconversions to be accomplished via a process known as site-directed mutagenesis. Many methods exist by which mutations can be introduced into a specific region of DNA sequence. Also, techniques are available to introduce specific mutations into a site where a restriction endonuclease can act. These are limited by a lack of specificity for changing a particular nucleotide and by the necessity for a restriction site (preferably unique) in the region one desires to change, respectively.

Oligonucleotide site-directed mutagenesis is certainly the most powerful tool available when considering sequence–function and sequence–structure relationships in proteins because it allows the investigator to introduce base substitutions, insertions, and deletions at will. In general, two methods are available to introduce mutations at specific sites. The first involves the use of the single-stranded bacteriophage m_{13}. The gene is first cloned into the polylinker region of the replicative form and it propagates to yield single-stranded circular ϕ DNA. From the gene sequence of the protein around the residue to be mutated a nucleotide sequence encompassing two or three amino acid residues either side of the altered codon is derived and the oligonucleotide chemically synthesized. The oligonucleotide containing the mismatch is allowed to anneal in vitro and act as a primer for DNA synthesis using the large (Klenow) fragment of E. coli DNA polymerase I, dNTPs, ligase, and ATP to seal the circle. The double-stranded circular DNA containing the mismatch is introduced into E. coli and the mutation "fixed" by the host. Either the mutant can be propagated and segregated by host replication machinery or the host can "repair" the mismatch either in favor of the wild-type or the mutant allele. Some approaches employ two primers, but the basic theory is the same.

Oligonucleotide site-directed mutagenesis can also be accomplished using double-stranded plasmid DNA in a variety of ways. In all cases, the gene to be mutated is cloned into a plasmid vector. One method involves introducing a single-stranded nick (most often chemically) into the covalently closed circular molecule. All or part of the nicked strand is then degraded using (usually) exonuclease III, leaving a single circular strand of DNA. The same approach is then taken as with the single-stranded phage mutagenesis. An alternative to this is to denature the double-stranded plasmid DNA in the presence of two synthetic oligonucleotides that are complementary to the same strand, lie about 250 base pairs apart, and one of which contains the desired mutation. Klenow, dNTPs, ATP, and ligase are added and the resultant plasmid is introduced into an appropriate E. coli host. The mismatch in the plasmid is handled in a similar manner as that in the single-stranded phage.

Whether single-stranded ϕ or plasmid DNA is used in these methods, mutant clones can be detected by taking advantage of the fact that short duplex DNA containing a single mismatch is more easily denatured than a perfect match. Therefore, using the mutant synthetic oligonucleotide end-labeled with ^{32}P and conditions that only allow perfect matches, one can screen a large number of colonies in a filter hybridization for clones containing the mutated DNA.

UNANSWERED QUESTIONS

The synthesis of small peptides using either the solution or the solid-phase support approach has become almost commonplace and has allowed tremendous advances to be made in understanding structure–function relationships of the peptide hormones. The advent of genetic manipulation of primary sequences will permit similar advances to be made at the level of large proteins. The major challenge of peptide synthesis has shifted away from its original goals. (Although the techniques are now available to answer these questions, they have been applied in very few instances, and much work remains to be done in this area). Peptide synthesis is now being used in a variety of ways that are increasingly important in examining various aspects of protein structure–function relationships. These range from the synthesis of defined short peptides which can then be crystallized to examine the influence of local primary structure on conformation, to the synthesis of peptides with the ability to serve as substrates for various post-translational modifications such as phosphorylation. The most esoteric challenge of peptide synthesis is perhaps the design and synthesis of a protein with predicted structure and function. From the academic standpoint this represents a new stage in the understanding of protein structure–function relationships and depends for its success on the principles of protein architecture described in Chaps. 9 to 11.

7

Chemical Modification:
Side-Chain Specific Reagents

INTRODUCTION

Although chemical modification can be defined as any alteration of the covalent structure of a protein via introduction or removal of a chemical group, we use a more restrictive definition in the context of this chapter. To the enzymologist, chemical modification is the specific (at least attempted) modification of one or more amino acid side chains in a polypeptide chain, usually accompanied by the introduction of a substituent into that group. This type of protein modification has found a wide number of uses in protein chemistry, which we examine together with the practical requirements of the modification reagents that might be helpful in each context. The uses include:

1. The identification of residues involved in the catalytic mechanism of an enzyme or in a binding site on a protein.
2. The introduction of reporter groups, which can be used to report on their environment, to indicate conformational changes or to act as points of reference in distance measurements. Such introduced groups are often fluorescent or have magnetic properties suitable for ESR or NMR, although in theory they may contain any group with distinct spectral properties.
3. The direct detection of conformations or conformational changes in proteins. The reactivity of amino acid side chains is governed by their environment, and the nature of their reaction with chemical modification reagents can reflect this environment.

4. The chemical cross-linking, either of different polypeptide chains within an oligomer or of residues within the same polypeptide chain. Such modifications require bifunctional reagents.

5. The preparation of enzyme derivatives that can be used in hybridization studies to establish subunit composition or to examine subunit–subunit interactions. Such reagents must alter the charge of the protein to allow separation from native molecules.

The general characteristics of the modification reagents that might be required for these various uses can differ depending on the use. Before considering some of the theoretical and practical aspects of chemical modification we define six terms that will be used in this chapter and Chap. 8, on site-specific modification reagents.

Specificity: Refers to the chemical nature of the amino acid side chains modified by a particular reagent. In some cases it is advantageous to use a reagent with high specificity: that is, one that reacts only with, for example, cysteine residues. Such a reagent is particularly useful when attempting to modify a specific side chain in a protein to obtain evidence for its involvement in some aspect of the protein's function. In other instances a high specificity is detrimental—for example, with site-directed (see the following definition) or cross-linking reagents it makes it less likely that the reagent will be of use: They are designed for a particular purpose, not for reaction with specific residues.

Site Directed: Refers to a chemical modification reagent that has incorporated into the molecule some moiety which "directs" it toward a specific binding site on the protein such that any side chains modified have a high probability of residing in or near that site. When it is the active site of an enzyme the term "active-site-directed irreversible inhibitor" is often applied to such a reagent. Since their aim is often to label residues residing within the specific site for later identification, specificity and selectivity (see the following definition) are not highly sought-after characteristics of such reagents.

Selectivity: Refers to the sensitivity a particular reagent shows for the environment surrounding the group or groups it can modify. One with a high degree of specificity is of little use if it shows no selectivity. Little is gained concerning an understanding of the role of a particular type of side chain in a protein's function if all the residues react equally with the reagent. Ideally, it reacts with a high degree of selectivity, modifying either only one of a specific side chain or several of that side chain but with very different rates of modification. Such reagents are particularly useful for examining the environment of certain side chains in a protein.

Reversibility: Most chemical modifications are not readily reversible—in fact, it would be a serious handicap if they were as it would make identification of the modified residue or residues quite difficult. However, in certain circumstances it is advantageous to be able to reversibly modify a protein. This is particularly true in cases where chemical modifications are to be used in the construction of hybrid molecules or where loss of biological activity as a result of a modification must be reacquired.

Photo-activatable: Refers to a group incorporated into a modification reagent which remains inactive until activated by light. Two types are commonly used: (1) so-called photo-affinity labels, where the photo-activatable moiety is the only chemical modification reagent in the molecule, the rest representing a site-directing moiety, and (2) hetero-bifunctional reagents, where the reagent contains two reactive moieties, one of which is reactive at all times and the other only after photo-activation. The advantages and uses of these types of reagents is discussed fully later in this chapter.

Suicide: Refers to reagents that *become* modification reagents after an enzymatic process has taken place that generates an active group on the reagent. They can be regarded as the ultimate in site-directed reagents.

Chemical modification experiments have taken many forms, but in general, reagent selection usually involves some form of trial and error. The pH dependence of kinetic parameters or ligand binding is often used to give a pK that can guide the design of a modification experiment. Once a modified protein is obtained it is often necessary to determine which amino acid residue (or residues) have been modified in terms of their location in the primary amino acid sequence of the protein. This involves the procedures discussed in Chap. 5.

As detailed at the start of this chapter, chemical modification experiments have been employed in a wide variety of ways. Because of this broad application and the detailed information that can be obtained, it is necessary to consider some of the pitfalls that await the experimentalist using chemical modification methods. The purpose is not to discourage the use of these approaches, but to give an awareness so that appropriate safeguards can be taken. Not all of the pitfalls listed here are always pitfalls—it depends on the nature of the information that the experiments are designed to provide.

1. Few reagents are specific to the extent that they react only with a particular type of side chain. Inferences regarding the functioning of specific side chains thus depend on the direct experimental demonstration that a particular type of side chain is being modified.

2. It is not possible to chemically modify an amino acid side chain without affecting the conformation of a protein. When chemical modification is used to introduce reporter groups, not only must the uniqueness of labeling be established but also the lack of *significant* (in terms of biological activity) effect on the protein's conformation must be confirmed before inferences concerning the potential role of measured conformational changes can be established. Reporter groups used to monitor the environment around a particular side chain *do not*: They reflect the environment around the reporter group, and since such groups are often bulky and hydrophobic, they can alter the environment of the side chain. Information concerning the side chain of a particular amino acid is best obtained by monitoring the

reactivity of that side chain, not the spectral properties of an introduced group after modification.

3. It is hard to predict the behavior of a particular reagent toward a protein—because it modifies a particular side chain in model compounds or in another protein does not mean that it will behave similarly in all proteins.

4. Interpretation of chemical modification data is not always straightforward. Except in the case of site-directed reagents it is difficult ruling out conformational effects as being the cause of loss of activity rather than the specific modification occurring in the appropriate specific site. Even ligand protection experiments do not allow an unequivocal determination to be made concerning the location of a specifically modified residue.

In this chapter we first examine various theoretical aspects of chemical modification experiments, followed by a discussion of different side-chain specific reagents. Finally, some applications of modification studies are considered.

THEORETICAL CONSIDERATIONS

There are two major factors affecting the reactivity of a specific amino acid side chain toward a reagent that under ideal conditions reacts with it. These are: (1) effects on the pK of the reacting group (which affects its nucleophilicity), and (2) steric effects.

pK Values of Reacting Groups

A number of factors influence the pK of an ionizable amino acid side chain in a protein, including the electrostatic field of the protein (especially the local charge distribution around the particular side chain), the solvation of the group and the hydrogen bonds, if any, that the group may be involved in. An examination of th. pK values commonly found in protein side chains shows a range from approximately 3.75 to >12. Table 7-1 shows pK values for amino acid side chains as determined in the free amino acids.

These pK values do not, however, necessarily represent what would be found in a protein for each side chain. Even in dipeptides the pK for histidine's imidazole

TABLE 7-1   Side-chain pK values in free amino acids

α-COOH	3.75
β,γ-COOH	4.6
Imidazole	7.0
α-NH$_2$	7.8
ε-NH$_2$	10.2
—SH	9.2
Phenolyic	9.6
Guanidyl	>12.0

TABLE 7-2 Imidazole pK values in small peptides

His-Gly	6.22
Gly-His	6.95
His-Lys	6.48
Gly-His-Gly	6.72

is quite variable. Table 7-2 shows imidazole pK values determined by NMR measurements for a series of peptides.

If we examine the experimentally determined pK values for histidines in several proteins, we find further evidence for the range of pK values the imidazole side chain can have as a result of its environment in a protein. Carbonic anhydrase and staphylococcal nuclease each have four histidine residues. The pK values of these residues (determined by NMR) are:

Carbonic anhydrase: 5.91, 6.04, 7.0, 7.23
Staphylococcal nuclease: 5.37, 5.71, 5.74, 6.5

Between these two proteins we have a range from 5.37 to 7.23, almost 2 full pH units, for the pK of histidine.

The effects of environment on pK can also be demonstrated by examining the pK of a single group under a number of different circumstances. The γ-COOH of glutamate-35 in lysozyme has a pK of 5.9 in the native enzyme, compared to a value of 4.6 for the free amino acid. When lysozyme is denatured the pK of glutamate-35 falls to a value of 4.6. When the native enzyme is allowed to bind the inhibitor tri-N-acetylglucosamine, the pK rises to 6.4. The native conformation, and conformational changes involving a particular residue, can affect the pK of a particular side chain.

Competitive Labeling Method. The pK values of reacting side chains can be determined via this method, which is based on the fact that in the presence of a trace amount of a radioactive modification reagent, the various groups in a protein that can be modified by the reagent will compete for the label. As a result, the amount of radioactivity incorporated into any group of the protein is determined by the pK of the group and the environment. Because only a trace amount of label is used, complexity as a result of one modification affecting subsequent modifications is avoided.

In practice, the protein and an internal reference standard are reacted with a limiting amount of radioactive label. The reaction is quenched by addition of an excess of unlabeled reagent. This not only stops the reaction, but generates a chemically homogeneous product which is heterogeneous with respect only to isotopic label, preventing problems due to different chemical modifications in subsequent peptide analysis. After quenching, the protein is subjected to this analysis using fragmentation methods which give peptides containing only one modified residue. The amount of label in the various peptides depends on the reactivity of the individual groups, and the rate of labeling of the various peptides relative to the rate of labeling of the internal standard is found from a time-dependence profile. The relative rate of modification is then determined at a series of pH values, and from a plot of the relative rate versus pH the pK of the reacting group can been calculated. Once the pK of the reacting group has been found, the rate constant of reaction with the group can be expressed by the pH-independent or specific rate constant given in the equation

$$K_{\text{obs}} = K_n \frac{K_i}{K_i + [\text{H}^+]} \tag{7-1}$$

where K_{obs} is the rate constant at a particular pH, K_n the pH-dependent rate constant, K_i the ionization constant of the group, and H^+ the hydrogen-ion concentration at the particular pH. The ratio of K_n to the pH-independent rate constant of modification of a standard, free residue (K_s) gives $K_r = K_n/K_s$. In practice, the *amount* of modification of each modifiable residue is taken as a ratio to the amount of modification of an added standard under the same conditions, to give a value of K_r. A Brønsted plot (see Fig. 7-1) of log K_r versus pK for all similar groups in the protein can be plotted. Points that lie on the line indicate groups that show normal reactivity, points that lie below indicate those that are subject to steric hindrance, while points that lie above indicate residues that are especially reactive, either as the result of a uniquely reactive orientation or of specific reagent binding in the proximity of the reacting group giving local concentration effects.

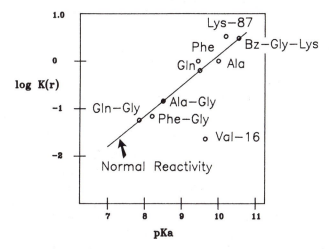

Figure 7-1 Brønsted plot for the reaction of acetic anhydride with various amines and with certain residues in native elastase. K_r is the ratio of the second-order velocity constant for the unprotonated amino group (K) to that for unprotonated phenylalanine (K_s).

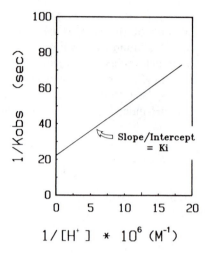

Figure 7-2 Plot of $1/K_{obs}$ versus $1/[H^+]$ to determine the ionization constant, K_i, for a reacting group.

Where a single group is known to react with a reagent, the pH dependence of the reaction can be expressed as

$$K_{obs} = \frac{K_{max}}{1 + K_i/[H^+])} \qquad (7\text{-}2)$$

where K_{max} is the intrinsic pseudo-first-order rate constant when the group is protonated. Equation (7-2) in double reciprocal form,

$$\frac{1}{K_{obs}} = \frac{1}{K_{max}} + \frac{K_i}{K_{max}(1/[H^+])} \qquad (7\text{-}3)$$

shows that a plot of $1/K_{obs}$ versus $1/[H^+]$ yields a straight line (Fig. 7-2) and that K_i can be obtained from the slope divided by the intercept.

Steric Effects

In a protein, the reactivity of a particular side chain to a modification reagent may well be affected by the accessibility of the reagent to the group. Such steric effects may slow reaction with the reagent (resulting in a below-the-line point in a Brønsted plot) or prevent reaction altogether. Reacting groups are sometimes classified as exposed or buried based on their reactivity with chemical modification reagents. Because of the reagents' varying natures, all of which react with similar specificity, some amino acid side chains may appear to be exposed to some reagents and buried to others. Classification of groups as exposed or buried on the basis of chemical reactivity alone may not be completely accurate. Generally, three classes are identified:

1. *Internal.* residues with completely buried side chains having no solvent accessibility

2. *Surface.* side chains partially buried, or accessible to solvent from one side only

3. *External.* side chains that project into the surrounding solvent and are freely accessible to modification reagents

QUANTITATION OF MODIFICATION AND DATA ANALYSIS

During a chemical modification experiment a number of experimental parameters can be determined and related to the number of residues being modified and their effects on activity. Here we examine some of the more usual methods of representing chemical modification data and discuss what information can be obtained from such analysis.

Time-Course Analysis

In most experimental situations there is a considerable excess of modification reagent to enzyme residues that can be modified, and the reaction can be regarded as pseudo-first order. The modification can be monitored either by following directly the modification of a particular type of side chain (which is quite easy if spectral changes result from the process), or by following the effect of modification on some enzymatic parameter of the protein (activity, regulation by an allosteric ligand, etc.).

The rate of inactivation, V_{inact}, is given by

$$V_{inact} = \frac{-d[E]}{dt} = K_{inact}[E] \qquad (7\text{-}4)$$

Rearrangement and integration between the limits 0 and t for time and E_0 and E_t for $[E]$ gives

$$\ln \frac{E_t}{E_0} = -K_{inact}t \qquad (7\text{-}5)$$

where E_t is the activity at time t and E_0 is the initial activity (these are alternatively represented as A_t and A_0, respectively, by some workers).

Since $\ln(E_t/E_0)$ is the log of the residual activity at any time t, a semilog plot of log residual activity versus time, as shown in Fig. 7-3, gives a value for the rate constant of inactivation, K_{inact}.

In plots of log residual activity versus time, straight-line plots are obtained unless two or more residues per protein molecule, each contributing to the activity but reacting with quite different rate constants, exist. In such a case the plot of log residual activity versus time is multiphasic (if the rate constants are sufficiently separated), and rate constants can only be obtained by fitting the data to two or more exponentials. In plots of log percent residues modified versus time, complex plots are obtained if more than one residue reacts and the rate constants are sufficiently separated.

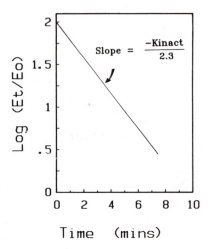

Figure 7-3   Semilog plot of log residual activity (E_t/E_0) versus time of incubation with reagent. The slope of this plot is $-K_{inact}$.

It is important to note that such analysis of the time dependence of inactivation assumes that the modification process is the only one utilizing the modification reagent. It is possible that the reagent may be subject to a competing process such as hydrolysis, and in such a case this process will contribute to a decrease in reagent concentration as a function of time. To obtain the true value of the apparent first-order rate constant for inactivation, a plot of log residual activity versus $(1 - e^{-k't})k'$ must be made, according to

$$\ln \frac{E_t}{E_0} = -\frac{k}{k'} I(1 - e^{-k't}) \tag{7-6}$$

where k' is the rate constant for hydrolysis of the reagent and I is the initial concentration of the reagent. k' must be experimentally determined independently. This problem is often encountered with reagents such as diethylpyrocarbonate, and Fig. 7-4 shows experimental data obtained with modification of S-adenosylhomocysteinase at a series of diethylpyrocarbonate concentrations. Also shown is a plot of K_{obs} versus [reagent].

The rate constant, K', for the decomposition of diethylpyrocarbonate is determined by incubating the reagent under the reaction conditions, withdrawing aliquots at appropriate time intervals, adding a large excess of imidazole and quantitating the formation of ethoxyformyl imidazole at 242 nm, using a molar extinction coefficient of 3200 cm^{-1}. In this way the quantity of reactive diethylpyrocarbonate at any time, t, is quantitated and the rate constant of decomposition is determined from a plot of log ([DEP]$_t$/[DEP]$_0$) versus time, as in Fig. 7-3.

An important property of K_{inact} is that except in cases of specific complex formation of the target protein with the modification reagent (MR), a plot of K_{inact} versus the concentration of the modification reagent should be linear and pass through the (0, 0) axis, as shown in Fig. 7-5.

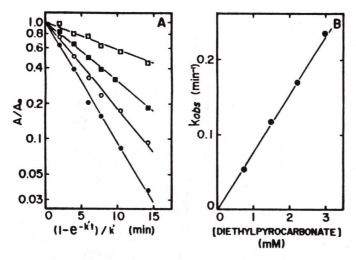

Figure 7-4 Inactivation of S-adenosylhomocysteinase by diethylpyrocarbonate. The enzyme (16 μM subunit) was incubated with 0.74 ($\square$), 1.48 ($\blacksquare$), 2.21 ($\bigcirc$), or 2.98 ($\bullet$) diethylpyrocarbonate in 0.1 M potassium phosphate buffer, pH 6.9, at 0°C. At time intervals, aliquots were removed for measurements of the residual enzyme activity. The first-order rate constant for decomposition of diethylpyrocarbonate (K') was determined separately as described in the text. Values on the abscissa are calculated with $K' = 7.5 \times 10^{-3}$ min^{-1}. S-Adenosylhomocysteinase incubated in the absence of diethylpyrocarbonate lost no activity under these conditions. (Reprinted with permission from: T. Gomi and M. Fujioka, *Biochemistry*, *22*, 137–143. Copyright 1983 American Chemical Society, Washington, D. C.)

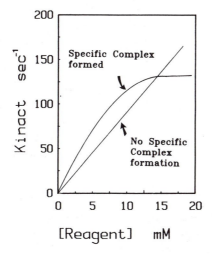

Figure 7-5 Plot of K_{inact} versus concentration of modification reagent.

In some instances a specific complex between the reagent (MR) and the protein may occur, and the modification process consists of two stages:

$$\text{protein} + \text{MR} \underset{}{\overset{K_d}{\rightleftharpoons}} \text{protein} - \text{MR} \overset{k}{\longrightarrow} \text{inactivation} \tag{7-7}$$

If such complex formation occurs, a plot of K_{obs} versus [reagent] concentration is hyperbolic and K_d, the dissociation constant for protein–reagent complex formation, can be obtained either by direct fitting of the hyperbola or from a double reciprocal plot of K_{obs} versus 1/[reagent].

Under conditions where the reagent concentration is not in large excess over the protein the assumption of pseudo-first-order kinetics is not valid, and the second-order rate constant for inactivation can be calculated from

$$\frac{2.303}{a - b} \log \frac{b(a - x)}{a(b - x)} = k_t \tag{7-8}$$

where a and b are the concentrations of reagent and enzyme at time $t = 0$, and x is the amount of enzyme modified at any time, t.

This analysis does not indicate how many of a particular residue must be modified for activity to be lost. The *minimal* order of the reaction is the slope obtained from a plot of log $1/t_{1/2}$ versus log [reagent]. This type of plot, although often used, has been criticized since it can lead to an underestimate of the number of residues necessary for activity in circumstances where a slow, rate-limiting reaction with one residue leads to a more rapid, kinetically unobservable modification of other residues that are essential for activity. *Estimates of the number of residues reacted from kinetic measurements are not a substitute for direct chemical evidence.*

Quantitation

As indicated in the last section, the number and types of groups that have been modified in a particular experiment must be directly determined by use of radioactive modification reagents, by amino acid analysis after modification, or by spectroscopic means when the modification results in the incorporation of a spectrally active group. There are two basic ways that quantitation can be achieved: The unreacted residues can be estimated by amino acid analysis and the number of residues reacted established by difference or **alternatively,** the reacted residues can be quantitated directly. In many ways the latter approach is preferable. With the first approach it is often a case of determining a difference between two quite large numbers, both of which have significant experimental uncertainty. Also, the modified residues may not separate sufficiently from the parent residue to allow unique determination of the unmodified residues. Direct quantitation can be achieved easily and accurately if radiolabeled reagents are available. Many reagents lead to the incorporation of chromophoric groups, which, after removal of unreacted reagent, may be quantitated spectrally. One particular problem with this appraoch is the environment of the group, which can affect the spectral properties. When environmentally sensitive chromophores are used, it is advisable to denature the protein after modification and quantitate the incorporation in such a uniform state where tertiary structure does not influence the spectral properties. Where quantitation can be carried out via spectral measurements, the utility of this approach is indicated for individual reagents discussed. Assuming that this can be achieved, plots of residual activity versus number of groups modified per mole of enzyme can be made.

We consider several types of chemical modifications, depending on the number of amino acid residues modified and the relationship between the rate of chemical modification and the rate of inactivation.

In the most simple case, the rate of modification equals the rate of inactivation and there is a single residue involved. The rate of modification, V_{mod}, is given by

$$V_{mod} = \frac{dx}{dt} = K_{mod}(x_m - x) \tag{7-9}$$

where x is the number of modified groups and x_m is the maximum number for modification.

$$\int_0^{x_1} \frac{dx}{x_m - x} = \int_0^1 k_{mod}\, dt \tag{7-10}$$

Gives

$$\ln \frac{x_m - x_1}{x_m} = -k_{mod}t \tag{7-11}$$

Since $K_{inact} = K_{mod}$, we get

$$\ln \frac{x_m - x_1}{x_m} = -k_{inact}t = \ln \frac{E_t}{E_0} \tag{7-12}$$

or

$$\frac{E_t}{E_0} = 1 - \frac{x_1}{x_m} \tag{7-13}$$

and a plot of E_t/E_0 as a function of x_t, the number of residues modified per molecule of enzyme, is linear with a negative slope of $1/x_m$. When $E_t/E_0 = 0$, $x_t = x_m$. Figure 7-6 shows such a case.

When the rates of modification and inactivation are not equal, we can define a ratio of the rate constants, r.

$$r = \frac{K_{inact}}{K_{mod}} \tag{7-14}$$

As before,

$$\ln \frac{E_t}{E_0} = -k_{inact}t \tag{7-15}$$

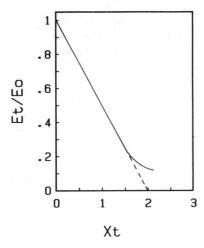

Figure 7-6   Plot of E_t/E_0 versus X_t for determining the maximum number of "essential" groups. E_0 is the initial enzyme activity at time t. The maximum number of "essential" groups in this case is interpreted to be two.

and

$$v_{\text{mod}} = \frac{dx}{dt} = k_{\text{mod}}(x_m - x) \tag{7-16}$$

which, in conjunction with Eq. (7-14), gives

$$r \ln \frac{x_m - x_1}{x_m} = -k_{\text{inact}}t \tag{7-17}$$

and we get

$$r \ln \frac{x_m - x_1}{x_m} = \ln \frac{E_t}{E_0} \tag{7-18}$$

which can be rearranged to give

$$\left(\frac{E_t}{E_0}\right)^{1/r} = 1 - \frac{x_t}{x_m} \tag{7-19}$$

In terms of E_t/E_0, we get

$$\left(\frac{E_t}{E_0}\right) = \left(1 - \frac{x_1}{x_m}\right)^r \tag{7-20}$$

which predicts the family of parabolic curves shown in Fig. 7-7, with the shape depending on the value of r. When $r = 1$ the plot resembles the previous one, but a variety of plots are possible.

Since it is inherently unlikely that $r > 1$, plots with $r < 1$ are most likely to be encountered. Mechanistically, such a plot indicates that inactivation is the result of a rate-limiting conformational change resulting from the modification event.

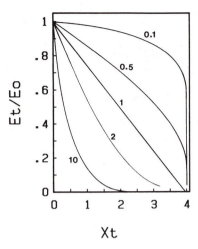

Figure 7-7   Effects of the value of r on the shape of plots of E_t/E_0 versus X_t.

A more likely situation is one in which there are several classes of a particular type of residue that can react and lead to loss of activity, each of which react with different rates, K_{mod_n}. The overall rate of modification, V_{mod}, is the sum of the rates of the individual classes:

$$v_{mod} = \sum^n \frac{d(x_n)}{dt} = \sum^n k_{mod_n}[(x_n)_m - (x_n)] \tag{7-21}$$

where $(x_n)_m$ is the maximum number of groups available in each class and x_n is the number of groups reacted. For each class of residue,

$$\ln \frac{(x_n)_m - (x_n)_t}{(x_n)_m} = -k_{mod_n}t \tag{7-22}$$

and the overall modification reaction is described by

$$\sum^n \ln \frac{(x_n)_m - (x_n)_t}{(x_n)_m} = -\sum^n k_{mod_n}t \tag{7-23}$$

$$\sum^n (x_n)_t = x_T \tag{7-24}$$

and is the quantity experimentally measured—that is, the sum of all modified residues at time t.

E_t/E_0 as a function of x is obtained as follows:

$$(x_1)_t = x_T - \sum_{n=2}^n (x_n)_t \tag{7-25}$$

and from previously,

$$(x_n)_t = (x_n)_m[1 - \exp(-k_{mod_n}t)] \tag{7-26}$$

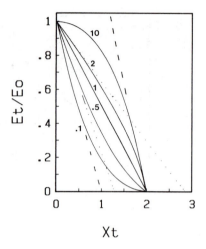

Figure 7-8   Plot of Eq. (7-27) for $n = 2$; $(X_1)_m = (X_2)_m = 1$; $K_{mod_1} = K_{inact}$; and K_{mod_2}/K_{mod_1} as indicated for each solid curve. ---, lower and upper limiting curves; $\cdots$, extrapolation of curves to $E_t/E_0 = 1$ and $E_t/E_0 = 0$.

Combining Eqs. (7-25) and (7-26) and the equation for E_t/E_0 when there is a single inactivation and modification rate from previously [Eq. (7-20)], we get

$$\frac{E_t}{E_0} = 1 + \frac{\sum\limits_{n=2}^{n} \{(x_n)_m[1 - \exp(-k_{mod_n}t)]\}}{(x_1)_m} - \frac{x_T}{(x_1)_m} \tag{7-27}$$

which predicts that a plot of E_t/E_0 versus x_t will generally be exponential since t is an independent variable. Plots of this equation are shown in Fig. 7-8 for a variety of combinations of $(x_n)_m$ and K_{mod_n} and n.

In the event that K_{mod} for one of the groups is equal to K_{inact} and the extent of modification can be determined independently of other modifications, we get

$$\ln \frac{(x_1)_m - (x_1)_t}{(x_1)_m} = -k_{inact}t = \ln \frac{E_t}{E_0} \tag{7-28}$$

Therefore,

$$\frac{E_t}{E_0} = 1 - \frac{(x_1)_t}{(x_1)_m} \tag{7-29}$$

which is analogous to the simplest case first considered.

Since this is the only situation that can give unique information, it is necessary, where simple E_t/E_0 versus extent of modification plots are not obtained, to experimentally determine the extent of modification of individual residues where more than one residue can be modified, and to correlate activity losses with such unique modifications.

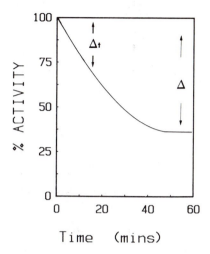

Figure 7-9 Plot of data indicating that modification of one or more residues leads to only *partial* loss of activity. The maximum loss of activity is Δ, and the activity loss at any intermediate time is Δt.

In the examples thus far considered we have assumed that maximal modification leads to *complete* loss of activity (whether it is catalytic activity or regulatory activity). It is quite possible (and often encountered) that modification leads to only partial loss of activity, as indicated in Fig. 7-9.

When such plots are constructed using varied time it is important that the same value of Δ be obtained at different concentrations of modification reagent. If increasing values of Δ are obtained as the concentration of the modification reagent is increased, two explanations are possible. In the first, the progress curves all behave as simple exponentials and the increased values of Δ indicate that an equilibrium between bound and unbound modification reagent is involved. In the second, the progress is multiphasic, indicating that more than one class of residues is being modified, each affecting the activity, possibly to different extents.

In the simple case where Δ does not change with increasing reagent concentration, the parameter $\Delta t/\Delta$ is used in place of E_t/E_0.

The Tsou Plot. When both the amount of modification per protein molecule *and* the percent residual activity can be determined as a function of time, information concerning the number of modified residues essential for activity can be obtained by application of an equation first developed by Tsou:

$$nx = pa^{1/i} + (n - p)a^{\alpha/i} \tag{7-30}$$

where n is the total number of residues in the protein of the type being modified, x the molar function of those residues remaining unmodified after a given time, a the residual activity, p the number of the residues that are modifiable, i the number

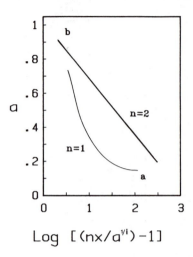

$$\text{Log } [(nx/a^{vi}) - 1]$$

Figure 7-10 Tsou plot for a hypothetical situation with $n = 7$ and $p = 3$. Line (a) results from $i = 1$, line (b) results from $i = 2$. The linearity of line (b) indicates that $i = 2$ in this hypothetical situation.

of those modifiable residues essential for activity and α the ratio of the rates of modification of nonessential versus essential residues. Equation (7-30) can be rewritten as

$$\log\left(\frac{nx}{a^{1/i}} - p\right) = \log\left(n - p\right) + \frac{\alpha - 1}{i}\log a \qquad (7\text{-}31)$$

A plot of $\log a$ versus $\log (nx/a^{1/i} - p)$ has a slope of $(a - 1)/i$.

Once the rate curves of the inactivation are determined, and assuming that n and p are known from independent experiments, the data are plotted according to Eq. (7-31) with a series of values of i. The correct value of i results in a linear plot, as shown in Fig. 7-10.

Protection Experiments

As will be discussed in Chaps. 11 and 17, chemical modification studies can be used in a variety of ways to study ligand binding or conformational changes induced by ligand binding. Further aspects of chemical modification studies in relation to these topics are dealt with in the appropriate chapters. Here we briefly discuss the concept of protection experiments and what information from them may mean with regard to the existence of certain amino acid residues in a particular binding site. One of the prime aims of chemical modification experiments is to attempt to elucidate what types of residues may be important in certain protein functions. Modification studies are often designed to establish important residues in substrate or ligand binding sites. Frequently, studies report that modification of a particular type of residue results in loss of ligand binding ability by the protein, and that the presence of a particular ligand prevents the modification and loss of activity. The inference of such experiments is often that these results indicate the presence of the particular type of residue being studied in or near the ligand binding site. The "or near" is added to cover the possibility that the ligand in its binding site may sterically block access of the modification reagent to a nearby reactive residue, and vice versa. Such conclusions are unwarranted since there is another equally probable explanation. Modification of a residue some distance removed from a particular ligand binding site may lead to a conformational change in the protein which prevents ligand from binding. In the presence of ligand, there is a ligand-induced conformational change which prevents or hinders the modification reagent from attacking its target group. Chemical modification experiments with side-chain specific reagents are not particularly useful in defining what types of residues may be present in or near a ligand binding site. These comments do not of course apply to site-specific reagents, and with these, as is discussed later, ligand protection experiments can provide useful information.

SIDE-CHAIN SPECIFIC REAGENTS

This section of the chapter is not designed to give a comprehensive account of the many reagents available for the chemical modification of amino acid side chains in proteins. The reagents discussed here have been selected on the basis of (1) their widespread use, and (2) the purpose behind the modification reaction (i.e., the introduction of a reporter group, the alteration of the charge of a residue, or their reversibility). For ease of discussion the reagents described are dealt with in terms of their prime specificity, although, as emphasized earlier, many are not particularly specific. In some cases specificity is improved under certain conditions, and some of these are mentioned. During the descriptions of these various reagents, comments regarding the quantitation of modification by spectral methods are made where appropriate.

Modification of Amino Groups

Modifications of amino groups can be categorized in terms of how they effect the charge of the side chain: the formal charge at neutral pH can be retained, eliminated, or reversed. Table 7-3 shows the average amino acid composition and fractional exposure of residues in proteolytic enzymes whose crystal structures have been examined. Lysine is clearly the most exposed residue in these proteins and makes up a fairly large proportion of the composition of the "average" protein. In fact, of the residues that can be readily modified, lysine is present in much higher proportions than the others. Although Table 7-3 is based on one class of proteins, they are in general of "average" size and compact structure and can be considered as representative of proteins in general. As a result, it is likely that most proteins have lysine residues that can be modified.

TABLE 7-3 Exposure and average percent composition of amino acids in proteolytic enzymes of known crystal structure

Residue	Fractional exposure	Percent composition
Lysine	0.58	7.0
Glutamate	0.49	5.3
Glutamine	0.43	4.1
Arginine	0.43	3.9
Serine	0.43	7.8
Asparagine	0.43	4.8
Aspartate	0.42	5.2
Proline	0.40	5.5
Threonine	0.37	6.5
Glycine	0.33	7.6
Alanine	0.32	8.1
Histidine	0.27	2.2
Tyrosine	0.25	3.4
Phenylalanine	0.17	3.5
Valine	0.16	6.9
Isoleucine	0.14	4.6
Cysteine	0.13	3.4
Methionine	0.11	1.6
Tryptophan	0.11	1.2
Leucine	0.10	7.3

Modification with Retention of Charge. The following are examples of reagents that leave an ionizable group capable of being positively charged on the lysine side chain. The charge is, however, located at a different position than it was originally.

Reductive Alkylation: Reaction with an aldehyde gives a Schiff's base which is subsequently reduced with sodium borohydride (Fig. 7-11). The aldehyde used can have different R groups, such as formaldehyde, acetaldehyde, propionaldehyde, and so on, which allows information to be obtained regarding the steric environment of the side chain.

$$\text{ENZ-NH}_2 + \text{RHCO} \underset{+H_2O}{\overset{-H_2O}{\rightleftharpoons}} \text{ENZ-N=CHR} \xrightarrow{[H]} \text{ENZ-NH-CH}_2\text{R}$$

Figure 7-11 Schiff's base formation and reduction with sodium borohydride.

Amidination with Methylacetimidate: The reaction (Fig. 7-12) gives a derivative that is quite stable at acid or neutral pH but labile at alkaline pH. The formal positive charge, which is retained, is delocalized over several atoms by resonance.

Modification with Loss of Charge. These reagents suppress the protonation of the lysine ε-amino group, and thus the derivatives can carry no charge.

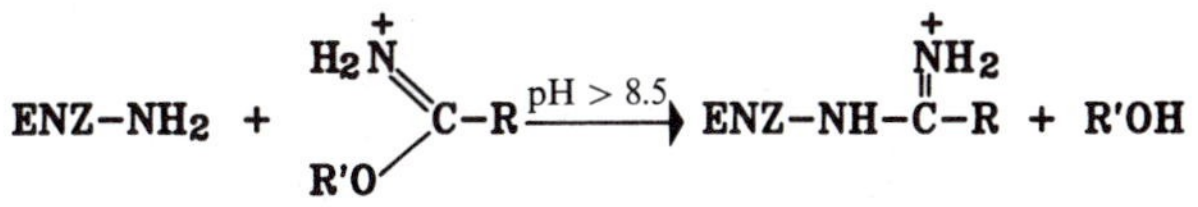

$$\text{ENZ-NH}_2 \ + \ \begin{matrix} \overset{+}{\text{H}_2\text{N}} \\ \diagdown \\ \text{R'O}^{\diagup} \end{matrix} \text{C-R} \xrightarrow{\text{pH} > 8.5} \text{ENZ-NH-}\overset{\overset{+}{\text{NH}_2}}{\underset{\|}{\text{C}}}\text{-R} \ + \ \text{R'OH}$$

Figure 7-12 Amidination of lysine residue.

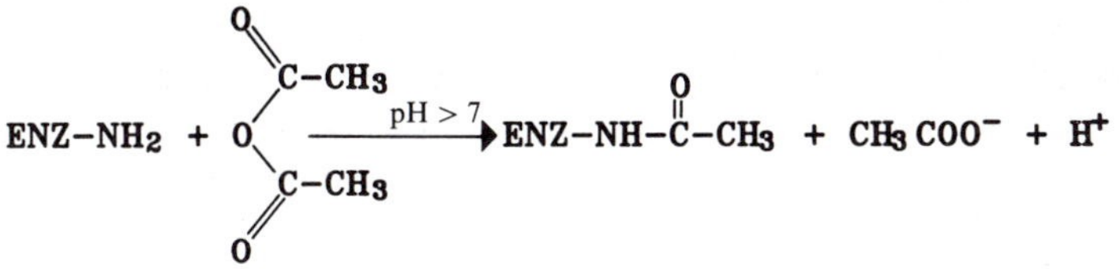

$$\text{ENZ-NH}_2 \ + \ \begin{matrix} \overset{O}{\diagdown}\text{C-CH}_3 \\ O \\ \diagup \\ \underset{O}{\text{C-CH}_3} \end{matrix} \xrightarrow{\text{pH} > 7} \text{ENZ-NH-}\overset{O}{\underset{\|}{\text{C}}}\text{-CH}_3 \ + \ \text{CH}_3\text{COO}^- \ + \ \text{H}^+$$

Figure 7-13 Reaction of amino group with acetic anhydride.

Acylation: Reaction of amino groups with acetic anhydride gives an acylated derivative (Fig. 7-13). Although acetic anhydride reacts readily with amino groups at neutral or slightly alkaline pH values, reaction with either sulfhydryl or imidazole groups can occur.

Carbamoylation: This reaction is analogous to that discussed in Chap. 5 for amino-terminal labeling. Reaction of amino groups with cyanate (Fig. 7-14) gives a quite stable derivative. As is also shown in Fig. 7-14, cyanate can react with sulfhydryl and phenolic groups, but at mildly alkaline pH values (approximately pH 8.0) derivatives formed by these reactions are not stable.

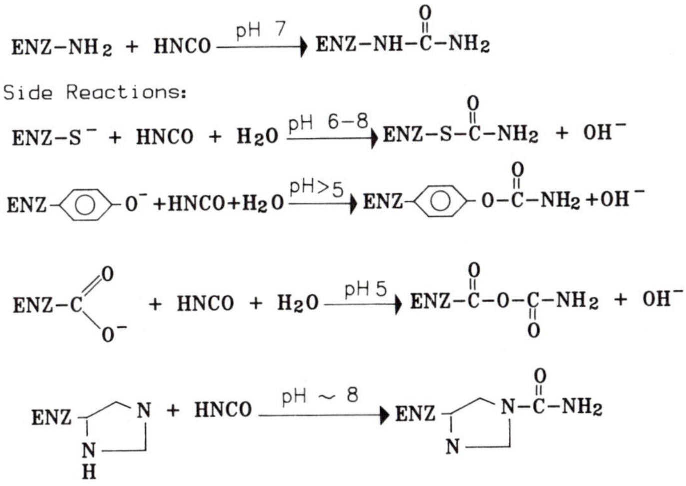

$$\text{ENZ-NH}_2 \ + \ \text{HNCO} \xrightarrow{\text{pH } 7} \text{ENZ-NH-}\overset{O}{\underset{\|}{\text{C}}}\text{-NH}_2$$

Side Reactions:

$$\text{ENZ-S}^- \ + \ \text{HNCO} \ + \ \text{H}_2\text{O} \xrightarrow{\text{pH } 6-8} \text{ENZ-S-}\overset{O}{\underset{\|}{\text{C}}}\text{-NH}_2 \ + \ \text{OH}^-$$

$$\text{ENZ-}\langle\bigcirc\rangle\text{-O}^- \ + \text{HNCO} + \text{H}_2\text{O} \xrightarrow{\text{pH}>5} \text{ENZ-}\langle\bigcirc\rangle\text{-O-}\overset{O}{\underset{\|}{\text{C}}}\text{-NH}_2 + \text{OH}^-$$

$$\text{ENZ-C}\begin{matrix}\diagup O \\ \diagdown \\ O^-\end{matrix} \ + \ \text{HNCO} \ + \ \text{H}_2\text{O} \xrightarrow{\text{pH } 5} \text{ENZ-}\overset{O}{\underset{\|}{\text{C}}}\text{-O-}\overset{O}{\underset{\|}{\text{C}}}\text{-NH}_2 \ + \ \text{OH}^-$$

$$\text{ENZ}\begin{matrix}\diagup\diagdown\text{N} \\ | \\ \text{N} \\ | \\ \text{H}\end{matrix} \ + \ \text{HNCO} \xrightarrow{\text{pH} \sim 8} \text{ENZ}\begin{matrix}\diagup\diagdown\text{N-}\overset{O}{\underset{\|}{\text{C}}}\text{-NH}_2 \\ | \\ \text{N}\end{matrix}$$

Figure 7-14 Carbamoylation of amino groups; also shown are various potential side reactions with other side chains.

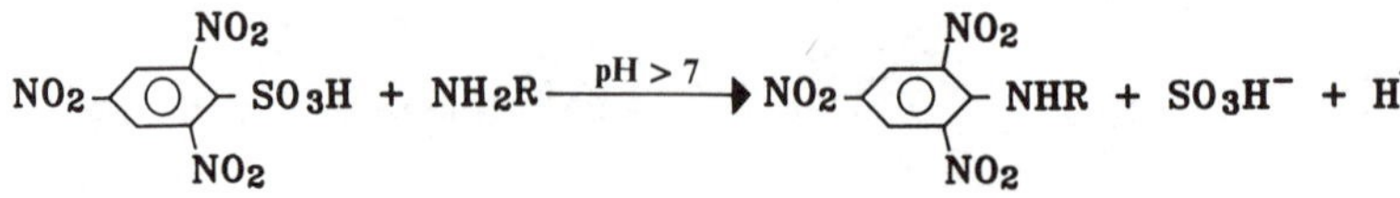

NO$_2$—⟨O⟩—SO$_3$H + NH$_2$R $\xrightarrow{\text{pH} > 7}$ NO$_2$—⟨O⟩—NHR + SO$_3$H$^-$ + H$^+$

(NO$_2$ substituents shown above and below each benzene ring)

Figure 7-15 Trinitrobenzylation of amino groups by TNBS.

2,4,6-Trinitrobenzene-1-sulfonic Acid (TNBS): TNBS reacts to give a trinitrobenzyl derivative of the amino group (Fig. 7-15). This has an absorption spectrum centered around 367 nm, and TNBS modification can be quantitated via absorbance measurements at this wavelength. The molar extinction coefficient of the derivative is 1.1×10^4 cm^{-1}, which at micromolar concentrations of protein means that even one modification per polypeptide can be reasonably quantitated. As with many of the reagents discussed here, reaction of TNBS with sulfhydryl groups can occur, although this problem has not been frequently reported.

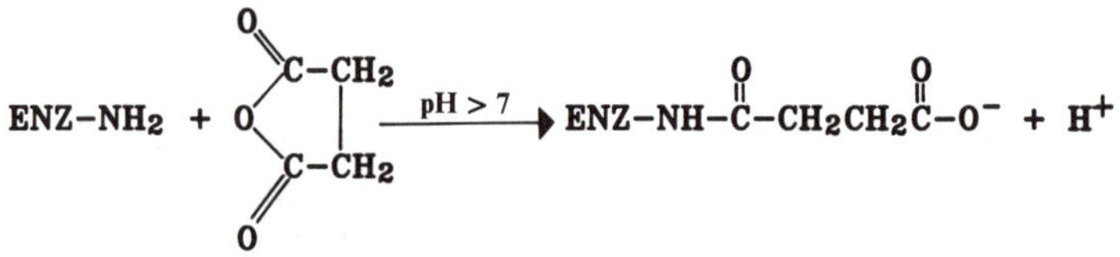

ENZ–NH$_2$ + [succinic anhydride] $\xrightarrow{\text{pH} > 7}$ ENZ–NH–C–CH$_2$CH$_2$C–O$^-$ + H$^+$

Figure 7-16 Acylation of amino group by succinic anhydride.

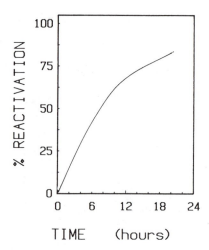

Figure 7-17 Reactivation of acylated protein by incubation at low pH.

Modification with Introduction of Negative Charge. As with modification reactions that retain a positive charge on the modified side chain, these reagents introduce a negative charge, but at a location spatially distinct from the original positive charge.

Acylation: This reaction is analogous to that discussed with acetic anhydride except that the anhydride carries a negative charge which results in the introduction of an overall negative charge into the side chain. Reagents such as succinic anhydride and tetrahydrophthalic anhydride have been successfully used for introducing a negative charge onto lysine side chains (Fig. 7-16). Succinic anhydride introduces a single negative charge per modified residue, while tetrahydrophthalic anhydride introduces a greater charge. Either modification can be completely reversed by incubation at low pH, and the rate of reversal is pH dependent (Fig. 7-17). As shown in Fig. 7-17, it may take up to 24 hours to achieve complete reversal. A potential problem is the stability of the protein at low pH for long periods. However, this approach to reversal has been used successfully in a number of cases, with full regain of native activity.

Pyridoxal-5'-Phosphate (PLP) Modification: Reaction of lysine residues with PLP at neutral pH results in a Schiff's base formation (Fig. 7-18). Subsequent reduction of the Schiff's base with sodium borohydride gives an irreversible derivative of

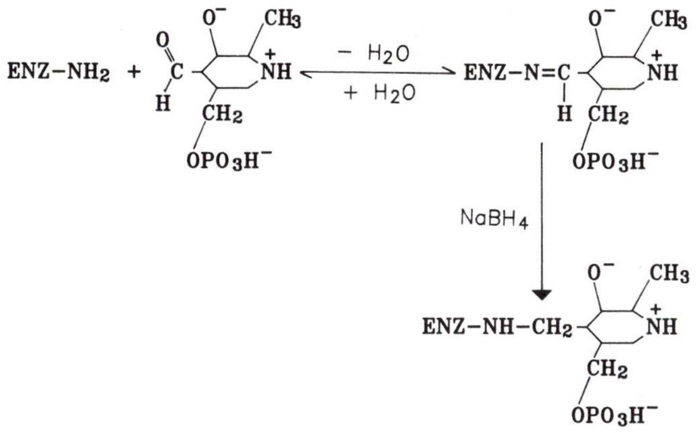

Figure 7-18   Pyridoxylation of amino groups.

the lysine. The introduced PLP group has a number of properties that make this an attractive modification for many purposes. The reaction product has an overall negative charge, as indicated in Fig. 7-18. The introduced group is also a chromophore, with a millimolar extinction coefficient at 316 nm of 8.5 cm^{-1}, allowing for quantitation. The chromophore is also fluorescent. The excitation and emission spectra of protein-bound PLP is shown in Fig. 7-19.

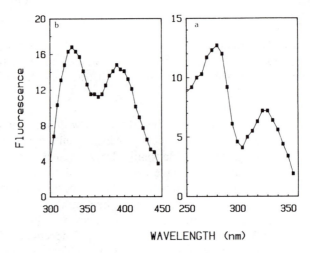

Figure 7-19 Fluorescence emission (A) and excitation (B) spectra of pyridoxylated glutamate dehydrogenase. In (A), excitation was at 280 nm. In (B), emission was at 400 nm.

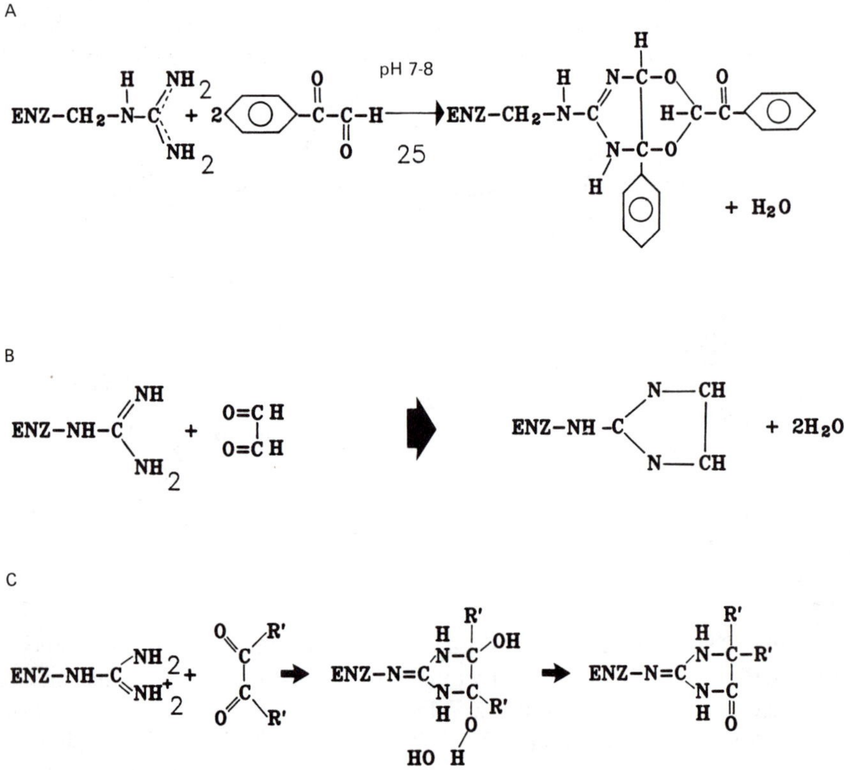

A

$$ENZ-CH_2-N-C \overset{NH_2}{\underset{NH_2}{}} + 2 \quad \bigcirc \text{—C—C—H} \quad \xrightarrow{\text{pH 7-8}}_{25} \quad ENZ-CH_2-N-C \quad + H_2O$$

B

$$ENZ-NH-C \overset{NH}{\underset{NH_2}{}} + \overset{O=CH}{\underset{O=CH}{}} \quad \blacktriangleright \quad ENZ-NH-C \overset{N—CH}{\underset{N—CH}{}} + 2H_2O$$

C

$$ENZ-NH-C \overset{NH_2}{\underset{NH_2^+}{}} + \overset{O}{\underset{O}{}}C\overset{R'}{\underset{R'}{}} \quad \blacktriangleright \quad ENZ-N=C \quad \blacktriangleright \quad ENZ-N=C$$

Figure 7-20 Reactions of arginine residues with (A) phenylglyoxal, (B) glyoxal, and (C) 2,3-butanedione.

154

Modification of the Guanidino Group of Arginine

The modification of arginine side chains is usually based on the formation of heterocyclic condensation products, with reagents having two adjacent or closely proximal carbonyl groups. Compounds with 1,2- or 1,3-dicarbonyl groups readily participate in such reactions because the spacing of the carbonyl groups closely matches that of the two unsubstituted nitrogens in the side chain. Although such compounds can react with amines or sulfhydryls, these reactions, at neutral pH, are very much slower than reaction with the guanidino group. Three reagents in particular have found widespread use: phenylglyoxal, glyoxal, and 2,3-butanedione. With phenylglyoxal and 2,3-butanedione the stoichiometry of reaction does not appear to be 1:1—products with two phenylglyoxal moieties are obtained and there is evidence that 2,3-butanedione reacts first to give a trimer which then reacts with the guanidino group. The modification reactions of these reagents are shown in Fig. 7-20. Modification of arginine residues with these reagents can be reversed, in the absence of excess reagent, at alkaline pH values, although the product obtained with phenylglyoxal modification has been shown to be slowly reversed at pH 7.0.

Modification of Carboxyl Groups

The primary means of modifying carboxyl groups in proteins involves carbodiimide activation via O-acylisourea formation at slightly acid pH (Fig. 7-21). The activated intermediate can either rearrange via an $O \rightarrow N$ acyl shift or react with an added nucleophile to give the corresponding amide. The nature of the added nucleophile can vary depending on the nature of the adduct desired, although glycine methyl ester is frequently used.

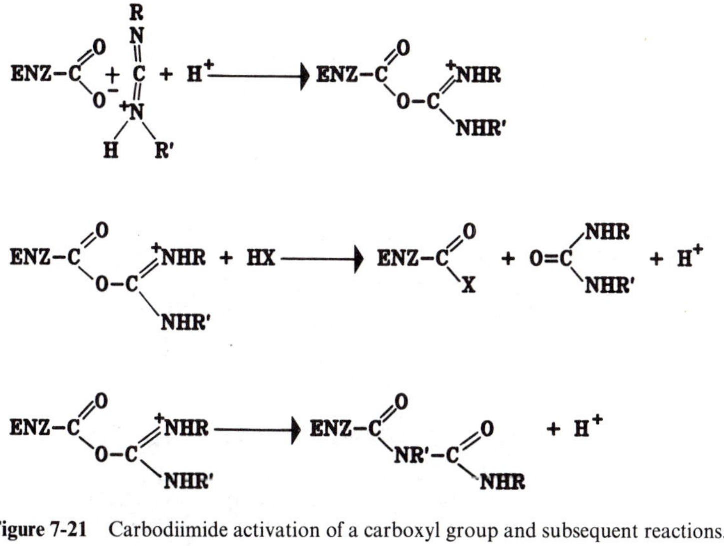

Figure 7-21 Carbodiimide activation of a carboxyl group and subsequent reactions.

Modification of Sulfhydryl Groups

A wide variety of reagents are available for the modification of sulfhydryl side chains in proteins. Many of the reagents are based on three reactions, involving modification by (1) iodoacetic acid or iodoacetamide, (2) maleimide or maleic anhydride, or (3) mercury derivatives such as *p*-chloromercuribenzoate (Fig. 7-22).

Iodoacetate derivatives are quite stable to various fragmentation or amino acid analysis conditions and are often used to alkylate cysteine residues during sequence

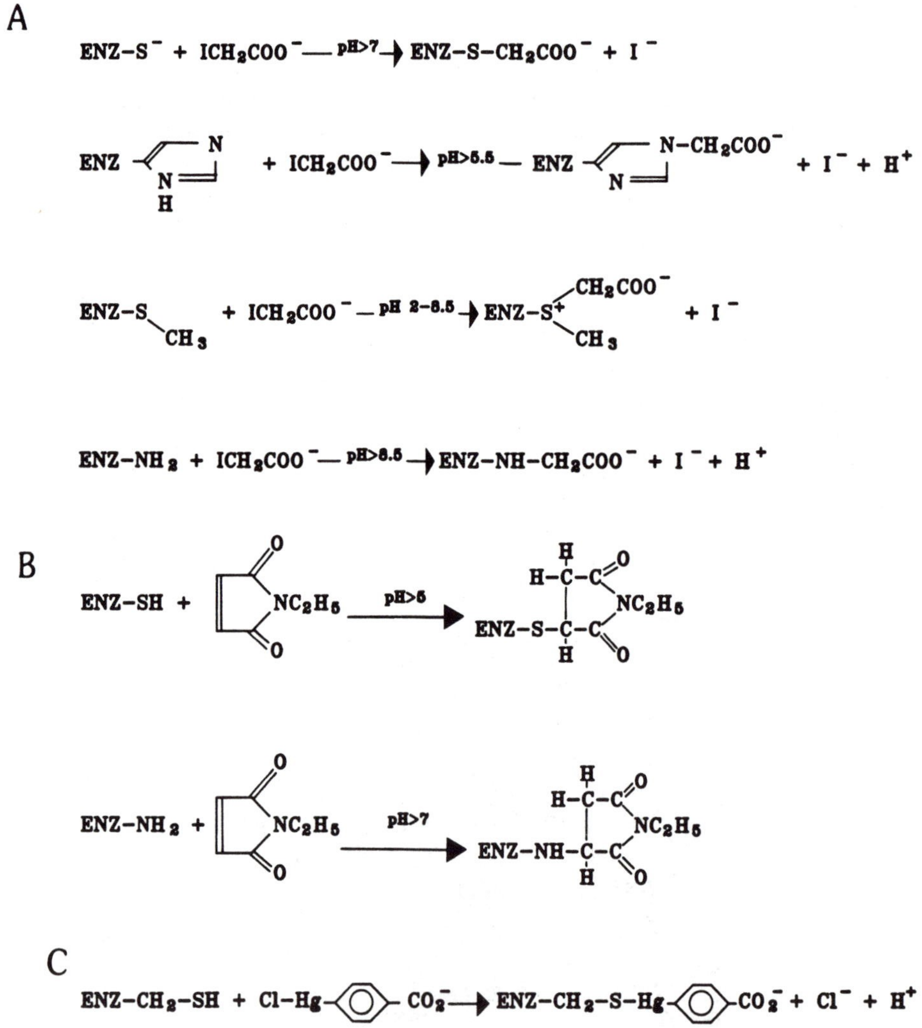

A

$$ENZ\text{-}S^- + ICH_2COO^- \xrightarrow{\ pH>7\ } ENZ\text{-}S\text{-}CH_2COO^- + I^-$$

$$ENZ\text{-}imidazole + ICH_2COO^- \xrightarrow{\ pH>5.5\ } ENZ\text{-}imidazole\text{-}N\text{-}CH_2COO^- + I^- + H^+$$

$$ENZ\text{-}S\text{-}CH_3 + ICH_2COO^- \xrightarrow{\ pH\ 2\text{-}5.5\ } ENZ\text{-}S^+(CH_2COO^-)(CH_3) + I^-$$

$$ENZ\text{-}NH_2 + ICH_2COO^- \xrightarrow{\ pH>5.5\ } ENZ\text{-}NH\text{-}CH_2COO^- + I^- + H^+$$

B

$$ENZ\text{-}SH + \text{maleimide-}NC_2H_5 \xrightarrow{\ pH>5\ } ENZ\text{-}S\text{-}succinimide\text{-}NC_2H_5$$

$$ENZ\text{-}NH_2 + \text{maleimide-}NC_2H_5 \xrightarrow{\ pH>7\ } ENZ\text{-}NH\text{-}succinimide\text{-}NC_2H_5$$

C

$$ENZ\text{-}CH_2\text{-}SH + Cl\text{-}Hg\text{-}C_6H_4\text{-}CO_2^- \longrightarrow ENZ\text{-}CH_2\text{-}S\text{-}Hg\text{-}C_6H_4\text{-}CO_2^- + Cl^- + H^+$$

Figure 7-22 Modification of sulfhydryl residues by (A) iodoacetic acid, (B) maleic anhydride, and (C) *p*-chloromercuribenzoate. Also shown are potential side reactions in parts (A) and (B).

TABLE 7-4   Spectrally active reagents based on iodoacetamide

N-Iodoacetylaminoethyl-5-naphthylamine-1-sulfonic acid	
1,5,1-AEDANS	350F495
1,8,1-AEDANS	350F495
7-Diethylamino-3-[(4'-iodoacetylamino)phenyl]-4-methyl coumarin	390F460
2-Anthraceneiodoacetamide	320F420

TABLE 7-5   Spectrally active reagents based on maleimide

N-Pyrene maleimide	386F405
N-1-Anilinonaphthyl-4-maleimide	355F488
7-Diethylamino-3-(4'-maleimidylphenyl)-4-methyl coumarin	390F460
N-[p-(2-Benzoxazoyl)phenyl]maleimide	310F375
4-Dimethylamino-4'-maleimido-stilbene	345F480

analysis to prevent disulfide bond formation. A wide series of fluorescent reagents based on iodoacetate have been developed and some of them are listed, together with their fluorescence properties, in Table 7-4. All of these alkylating reagents can react with other amino acid side chains, such as methionine, lysine, or histidine.

Maleic anhydride- or maleimide-based reagents react with sulfhydryls to give acid-stable derivatives. Reactions with other groups may occur, but these derivatives tend to be acid labile. The derivative produced by modification with N-ethylmaleimide has an absorption maximum at 300 nm, with a molar extinction coefficient of 620 cm^{-1}, which is too low to allow reasonable quantitation except at high protein concentrations or high degrees of modification. As with the iodoacetate-based reagents, a variety of spectrally active reagents are available based on maleimide, some of which are given in Table 7-5.

The mercury-based reagents are the most specific for sulfhydryls. The p-mercuribenzoate derivative of a sulfhydryl has an extinction coefficient at 250 nm of 7500 M^{-1} cm^{-1} at pH 7.0, which allows reasonable quantitation spectrophotometrically even at low (i.e., μM) protein concentrations. A particularly useful derivative of pMB is S-mercuric-N-dansyl cysteine, whose reaction with sulfhydryls (Fig. 7-23)

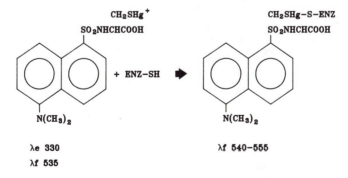

Figure 7-23   Reaction of S-mercuric-N-dansyl cysteine with sulfhydryl residues.

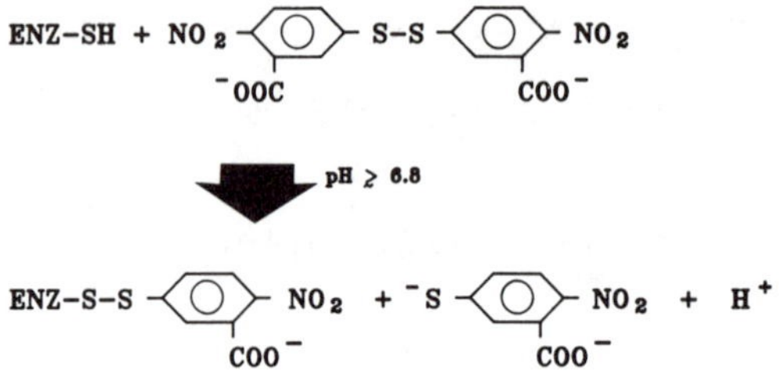

Figure 7-24   Reaction of sulfhydryl group with DTNB.

gives a derivative that can be used to estimate sulfhydryl groups by fluorescence titration, or to introduce a fluorophore onto a reactive sulfhydryl.

Iodoacetic acid and maleic anhydride can both react with other residues in addition to sulfhydryl moieties, and some of these potential side reactions are shown in Fig. 7-22.

The final reagent that is widely used with sulfhydryls is 5,5'-dithiobis(2-nitrobenzoic acid) (DTNB). This forms a mixed disulfide with cysteine (Fig. 7-24) and releases the thionitrobenzoate anion, which can be quantitated spectrophotomerically by absorbance measurements at 412 nm. At pH 8 in aqueous buffers the thionitrobenzoate anion has a molar extinction coefficient of 13,600 cm^{-1}. This coefficient, which is quite dependent on other compounds in the buffer, allows easy and rapid titration of sulfhydryl groups in either native or denatured proteins. It is advisable to determine the extinction coefficient under individual conditions by titration of cysteine with DTNB in control experiments. The derivatives produced by DTNB modification are easily reversible by dithiothreitol or mercaptoethanol.

Modification of Histidine Residues

As was indicated, iodoacetate alkylates histidine residues and both mono- and disubstituted derivatives can be obtained (Fig. 7-25). A particularly useful reagent for histidine modification is diethylpyrocarbonate. The reaction proceeds (Fig. 7-26)

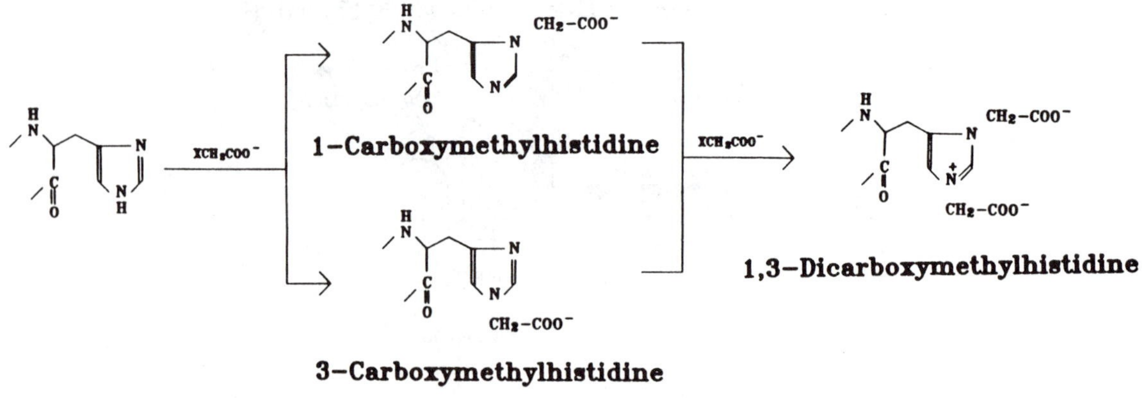

Figure 7-25   Modification of histidine with iodoacetic acid derivatives.

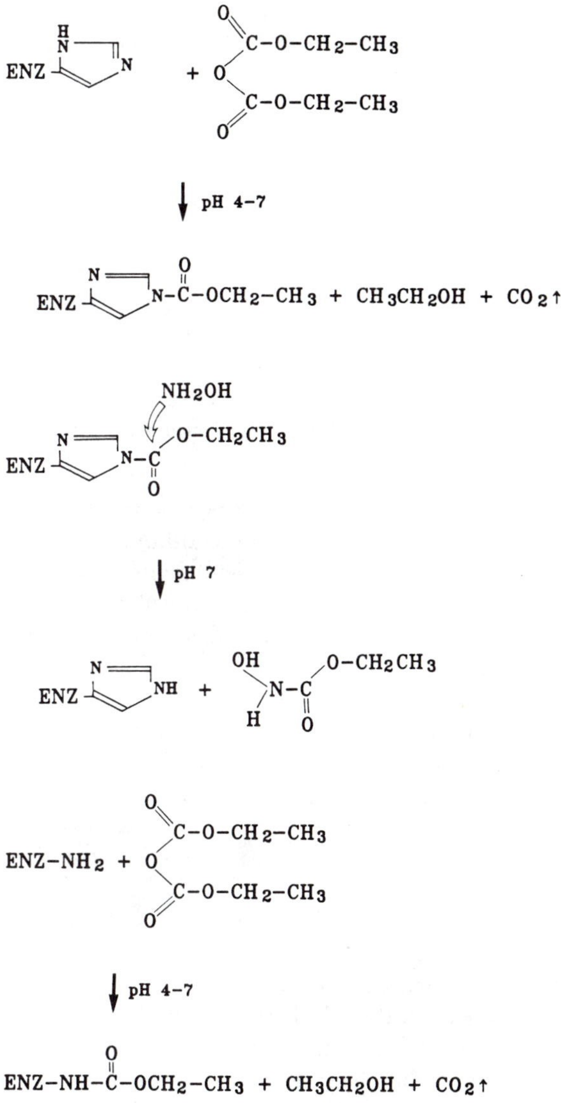

Figure 7-26 Modification of histidine by diethylpyrocarbonate and reversal by hydroxylamine. Also shown is the side reaction with lysine residues.

to produce an N-carbethoxy derivative, which has an absorbance band between 230 and 250 nm. At 240 nm, a molar extinction coefficient for this derivative of 3200 cm^{-1} allows the reaction to be quantitated and followed spectrophotometrically. Diethylpyrocarbonate is subject to a rapid hydrolysis, to carbon dioxide and ethanol, and as described earlier, time-course studies must take this into account. In addition to being easily monitored, modification of histidine by diethylpyrocarbonate is readily

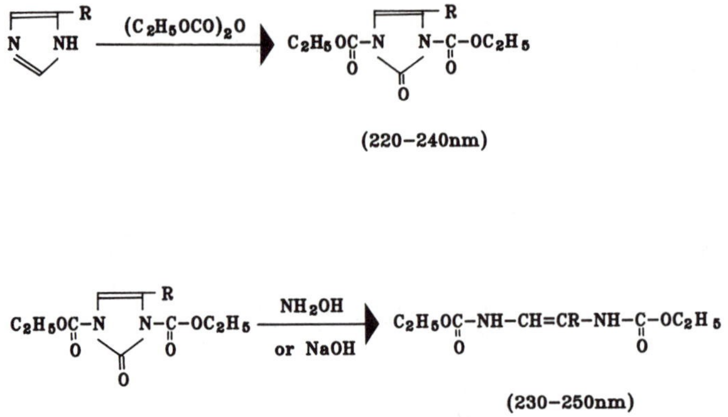

(220–240nm)

(230–250nm)

Figure 7-27   Formation of disubstituted imidazole ring.

reversed by incubation with hydroxylamine. Since diethylpyrocarbonate can ethoxy-formylate tyrosine, lysine, or sulfhydryl residues (although reaction is minimal at pH 6, where histidine reacts readily), reversal of histidine modification by hydroxylamine is important, as it does not reverse modification of sulfhydryls or lysine side chains.

A point of caution concerning the use of diethylpyrocarbonate must be raised. At very high excesses of diethylpyrocarbonate to histidine it is possible to get modification at both imidazole nitrogens, giving a derivative with a higher extinction at 240 nm, which can lead to misquantitation of the number of histidine residues modified. The disubstituted histidyl derivative (Fig. 7-27) reacts with hydroxlamine, not to give reversal of the modification, but to give cleavage of the imidazole ring and a derivative that still absorbs at 230 to 250 nm. Thus if the disubstituted derivative is formed, it may appear on the basis of hydroxylamine reversal that lysine or cysteine has been modified.

Both iodoacetic acid and diethylpyrocarbonate have been shown to modify *either* N atom in the imidazole ring of histidine. In the case of iodoacetic acid it is possible to isolate 1- and 3-substituted carboxymethyl derivatives and relate activity effects to which atom is modified. Although diethylpyrocarbonate can react with both N atoms to give stable disubstituted histidines, in the monosubstituted product the N-1 substituent is much more stable than the N-3 substituent and diethylpyrocarbonate can effectively be considered as specific for N-1 in monosubstitution reactions.

N-1 modification by diethylpyrocarbonate of an N-3 carboxylmethyl histidine residue in bovine α-lactalbumin has been shown to produce complete inactivation, while the N-3 substituent retains activity.

Modification of Tryptophan Residues

In Chap. 5 we examined the use of brominating reagents such as *N*-bromosuccinimide to cleave polypeptide chains at tryptophan residues. Under mild conditions this reagent oxidizes tryptophan side chains, resulting in a loss of the absorbance

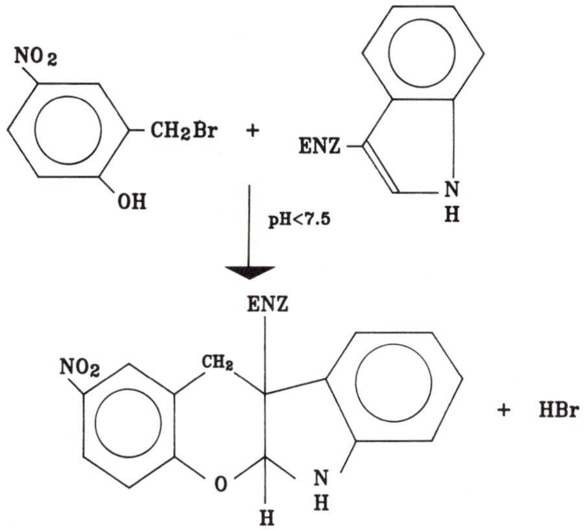

Figure 7-28   Modification of tryptophan with 2-hydroxy-5-nitrobenzylbromide.

properties in the region 270 to 295 nm. Although it has been used successfully to modify tryptophan in some proteins, the conditions must be carefully monitored since too high a reagent concentration leads to peptide-bond cleavage at the modified tryptophan.

Various benzyl halides have been used to alkylate the indole ring of tryptophan. The most popular is 2-hydroxy-5-nitrobenzyl bromide, first proposed as a tryptophan modification reagent by Koshland. Because of solubility problems the analogous dimethyl(2-hydroxy-5-nitrobenzyl) sulfonium salt is often used (Fig. 7-28). In the absence of cysteine residues these reagents are quite specific for tryptophan.

Because of product heterogeneity this reagent is often passed over in favor of various sulfenyl halides, which react with similar specificity but give a single product (Fig. 7-29). When sulfenyl halides with a nitrophenyl substituent are used the chromophoric properties of the nitrophenyl group allow quantitation of accessible tryptophan residues.

Modification of Tyrosine Residues

Tetranitromethane has been extensively used to modify tyrosine residues. The reaction, which can be conducted at mildly alkaline pH, proceeds via a proposed free-radical mechanism (Fig. 7-30) to give a 3-nitrotyrosine derivative that has a pK value of approximately 7.0.

The nitrophenoxide ion has an intense visible absorption. At low pH an absorption maximum of 360 nm with a molar extinction of 2790 cm^{-1} is found, while at higher pH the absorption maximum shifts to 428 nm and the extinction coefficient

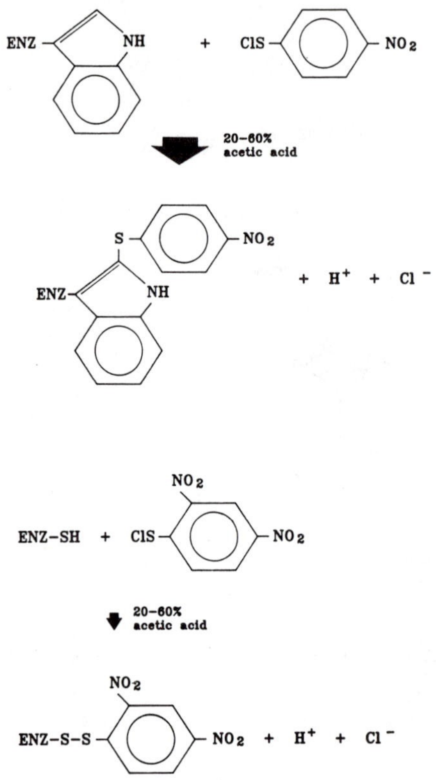

Figure 7-29   Modification of tryptophan with sulfenyl halides. Also shown is a competing reaction with sulfhydryl residues.

increases to 4200 M^{-1} cm^{-1}. This allows for a direct determination of the pK of the modified tyrosine residue. The nitrotyrosine is readily reduced to 3-aminotyrosine by sodium hydrosulfate. The pK value of this aromatic amino group is about 4.8, making it considerably lower than the other amino groups in the protein. This makes 3-aminoytrosine a particularly reactive target of amino-group-specific reagents such as 1-fluoro-2,4-dinitrobenzene (Chap. 5) at low pH, where the reactivity of other

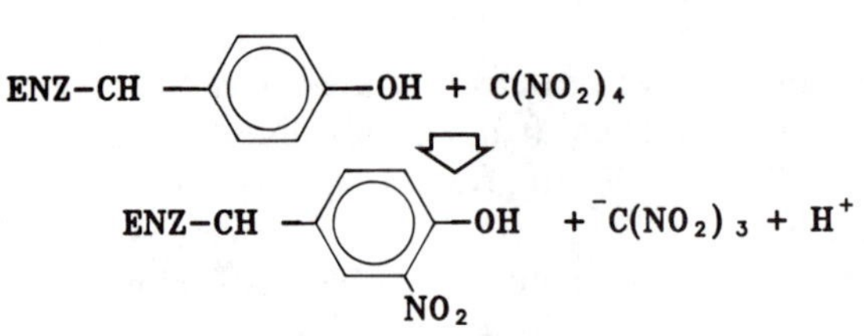

Figure 7-30   Nitration of tyrosine side chains by tetranitromethane.

TABLE 7-6   Specificity of commonly used modification reagents

Reagent	Side chain modified								
	Lys	Glu	Cys	Arg	Ser	His	Tyr	Trp	Met
Acetic anhydride	×		×		×		×		
Acyl anhydride	×		×				×		
Aldehydes	×		×	×		×			
N-Bromosuccinimide			×			×	×	×	
Carbodiimides		×							
Phenylisothiocyanate	×		×				×		
Tetranitromethane			×				×		
Trinitrobenzenesulfonate	×								
Diethylpyrocarbonate	×					×			
Iodoacetate	×		×			×			×
Maleimide	×		×						

amino groups is considerable lower. Where tyrosine can be specifically nitrated, subsequent reduction can allow for specific introduction of a modification with amino-group reagents.

Although tetranitromethane can modify cysteine, methionine, and tryptophan residues, the major problem encountered is the possibility of either intra- or inter-molecular cross-linking as a result of side reactions of the proposed tyrosine free-radical intermediate. As discussed in a later section, tetranitromethane is sometimes used as a cross-linking reagent.

As has been emphasized in this section, many of the reagents discussed are not uniquely specific. Table 7-6 gives the specificities of many commonly used modification reagents.

CROSS-LINKING REAGENTS

In the recent past a tremendous amount of work has gone into the synthesis and use of a wide variety of chemical cross-linking reagents. These reagents, which contain two reactive moieties, can be used to cross-link residues within a polypeptide chain, between polypeptide chains in an oligomer, or between protein molecules which may, for whatever reason, associate with one another. The identification of residues (or proteins) that can be cross-linked together can give much important information in the areas of protein conformation and protein–protein interactions. A number of considerations go into the design or selection of a cross-linking reagent, and prior to considering the reagents themselves we examine briefly some of these considerations and their implications.

1. *Reaction Specificity.* As with most chemical modification reagents, reaction specificity and conditions of reaction are important. The appropriate groups the

reagent reacts with must be present and, if structural information is to be obtained, they must react under "native" conditions. Most of the cross-linking reagents available react with amino groups or sulfhydryl groups or either. Cross-linking reagents fall into two groups based on their types of reactivity. When both reactive moieties are the same, the reagent is *homo-bifunctional.* When one of the reactive moieties has a different specificity from the other, the reagent is *hetero-bifunctional.* Hetero-bifunctional reagents often react with amino groups through one of their reactive moieties and a different group through the other. A special class of hetero-bifunctional reagents are the ones that include a *photo-activable* moiety as one of their reactive groups. In many situations such reagents have a distinct advantage over other cross-linking reagents, as once reaction via the non-photo-activable moiety has occurred, cross-linking via the photo-generated free-radical reacting is almost assured and does not depend on the availability of a particular amino acid side chain to react with. This can in some circumstances lead to random cross-linking events; however, such an occurrence may be desirable.

The availability of groups to react to give cross-linking leads to a consideration of the second criterion in selecting a cross-linking reagent.

2. Cross-Linking Distance. From a number of standpoints cross-linking distance is an important parameter. As the length between the two functional groups increases, there is an increased probability that a second group on the protein exists which can react with the cross-linker to also give cross-linking. Cross-linking with a reagent that spans, for example, 10 Å can occur only if a second reactive group is within 10 Å of the first. Because a cross-linking reagent *can* span 10 Å does not mean that the second group is 10 Å from the first, as many of the reagents are quite flexible and can react with groups located at distances less than the maximum span length. If a series of cross-linking reagents can be used with similar reaction specificities but varying span length, considerable spatial information concerning groups can be obtained.

Depending on the type of cross-linking that is being attempted (i.e., intra-peptide or inter-peptide), one might select reagents with either short cross-linking distances or long ones. In many instances where inter-peptide cross-linking is attempted, none is found until reagents of a certain minimum length are used.

The third characteristic that must be considered is useful both in the identification of cross-linked peptides and in controlling that any effects observed after cross-linking are due to the cross-linking itself rather than to the chemical modification event.

3. Cleavability. The development of cleavable cross-linking reagents has greatly assisted the isolation and characterization of cross-linked fragments. These reagents contain a moiety between the two functional groups which can, after cross-linking, be cleaved by a reagent, optimally under conditions that are not deleterious to the protein itself. Combined with two-dimensional gel electrophoresis, cleavable cross-linking reagents considerably ease the task of identification and isolation of cross-linked components.

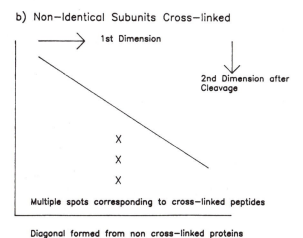

Figure 7-31   Schematic representation of two-dimensional PAGE results for homo- and hetero-polymers cross-linked with cleavable reagents.

Two situations can be envisaged involving multi-subunit proteins that give different types of experimental observations in two-dimensional gels with cleavage between the dimensions. They are summarized in Fig. 7-31.

In addition to this, cleavable cross-linking reagents can assist in control experiments to show whether or not any altered biological activity is the result of the cross-linking event or simply the chemical modification event. In the former case, the biological activity would be regained after cleavage of the cross-link (hence the importance of a cleavage procedure that does not harm the protein's integrity), whereas if the activity change resulted from the modification event, cleavage of the cross-link would not regenerate activity. A second way to examine this question is to use mono-functional reagents with the same reaction specificity as the cross-linking reagent. In this case, cross-linking cannot occur and the question of whether the modification itself causes the loss of activity can be examined. It must be emphasized that simply because a cross-linking reagent is being used it cannot be inferred that cross-linking is the only type of chemical modification taking place. It is quite probable that, in addition, monofunctional modification may be occurring. This monofunctional mod-ification can arise from two causes: (a) a lack of an available second amino acid side chain for cross-linking, and (b) degradation of one of the reactive moieties of the cross-linking reagent prior to cross-link formation, essentially converting the reagent into a monofunctional reagent.

4. *Reactive Moiety*: Finally, we can consider the reaction characteristics of the reactive moiety of the cross-linking reagent. As with many chemical modification reagents, the site of reaction can be controlled to a certain extent. It is possible to direct the reaction by considering the hydrophobicity, hydrophilicity, or charge of the reactive moiety. In addition, it may be possible to incorporate aspects of a site-directed irreversible inhibitor into one moiety of a hetero-bifunctional reagent, al-lowing reaction of one end of the cross-linker to be directed toward a particular binding site on the protein.

Types of Cross-Linking Reagents

A wide variety of cross-linking reagents have been developed in recent years with varying specificity, cleavability, and span lengths. They can be categorized into three basic classes, and within each class can be either cleavable or noncleavable. In many of the types of reagents used, a series of analogous reagents having differing span lengths are obtained by using a suitable spacer group between the reactive moieties. The three types of cross-linking reagents that we consider are summarized in Fig. 7-32.

Homo-bifunctional Reagents. This first class of reagents have the same reactive moiety at either reactive center, and their specificity is determined by this moiety. Typical of this class of reagent are the bisimidates, a series of bifunctional imido esters having $(CH_2)_n$ spacer groups with $n = 1$ (malonimidate) to $n = 6$ (suberimidate), spanning 5 to 11 Å and reacting with amino groups. Such reagents can be converted to cleavable homo-bifunctional reagents by inclusion of a disulfide bridge, as in dimethyl-3-3′-dithiobispropionimidate, which has a span length of 12 Å and is cleav-able by mercaptans. Although imidoesters are often used in homo-bifunctional reagents, other amino-specific reagents such as *N*-hydroxysuccinimide esters or 1,5-difluoro-2,4-dinitrobenzene have also been employed.

Essentially all homo-bifunctional reagents are amino-group specific, with the exception of reagents such as glutaraldehyde and formaldehyde. Although these aldehyde reagents have been used in cross-linking studies, their uncharacterized and

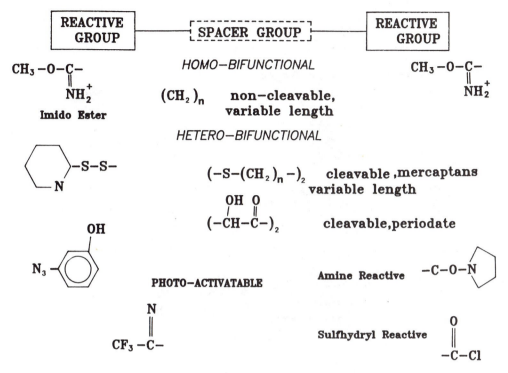

Figure 7-32 Types of cross-linking reagents.

nonspecific reactivity makes them reagents of last resort, especially when limited extents of cross-linking are required.

Hetero-bifunctional Reagents. In all cases, one of the reactive moieties in this class of reagents reacts with an amino group, and the other reactive moiety usually reacts with a sulfhydryl side chain, although in the case of the carbodiimides the second reacting group is a carboxyl group. Except for the various carbodiimides these reagents usually contain the amino-reactive moiety *N*-hydroxysuccinimide and often a sulfhydryl-reactive maleimido or dithio moiety.

Photo-activable Reagents. This group of reagents is really an extension of the hetero-bifunctional reagents but has a more widespread utility. This comes from the fact that one of the reactive moieties is generated in situ after reaction of the first reactive moiety with the protein, which helps reduce the potential for monofunctional modification, and from the fact that the generated second reactive moiety is either a carbene (from diazo reagents) or a nitrene (from azido reagents) free radical that is highly reactive but quite nonspecific. As a result, there is little requirement for certain types of residues to be located close to one another to allow for cross-linking.

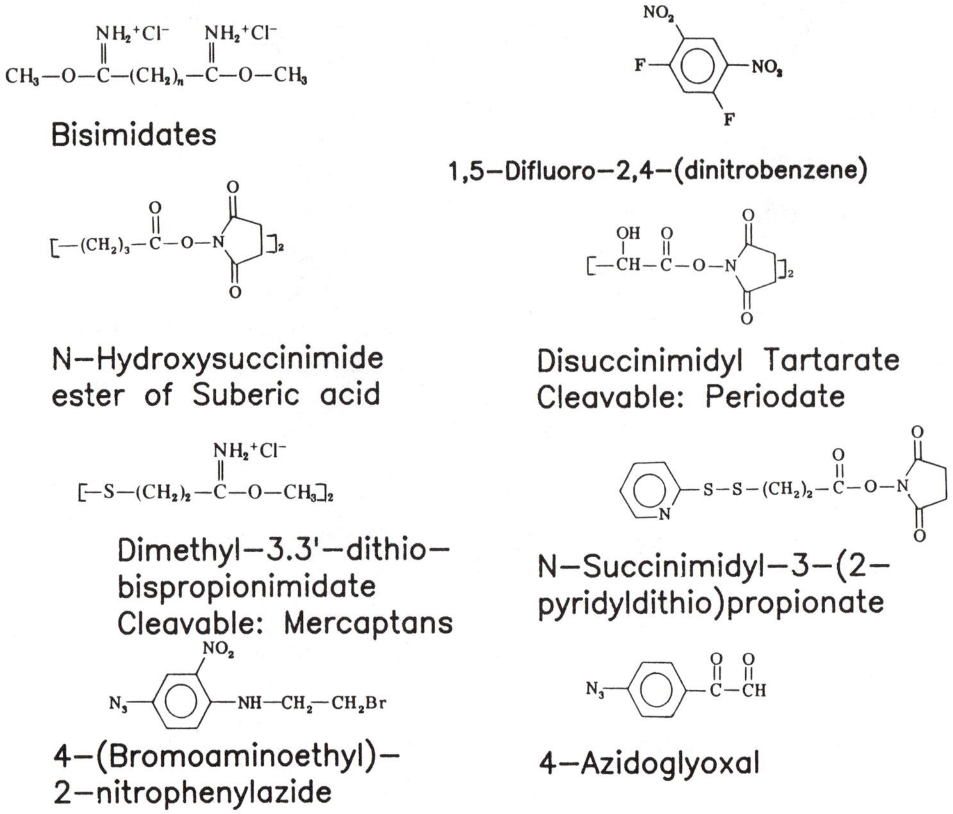

Bisimidates

1,5—Difluoro—2,4—(dinitrobenzene)

N—Hydroxysuccinimide ester of Suberic acid

Disuccinimidyl Tartarate Cleavable: Periodate

Dimethyl—3.3'—dithio-bispropionimidate Cleavable: Mercaptans

N—Succinimidyl—3—(2—pyridyldithio)propionate

4—(Bromoaminoethyl)—2—nitrophenylazide

4—Azidoglyoxal

Figure 7-33 Structures of some commonly used cross-linking reagents.

The initial reactive moiety is either amino or sulfhydryl reactive as with other hetero-bifunctional reagents, and as with other reagents discussed, various span lengths or cleavability can be built into the reagents.

The structures of some of the more popular cross-linking reagents that have been used are shown in Fig. 7-33, together with the basis of their cleavability where appropriate.

Examples of the Uses of Cross-Linking Reagents

Determination of Interlysyl Distances in Glycogen Phosphorylase b

Reference: J. Hadju, et al., *Biochemistry, 18,* 4037–4041 (1979).

This enzyme undergoes a dimer–tetramer association reaction that appears to be related to the allosteric properties of the enzyme. Experiments involved cross-linking the enzyme with a series of imido esters whose cross-linking span length ranges from 3.7 to 14.5 Å, electrophoresing the cross-linked products in an SDS–PAGE system, and quantitating the amount of each species, after staining, by densitometry. From such data two parameters are obtained: r_k, the rate-constant ratio of cross-linking, and C_d, the percent of cross-linked parameter, which are defined in Eqs. (7-32) and (7-33), respectively,

$$r_k = \frac{k_L}{k_0} \tag{7-32}$$

where k_L and k_0 are the apparent first-order rate constants for the disappearance of the monomer bound in the presence or absence of an added ligand, respectively.

$$C_d = \frac{[\text{trimer} + \text{tetramer}]}{[\text{total}]} \times 100 \tag{7-33}$$

Since all samples can be cross-linked under identical conditions except the presence of added ligand, both parameters can be calculated from densitometer scans

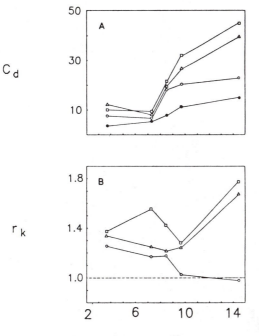

Maximal Effective Reagent Length, A

Figure 7-34 Effects of span length and AMP concentration on cross-linking parameters: cross-linking was performed with diimidates varying from 3.7 to 14.5 Å in span length in the absence of AMP (●) or in the presence of AMP: 0.1 mM (○); 0.3 mM (□); and 1 mM (△). (Reprinted with permission from: J. Hajdu, V. Dombradi, G. Bot, and P. Friedrich, *Biochemistry*, *18*, 4037–4041. Copyright 1979 American Chemical Society, Washington, D. C.)

of enzyme cross-linked in the absence or presence of ligand, and the parameter r_k is calculated from

$$r_k = \frac{\ln (\text{monomer/total})_L}{\ln (\text{monomer/total})_0} \qquad (7\text{-}34)$$

The effects of either cross-linker length or of an added allosteric ligand, AMP, on these two parameters is shown in Fig. 7-34. The amount of trimer and tetramer formed, as shown by C_d, increases with a span length greater than 8 Å, and this increase is amplified by the presence of AMP (Fig. 7-34A), indicating that for effective cross-linking to give tetramer there must exist two lysine residues approximately 8 Å apart. AMP clearly functions to increase the amount of tetramer present rather than bringing the lysines closer together in the tetramer, since even in the presence of saturating AMP concentrations, no significant tetramer is formed with cross-linking reagents having a span of <8.5 Å.

Covalent Cross-Linking of the Active Sites of
Vesicle-Bound Cytochrome b5 and NADH–Cytochrome
b5 Reductase

Reference: C. S. Hackett and P. Strittmatter, *J. Biol. Chem.,*
259, 3275–3282 (1984).

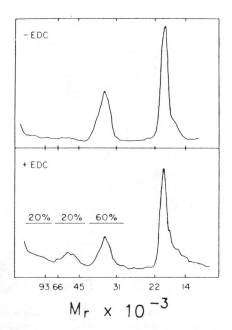

Figure 7-35 SDS-gel electrophoresis of cytochrome b_5–cytochrome b_5 reductase vesicles before and after cross-linking with EDC. [From *J. Biol. Chem., 259,* 3275–3282 (1984). Reprinted with permission of the copyright owner, The American Society of Biological Chemists, Inc., Bethesda, Md.]

The water-soluble carbodiimide 1-ethyl-3-(2-dimethylaminopropyl)carbodiimide hydrochloride (EDC) was used to cross-link cytochrome b_5 and NADH–cytochrome b_5 reductase. EDC is an interesting cross-linking reagent, being a so-called "zero-length" cross-linker. It acts by promoting the formation of an amide linkage between a carboxyl group on cytochrome b_5 and a lysyl residue on the reductase. Figure 7-35 shows a gel scan of cytochrome b_5 and the reductase electrophoresed with or without cross-linking by EDC. The cross-linked complex can then be incorporated into vesicles to study the effects of interaction of the two proteins during function.

Cross-Linked Galactosyltransferase and α-Lactalbumin:
Use in Site–Site Distance Estimates by Fluorescence
Resonance Energy Transfer Measurements

Reference: E. T. O'Keeffe, T. Mordick, and J. E. Bell, *Biochemistry*, *19*, 4962–4966 (1980).

A problem frequently encountered in physical measurements on the interaction of two proteins is the weak nature of such interactions. This is particularly difficult in fluorescence measurements where one component is fluorescent and the other can act as a resonance energy transfer acceptor of that fluorescence. This was the case in studies of the interaction of α-lactalbumin with galactosyltransferase, where dansylated α-lactalbumin was used to estimate a distance from the dansyl to cobalt bound to the metal-ion binding site of the transferase. Under normal circumstances an approximately 1000-fold excess of α-lactalbumin is required to bind the transferase completely. Since significant amounts of the 1:1 complex are required for the fluorescence measurements this would have led to an extremely unfavorable signal-to-background ratio. This problem was overcome by chemically cross-linking the dansyl α-lactalbumin to the transferase in a 1:1 complex and performing the fluorescence measurements on this complex. SDS-PAGE shows that cross-linking of α-lactalbumin to galactosyltransferase occurs. Since the cross-linked complex is not much larger than the un-cross-linked enzyme, it is difficult to separate the cross-linked from residual un-cross-linked activity by gel filtration. However, they can easily be separated using an α-lactalbumin affinity column. The resulting complex can be titrated with cobalt to estimate the quenching due to resonance energy transfer between the dansyl group on the α-lactalbumin (shown to be uniquely labeled at the amino terminal) and the cobalt metal binding site of the transferase. From the quenching a distance of 32 Å between the metal site and the dansyl group on the α-lactalbumin was calculated.

Possible Quaternary Structures for the Hexamer Glutamate
Dehydrogenase Established by Cross-Linking Studies

Reference: T. J. Smith and J. E. Bell, *Arch. Biochem. Biophys.*, *239*, 63–73 (1985).

Cross-linking studies can be used to distinguish between possible quaternary structural arrangements of subunits. This topic is dealt with in detail in Chap. 11, but a brief outline of the results obtained with glutamate dehydrogenase is given

SDS—PAGE of Cross—Linked GDH

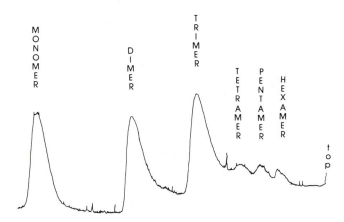

Figure 7-36 Densitometer scan of glutamate dehydrogenase cross-linked with dimethylpimilimidate.

here. The six subunits can be arranged either with cyclic symmetry (C-6) or with dihedral symmetry (D-3). They predict quite different cross-linking distributions of monomer, dimer, trimer, tetramer, pentamer, and hexamer obtained as a result of intersubunit cross-linking. Figure 7-36 shows a densitometer gel scan of glutamate dehydrogenase cross-linked with dimethylpimelimidate. From such data the amounts of each cross-linked species can be quantitated and compared with predicted stoichiometries. In this case the results eliminated cyclic symmetry from further consideration.

Identification of Intrapeptide Cross-Linking in the Monomer
*of Glutamate Dehydrogenase by **Dimethylpimelimidate***

Reference: T. J. Smith and J. E. Bell, unpublished results.

In attempts to predict tertiary structure on the basis of sequence information (see Chap. 10) the major problem encountered is the number of possible conformations a polypeptide may have. If this number can be limited by direct experimental information, it becomes feasible to attempt to predict tertiary structures. Chemical cross-linking of nearby residues and identification of the residues involved represents an attractive approach to obtaining information that can limit the number of conformations needing consideration. Clearly, experiments with cross-linking reagents having different span lengths can provide a wealth of information when intra-peptide cross-links can be identified. Experimentally, this is quite simple with a monomeric protein; however, with an oligomeric protein such as glutamate dehydrogenase the possibility of inter-polypeptide cross-links as well as intra-polypeptide cross-links presents difficulties. One approach is to use cross-linking reagents with short span lengths that cannot cross-link between subunits. This represents, however, an unnecessary limit on the possible information that can be obtained in such experiments.

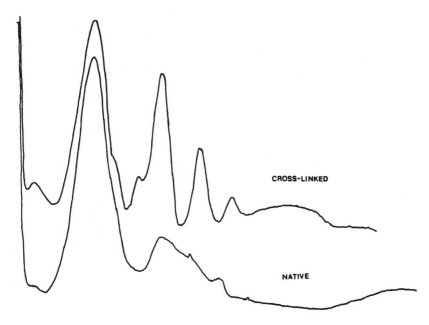

Figure 7-37   Densitometer scans of cyanogen bromide peptides obtained from native GDH and cross-linked trimer GDH.

In an alternative approach, cross-linked enzyme is electrophoretically separated as in the previous example, and the separated monomer, dimer, trimer, and so on, subjected to cleavage in situ in the gel. The cleaved monomer, dimer, trimer, and so on, are then electrophoresed in a second dimension together with a sample of un-cross-linked enzyme that has been electrophoresed and also cleaved in situ in the gel. Figure 7-37 shows a comparison of densitometer scans of native, un-cross-linked enzyme and cross-linked monomer. Clearly, a number of new peptides arising from intra-peptide cross-links are seen in the second dimension of the cross-linked monomer, which can be excised, further purified, and identified to give the desired information.

EXAMPLES OF MODIFICATION STUDIES

Uses of Chemical Modification in Hybridization Experiments

Reference: I. Gibbons and H. K. Schachman, *Biochemistry*, *15*, 52–60 (1976).

A particularly interesting example of some uses of chemical modification is provided by the work on aspartate trancarbamoylase. This enzyme consists of six catalytic peptides arranged in two trimers and six regulatory peptides arranged in three

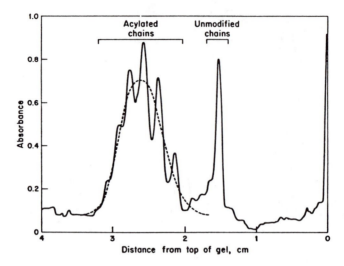

Figure 7-38   Electrophoresis patterns of polypeptide chains of catalytic subunit, the acylated derivative, and the deacylated product. Electrophoresis in polyacrylamide gels containing 8 m*M* urea was used. (Reprinted with permission from: I. Gibbons and H. K. Schachman, *Biochemistry*, *15*, 52–60. Copyright 1976 American Chemical Society, Washington, D. C.)

dimers. The regulatory dimers appear to act as a "bridge" or "cross-link" between the two catalytic trimers, which have no direct contact. The catalytic trimers and the regulatory dimers can be separated and purified.

The catalytic subunits can be nitrated with tetranitromethane in the presence of the substrate carbamoyl phosphate and the analog succinate to give nitrated subunits with approximately 0.8 nitrotyrosine per polypeptide chain but 80 to 90% activity. These active, nitrated catalytic subunits can be pyridoxylated at one lysine per polypeptide chain with 90% loss of activity.

Nitrated, pyridoxylated catalytic subunits can be made more negative by modification with 3,4,5,6-tetrahydrophthalic anhydride. As described earlier, acylation can be reversed by incubation at low pH to give the nitrated, pyridoxylated catalytic subunits. Figure 7-38 shows the electrophoretic separation of acylated from nonacylated catalytic subunits.

Reconstituted enzyme can be made by mixing catalytic and regulatory subunits. If native catalytic subunits, C_n, and inactive, nitrated, pyridoxylated, acetylated (negatively charged) subunits, C_t, are mixed with an excess of regulatory subunits (R), a series of reconstituted molecules will be formed: C_n-R-C_n, C_n-R-C_t, and C_t-R-C_t. These hybrid molecules can be separated by ion-exchange chromatography, as shown in Fig. 7-39. Once separated the acylation can be reversed to give a variety of normally charged but chemically modified hybrids.

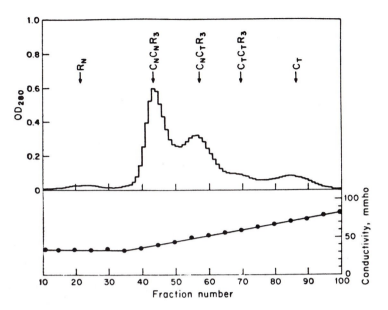

Figure 7-39 Chromatographic fractionation on DEAE-Sephadex of hybrid set formed by reconstitution of ATCase-like molecules from C_n, C_t, and R. (Reprinted with permission from: I. Gibbons and H. K. Schachman, *Biochemistry*, *15*, 52–60. Copyright 1976 American Chemical Society, Washington, D. C.)

Clearly, the possibilities of this type of manipulation with oligomeric proteins are almost unlimited, and some of the uses of this type of hetero-oligomer construction are examined in relation to allosteric proteins in a later chapter. Generation of differently charged subunits can also be of use in establishing quaternary structure, as detailed in Chap. 11.

Demonstration of Two Pyridoxal-5′-Phosphate (PLP)
Reactive Lysines in Glutamate Dehydrogenase

Reference: J. C. Talbot et al., *Biochim. Biophys. Acta*, *494*, 19–32 (1977).

The time course of Schiff's base formation by PLP binding to glutamate dehydrogenase is biphasic (Fig. 7-40), indicating the probable existence of at least two classes of lysine residues per polypeptide chain.

Examination of the time course of inactivation of the enzyme by PLP showed approximately 95% loss of activity in the absence of protecting ligands and no loss of activity in the presence of NADPH and 2-oxoglutarate (Fig. 7-41).

Tryptic mapping of enzyme modification with PLP after reduction with sodium borohydride in the absence or presence of protection showed two modified peptides

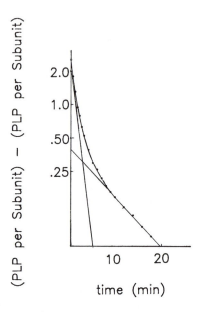

Figure 7-40 Time course of PLP binding to glutamate dehydrogenase. [Reprinted with permission from: J. C. Talbot, C. Gros, M. P. Cosson, and D. Pantaloni, *Biochim. Biophys. Acta, 494*, 19–32 (1977).]

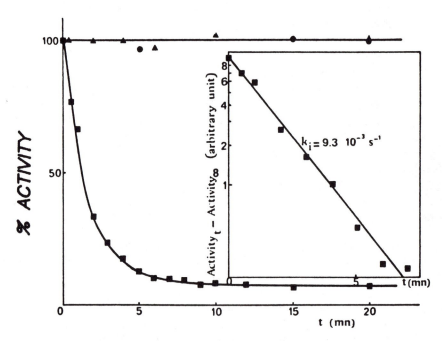

Figure 7-41 The inset shows a plot of log versus time and gives a rate constant of $9.3 \times 10^{-3} \text{ sec}^{-1}$. [Reprinted with permission from: J. C. Talbot, C. Gros, M. P. Cosson, and D. Pantaloni, *Biochim. Biophys. Acta, 494*, 19–32 (1977).]

176

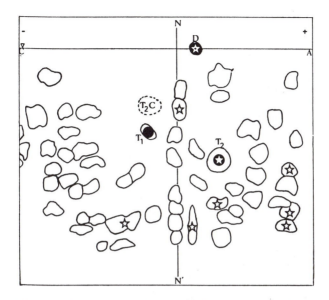

Figure 7-42 Peptide map of tryptic digest of [^{14}C]carboxymethylated reduced pyridoxal-P enzyme: T_1, peptide containing reduced pyridoxal-P-Lys (Lys-333); T_2, peptide containing reduced pyridoxal-P-Lys (Lys-126); T_2C, peptide containing reduced pyridoxal-P-peptide deriving from tryptic and chymotryptic digestion of pyridoxal-P enzyme, or from chymotryptic digestion of T_2. ●, Fluorescent-reduced pyridoxal-P-peptide spots under 300- to 350-nm ultraviolet light; ☆, radioactive spots; △, electrophoretic mobility of lysine. [Reprinted with permission from: J. C. Talbot, C. Gros, M. P. Cosson, and D. Pantaloni, *Biochim. Biophys. Acta*, *494*, 19–32 (1977).]

in the former case, corresponding to lysine-126 and lysine-333, but only a single modified peptide, corresponding to lysine-333, in the latter case (Fig. 7-42).

   Subsequent studies indicated that enzyme with only lysine-333 pyridoxylated no longer underwent the concentration dependent aggregation the enzyme is known to show, but otherwise appeared normal.

Attachment of Metal Chelating Groups to Macromolecules
Using "Bifunctional" Chelating Agents

Reference: C. Leung and C. F. Meares, *Biochem. Biophys. Res. Commun.*,
75, 149–155 (1977).

   In a novel approach to using chemical modification, diazonium chelating agents were used to derivatize proteins, mainly through reaction with lysine and histidine residues.

   Once derivatized the proteins could be titrated with the fluorescent lanthanide europium (Eu), which indicated excellent agreement with the molar incorporation of EDTA groups (Fig. 7-43).

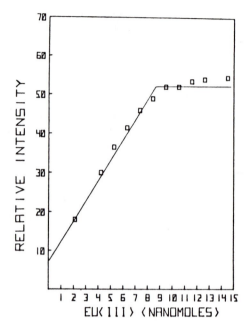

Figure 7-43   Fluorescence titration of albumin-bound EDTA groups (9 nmol) with Eu(III) in 0.1 *M* sodium citrate, pH 6.5. Fluorescence of Eu(III) excited at 310 nm and monitored at 618 nm. (Reprinted with permission from: C. S.-H. Leung and C. F. Meares, *Biochem. Biophys. Res. Commun.*, *75*, 149–155. Copyright 1977 Academic Press, Inc., New York.)

To be useful as a procedure for introducing fluorescent molecules onto a protein, such modifications must be shown to be unique if use in resonance energy transfer is anticipated.

Chemical Modification of Histidine Residues in p-Hydroxybenzoate Hydroxylase

Reference: R. A. Wijnands and F. Muller, *Biochemistry*, *21*, 6639–6646 (1982).

Figure 7-44 shows difference spectra of enzyme modified with diethylpyrocarbonate at pH 5.6 and at pH 8.0. The spectrum obtained at pH 5.6 is typical of histidine modification, and from the absorbance at 240 nm it was calculated that four histidine residues were modified, using an extinction coefficient of $3200\ M^{-1}\ cm^{-1}$. As shown in Table 7-7, essentially all the activity could be recovered after incubation of enzyme at pH 6, indicating the probability that only histidine residues are modified under these conditions.

The difference spectrum obtained by modification at pH 8.0 shows a distinctive region between 270 and 280 nm, suggesting that under these conditions tyrosine as well as histidine is being modified.

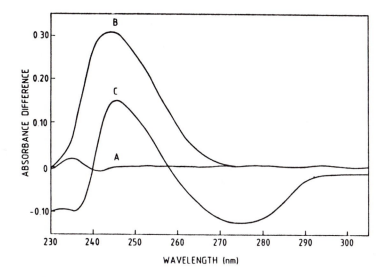

Figure 7-44 Ultraviolet difference spectra of hydroxybenzoate hydroxylase modified with DEP at pH 5.6 (B) or pH 8.0 (C). Line A is the baseline. (Reprinted with permission from: R. A. Wijnands and F. Muller, *Biochemistry*, *21*, 6639–6646. Copyright 1982 American Chemical Society, Washington, D. C.)

Figure 7-45 shows plots of percent activity remaining versus number of histidine residues modified in the absence of protecting ligand and in the presence of *p*-fluorobenzoate in part (A). Part (B) shows Tsou plots of the data, indicating that in the absence of *p*-fluorobenzoate, two of the four modifiable histidine residues are essential for activity, while in the presence of *p*-fluorobenzoate, one of the two modifiable residues is essential for activity.

Table 7-7 Time course of the reactivation of modified enzyme by hydroxylamine[a]

Time (hr)	Percent activity
0	2
0.5	9
1.0	14
1.5	75
2.0	86
2.5	92

[a] Enzyme was incubated with 116 mM hydroxylamine at pH 7.0.

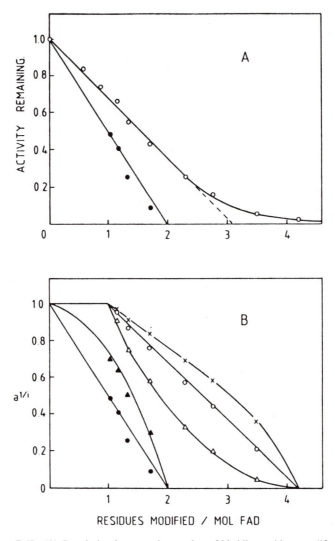

Figure 7-45 (A) Correlation between the number of histidine residues modified by diethylpyrocarbonate and the fractional activity remaining. The closed symbols represent the data for the enzyme–p-fluorobenzoate complex. (B) The data of part (A) are presented in the form of a Tsou plot for $i = 1$ ($\triangle$), $i = 2$ ($\bigcirc$), and $i = 3$ (X) (free enzyme), and for $i = 1$ ($\bullet$) and $i = 2$ ($\blacktriangle$) (p-fluorobenzoate-complexed enzyme). (Reprinted with permission from: R. A. Wijnands and F. Muller, *Biochemistry, 21*, 6639–6646. Copyright 1982 American Chemical Society, Washington, D. C.)

Modification of Arginine-148 in the β Subunit of
Tryptophan Synthase

Reference: K. Tanizawa and E. W. Miles, *Biochemistry*, 22, 3594–3603 (1983).

Reaction of apo-β-2 from tryptophan synthase with phenylglyoxal (Fig. 7-46) resulted in loss of the serine deaminase activity of the isolated subunit. Reciprocal pseudo-first-order rate constants calculated from the data in part A were plotted versus 1/[reagent] in part B, and the resulting plot indicated that the phenylglyoxal

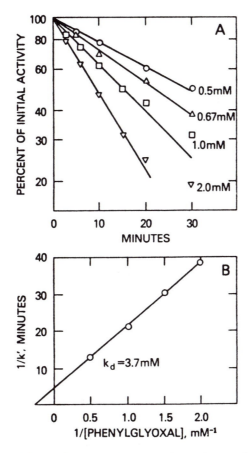

Figure 7-46 (A) Effect of phenylglyoxal concentration on the rate of inactivation of the apo-β-2 subunit of tryptophan synthase; (B) reciprocal first-order rate constants (k′) calculated from part (A) are plotted versus the reciprocal concentration of phenyl-glyoxal. The dissociation constant (K_d) for phenylglyoxal calculated from the inter-cept on the abscissa (1/K_d) is 3.7 mM. (Reprinted with permission from: K. Tanizawa and E. W. Miles, *Biochemistry*, 22, 3594–3606. Copyright 1983 American Chemical Society, Washington, D. C.)

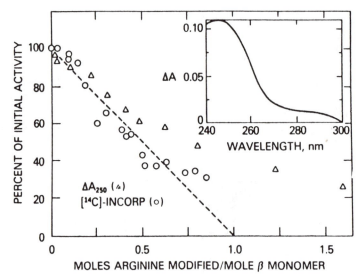

Figure 7-47   Effect of the extent of modification of the apo-β-2 subunit by phenyl-glyoxal upon activity. The extent of modification of arginyl residues determined from the change in absorbance at 250 nm after modification by phenylglyoxal in phosphate buffer ($\triangle$) or from the incorporation of radioactivity after modification by phenyl-[2-^{14}C]glyoxal in borate buffer ($\bigcirc$) is plotted versus the percent of initial activity. (Reprinted with permission from: K. Tanizawa and E. W. Miles, *Biochemistry, 22*, 3594–3606. Copyright 1983 American Chemical Society, Washington, D. C.)

formed a reversible complex prior to inactivation occurring, with a calculated dissociation constant of 3.7 mM.

When percent of residual activity was plotted versus moles of arginine modified per subunit, determined using radioactive phenylglyoxal, a single site per subunit was modified (Fig. 7-47). The inset shows a difference spectrum of the modified enzyme. From the absorbance at 250 nm, approximately 0.96 modified arginine per subunit was calculated using an extinction coefficient of 11,000 M^{-1} cm^{-1} for the diphenyl-glyoxal adduct of arginine, in good agreement with the incorporation determined by radioactivity. Cyanogen bromide fragmentation and isolation of the radioactively labeled fragment gave a peptide containing two arginine residues on amino acid analysis. Pepsin digestion of this cyanogen bromide fragment gave a single labeled fragment, which from amino acid analysis was identified as containing arginine-148.

As can be seen from these examples, chemical modification experiments have been used in numerous ways to answer many important questions in protein chemistry. The reagents we have discussed are meant only to illustrate the types of considerations that can be addressed in the design and execution of a modification experiment. There are many more reagents available.

In the introduction to this chapter we mentioned potential uses of chemical modification experiments. First was the identification of residues involved in the catalytic mechanism of an enzyme or in the binding site of a protein. Due to the possibility of a conformational change in the protein as a result of a chemical modification event, it is not possible to state unequivocally that because modification of a single amino acid residue in a protein blocks activity or binding that the residue is present in the catalytic or binding site. As discussed, even protection experiments do not allow this conclusion to be reached. The many other uses of chemical modification are not affected by this point, but where the purpose is to identify residues in or near particular binding sites of a protein, alternative approaches must be taken.

8

Chemical Modification:
Affinity Reagents

INTRODUCTION

A problem that is often encountered in chemical modification studies of the type described in Chap. 7 is limiting covalent modification to a few of the available side chains. An exception is where an unusually reactive side chain exists that can often be fairly specifically modified by reason of its more rapid reaction.

A second problem, also discussed in Chap. 7, is that side-chain-specific reagents may alter function by reaction at a site distant to the binding site or catalytic site being examined, leading to the erroneous conclusion that a certain residue is in or near that site.

The concept of site-specific labeling arose to circumvent some of these problems. The principle is quite simple: A reagent is synthesized that has characteristics of the ligand which usually binds to the site, but includes a reactive moiety that can react with residues in the binding site. Care is taken in the design of the reagent not to alter those ligand features that are essential for binding. Since such reagents have affinity for a specific site on the protein, they are usually referred to as "affinity reagents."

Once synthesized, the reagent (I^*), to be successful, must satisfy these requirements:

1. It must undergo reversible EI^* formation prior to irreversible inactivation. The existence of such complex formation is indicated by a saturable dependence of the rate constant for inactivation on the reagent concentration.

2. The extent of modification is limited relative to an unrelated reagent containing the same functional group when that group reacts readily with the protein. This is a result of effective concentration of the reagent. The affinity for the binding site gives a local concentration much higher than that in bulk solution, resulting in a more rapid modification of residues in the binding site than of similar residues elsewhere in the protein.

3. The modification can be protected by the presence of a competing, nonreactive ligand.

Before considering in more depth details of particular affinity reagents, we should complete our list of desirable characteristics of site-specific labels. As indicated, they should be structurally close to the parent compound and contain an active functional group that preferably reacts with a variety of different amino acid chains, and they should not be too bulky, so that steric problems do not arise. In addition to these three characteristics, site-specific reagents should produce stable derivatives of the modified protein in order that identification and quantitation of derivatized amino acid side chains is easy. As with many chemical modification reagents, availability of radio-labeled derivatives assists in quantitation and isolation of peptides. Finally, in the "ideal" reagent, one looks for such minor considerations as stability, solubility, and ease of synthesis or availability.

Affinity-labeling reagents have found increasing use in establishing the chemical nature of ligand binding sites in proteins. As we will see, in some cases it is possible to obtain information about a variety of amino acid residues in a particular site. This is achieved by the use of an affinity reagent with multiple reactivity or the use of a series of reagents with the reactive moiety at different positions in the molecule.

THEORETICAL CONSIDERATIONS

The formation of a reversible complex between the affinity reagent I* and its target protein is represented by

$$E + I^* \longleftrightarrow EI^* \longrightarrow E'I^* \tag{8-1}$$

where $K_d = [EI^*]/[E][I^*]$ and k is the rate of inactivation. The observed rate of inactivation at a given concentration of I*, K_{obs}, is given by

$$K_{obs} = \frac{k}{1 + K_d/[I^*]} \tag{8-2}$$

Therefore,

$$\frac{1}{K_{obs}} = \frac{1}{k} + \frac{K_d}{k} \frac{1}{[I^*]} \tag{8-3}$$

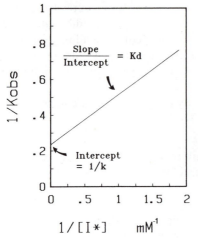

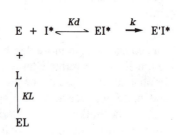

Figure 8-1   Determination of K_d and k for a functional site-specific reagent.

Figure 8-2   Scheme of competition between I* and L for protein binding sites.

indicating that a double-reciprocal plot, as in Fig. 8-1, allows determination of K_d and k.

One of the consequences of the formation of a reversible EI* complex is that protection against modification is afforded by analogs of I* that lack the reactive moiety. For such an analog (L), which forms a reversible EL complex as shown in Fig. 8-2, the dissociation constant K_L can be obtained from protection experiments.

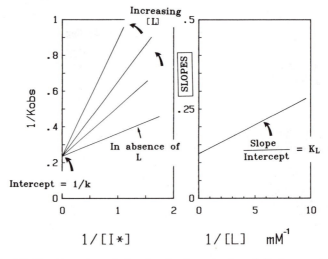

Figure 8-3   Double reciprocal plots for the determination of K_L from protection experiments.

By analogy with a simple competitive inhibitor we get

$$K_{obs} = \frac{k}{1 + (K_d/[I^*])}\left(1 + \frac{[L]}{K_L}\right)$$

(8-4)

and K_L as well as K_d, and k can be obtained from a series of determinations of the dependence of K_{obs} on $[I^*]$ at different concentrations of L, as shown in Fig. 8-3.

AFFINITY REAGENTS

Over the years many different types of site-specific reagents have been developed. Rather than attempt a comprehensive list, we examine one type in some detail since the considerations that go into their design and the problems encountered in their use are common to most site-specific reagents. The group we examine, because of their wide and varied uses, are the purine nucleotide analogs.

There are two classes of affinity reagents that can be considered. Some ligands, as part of their normal structure, contain a moiety that can be chemically converted to the required reactive group without a major perturbation of the structure of the ligand. We refer to them as endo-affinity reagents. In many cases, however, it is not possible to convert some part of the normal ligand to produce an affinity reagent, and a reactive moiety must be added to give an exo-affinity reagent. With exo-affinity reagents, particular attention must be paid to whether or not the additional moiety may interfere with or alter the normal binding specificity of the ligand.

Endo-affinity Reagents

The first class of purine nucleotide analogs that we consider are those arising from periodate oxidation. If we consider ATP as the parent compound, periodate cleavage results in oxidation of the ribose to give a 2′,3′-*cis*-dialdehyde (O-ATP, Fig. 8-4). This derivative, via the aldehyde moieties, will react primarily with lysine (or possibly N-terminal) amino groups, although a possible reaction with the sulfhydryl of cysteine has been suggested (Fig. 8-5).

If the amino group reacts to give a Schiff's base, reduction with sodium boro-hydride will give irreversible inactivation. Reduction with borotritiide will result in tritium labeling of the derivative. Such a scheme and results have been found in many enzymes with periodate-oxidized ATP, ADP, GTP, and even NADP (which can be enzymatically reduced to give O-NADPH).

In some enzymes, however [e.g., isocitrate dehydrogenase (ICDH)], labeling patterns obtained with radio-labeled reagents are inconsistent with the Schiff's base mechanism just mentioned. When ICDH is labeled with either O-ADP containing ^{14}C at the C-8 position, or with ^{32}P in the phosphates, one gets enzyme derivatives with different amounts of ^{14}C or ^{32}P labeling. In addition, reduction with sodium borotritiide results in *no* significant ^{3}H labeling. Such results have been explained in terms of the formation of a 4′,5′-didehydro-2′,3′-dihydroxymorpholino derivative after

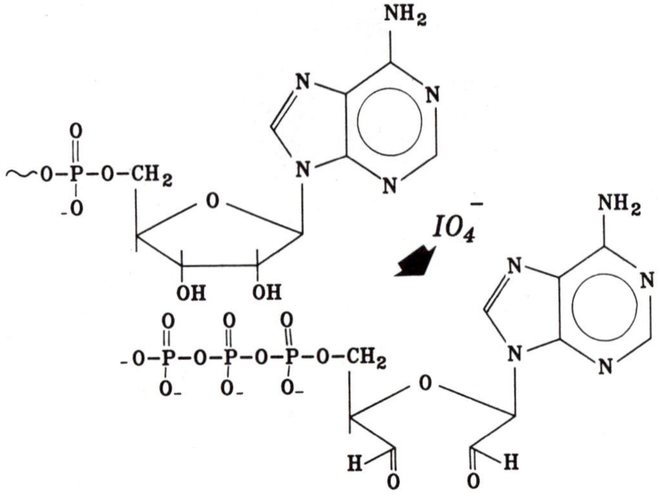

Figure 8-4 Periodate-oxidized ATP (*O*-ATP).

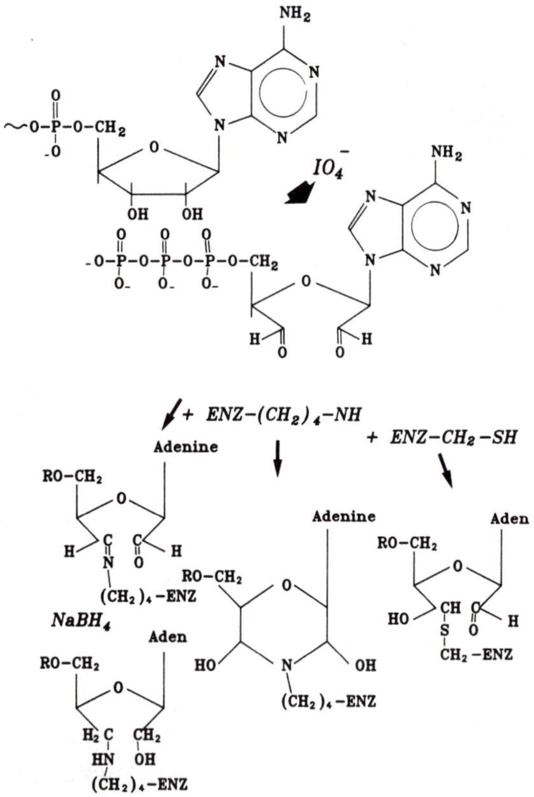

Figure 8-5 Possible reactions of periodate oxidized ATP with amino acid side chains.

188

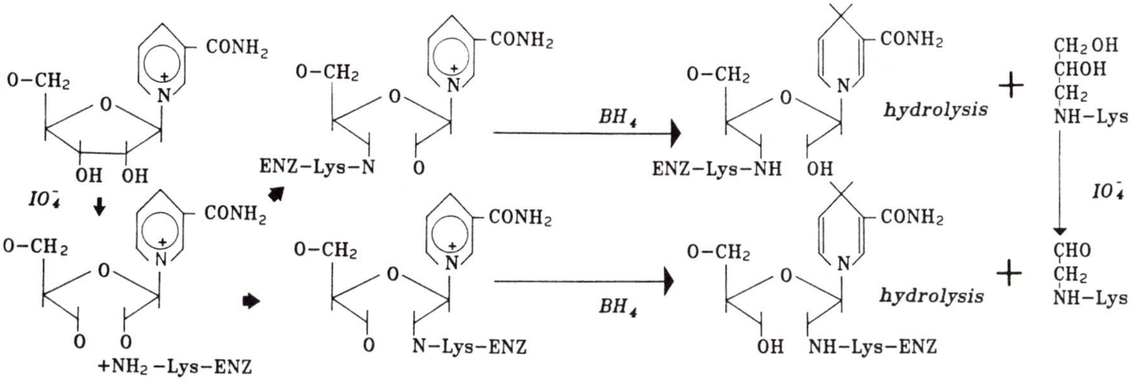

Figure 8-6 Reaction scheme for *O*-NADP.

reaction with the lysine amino group. Such products tend to be unstable to acid hydrolysis and subsequent amino acid analysis. The Schiff's base derivatives, however, are stable to hydrolysis to the extent that they give modified lysine derivatives. Shown in Fig. 8-6 is a reaction scheme for the inactivation and isolation of derivatized lysine for an enzyme modified by *O*-NADP. As can be seen, depending on which aldehyde in the *O*-NADP the Schiff's base is formed with, two derivatives of lysine can be obtained, which after periodate oxidation of the amino acid hydrolysate give only derivative II.

If we consider how periodate-cleaved purines rate as site-specific reagents in terms of our earlier criteria, we find that for structural considerations they rate well since only the ribose is modified and the modification does not involve large substituents. Many proteins that bind purines will tolerate alterations in the ribose ring. However, these derivatives do have drawbacks. The 2′,3′-dialdehyde reacts almost exclusively with lysine side chains rather than the more desirable case with wider side-chain specificity. In addition, when the reaction does not involve a Schiff's base the derivatives are somewhat unstable to acid hydrolysis or some of the harsher procedures that might, of necessity, be used in fragmentation.

Certain *alkyl halide derivatives* can also be considered endo-affinity reagents. This group can be regarded as having iodoacetamide as the parent compound, which can react with a variety of nucleophilic side chains, including cysteine, histidine, lysine, methionine, and to a minor extent, glutamate and aspartate. The most common derivatives are linked through the phosphates, as in the case of adenosine-5′-chloromethanepyrophosphonate and adenosine-5′-(β-chloroethylphosphate), whose structures, in relation to ATP, are shown in Fig. 8-7.

Not all alkyl halide derivatives of purines can be considered endo-affinity reagents, but with these derivatives of ATP the structure sufficiently resembles that of ATP for them to be considered as such. These derivatives can be very useful for labeling ATP binding sites if the purine part of the molecule provides the specificity. This type of derivative is not possible with NAD(P), of course.

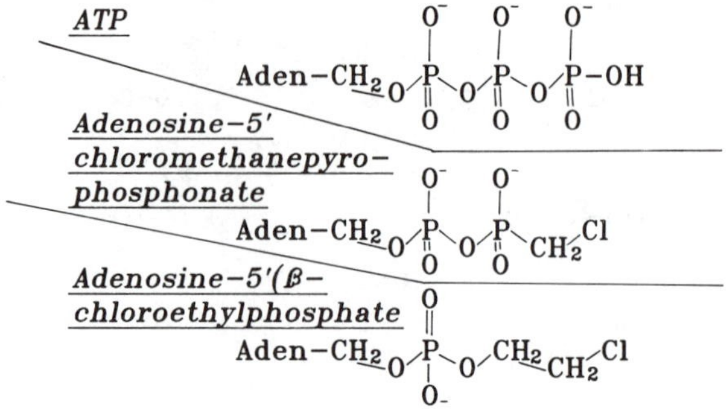

ATP

Aden$-CH_2$O$-\overset{O^-}{\underset{O}{P}}O-\overset{O^-}{\underset{O}{P}}O-\overset{O^-}{\underset{O}{P}}$-OH

$Adenosine-5'$
$chloromethanepyro-$
$phosphonate$

Aden$-CH_2$O$-\overset{O^-}{\underset{O}{P}}O-\overset{O^-}{\underset{O}{P}}CH_2$Cl

$Adenosine-5'(\beta-$
$chloroethylphosphate$

Aden$-CH_2$O$-\overset{O}{\underset{O_-}{P}}O-CH_2CH_2$Cl

Figure 8-7   Alkyl halide derivatives of ATP.

Exo-affinity Reagents

The alkyl halide moiety is usually considered as producing an exo-affinity reagent since it is normally added to the ligand in such a way as not to interfere with binding. An example of such a use is the linkage of an alkyl halide derivative via the N-6 of the adenine, as in the case of *N*-6-*p*-bromoacetamidobenzyl ADP (Fig. 8-8).

In a most elegant extension of the ideas of site-specific labels, a series of new affinity labels that are based in part on the foregoing alkyl halide derivatives have been synthesized and shown to be effective. As already mentioned, a problem often encountered with affinity labels is a too stringent side-chain specificity. This has been partly overcome by the synthesis of purine nucleotide derivatives with *two* reactive moieties instead of the more usual one. These derivatives, whose structures are shown in Fig. 8-9, have the alkyl halide moiety *and* a dioxobutyl moiety built into the same six-position substituent, which for ease of synthesis has been replaced by a thio derivative.

These novel reagents have specificity for arginine in addition to the range of side chains with which the alkyl halides react. The halide reactivity is also increased by

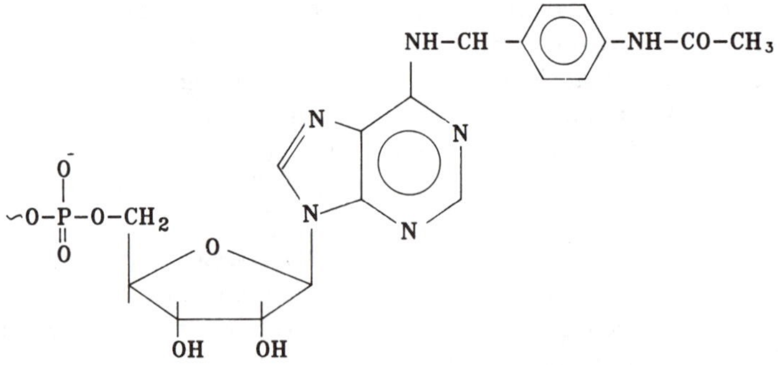

NH$-$CH $-$⟨◯⟩$-$NH$-$CO$-$CH$_3$

$\sim$O$-\overset{O^-}{\underset{O}{P}}O-CH_2$

OH   OH

Figure 8-8   Structure of *N*-6-*p*-bromoacetamido-benzyl-ADP.

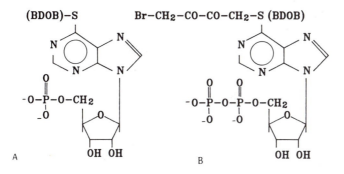

Figure 8-9 Structures of adenine nucleotide analogs: (A) 6-[(4-bromo-2,3-dioxobutyl)thio]-6-deaminoadenosine-5'-monophosphate; (B) 6-[(4-bromo-2,3-dioxobutyl)thio]-6-deaminoadenosine-5'-diphosphate.

having it in the form of a bromo-keto derivative, which helps overcome one inherent problem of the alkyl derivatives—low reactivity.

A different type of reactive group, which has been incorporated via the phosphate in both adenine and guanine nucleotides, is the fluorosulphonylbenzoyl (FSB) group found in adenosine 5'-(p-fluorosulphonylbenzoylphosphate), whose structure is shown in Fig. 8-10. These derivatives are particularly reactive toward lysine and tyrosine side chains. The FSB derivatives of ethanoadenosine have also been used to introduce a fluorescent group into proteins.

A special class of exo-affinity reagents are the photoaffinity reagents. In general, they have several distinct advantages over the other types of site-specific reagents:

1. They remain chemically inactive until they are activated (usually by light), which means that a wide variety of enzymic studies are possible without chemical modification occurring. In many cases the nonactivated derivatives are biologically active, allowing kinetic or regulatory properties of the derivatives to be established.

2. The generated group (diazo derivatives give a carbene, azido derivatives a nitrene: see Fig. 8-11) reacts fairly indiscriminately with a wide variety of side chains and thus does not suffer from being too side-chain specific.

There are, of course, problems, the most important of which is that the active species often has a quite long lifetime and as a result may dissociate from its binding site after activation but prior to labeling a residue in the binding site. Once in solution it may react nonspecifically with various side chains in the protein. Where such

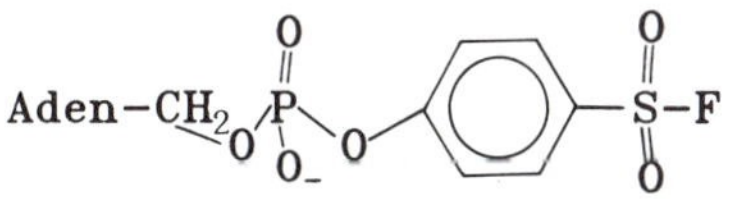

Aden—CH$_2$—$\overset{O}{\underset{O_-}{P}}$—O—⟨benzene⟩—$\overset{O}{\underset{O}{S}}$—F

Figure 8-10   Structure of fluorosulfonylphenyl group.

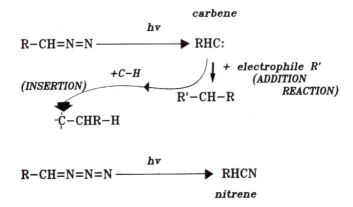

Figure 8-11   Generation of carbene or nitrene radicals in photo-affinity labels.

nonspecific labeling may present a problem (usually with long-lived active species that bind weakly to their specific binding sites), protection studies can distinguish nonspecific from specific labeling.

8-Azido derivatives (Fig. 8-12) have been used extensively; however, one problem that may be encountered is their tendency to adopt a syn conformation while the parent nucleotides exist predominantly in the anti conformation. Other derivatives, such as arylazido-β-alanine ATP or arylazido-β-alanine NAD(P), where the photo-affinity substituent is linked via ribose hydroxyls (Fig. 8-13), have been used successfully.

While the azido derivatives might be considered as endo-affinity reagents, the arylazido derivatives are clearly exo-affinity. These reagents have extremely bulky substituents, however, which can lead to two problems: (1) the substituent may interfere with binding—either preventing or distorting it, and (2) because of the distance between the reactive azido group and the major areas of the molecule involved in binding to its specific site, labeling may occur at adjacent residues rather than directly in the binding site.

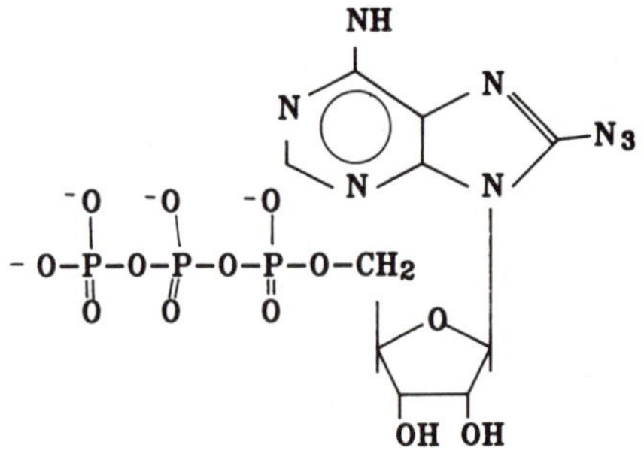

Figure 8-12   8-Azido adenosine triphosphate.

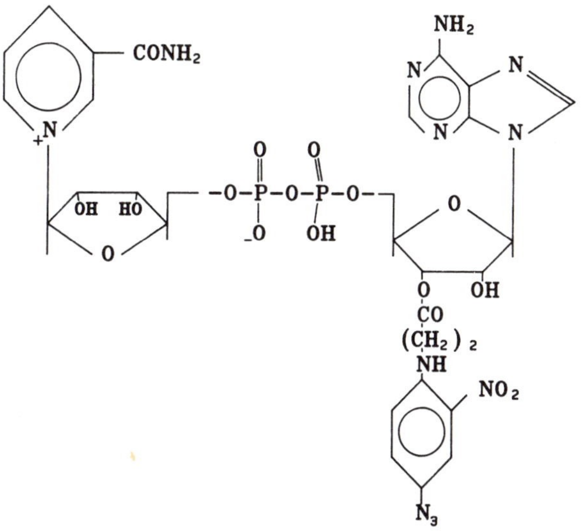

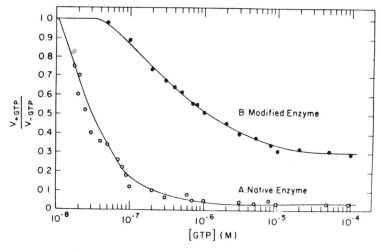

Figure 8-13   Structure of arylazido-β-alanine NAD.

This particular effect results in the type of data shown in Fig. 8-14 for modification of glutamate dehydrogenase by FSB-ethano-A, a supposed GTP-specific site reagent. The modified enzyme clearly is still responsive to the allosteric inhibitor GTP. The modification has occurred at some site adjacent to, rather than directly in, the GTP site, and caused decreased affinity rather than blocking GTP binding.

Figure 8-14   Inhibition of native and GTP-site "modified" glutamate dehydrogenase by GTP. (Reprinted with permission from: M. A. Jacobson and R. F. Colman, *Biochemistry, 21,* 2177–2186. Copyright 1982 American Chemical Society, Washington D. C.)

TABLE 8-1   Selected synthesized affinity reagents

Reagent	Target enzyme
4,5-Dimethoxy-2-nitrobenzyl esters of cAMP or cGMP	Cyclic nucleotide binding proteins
2-Azido-ADP	Chloroplast coupling factor 1
Procion Blue MX-R	Nucleotide binding sites
6-Diazo-5-oxo-L-norleucine	γ-Glutamyl transpeptidase
2-Diazoestrone sulfate	Steroid binding sites
Bromoacetylpyridoxamine	PLP binding sites
β-D-Galactopyranosylmethyl-p-Nitrophenyl triazine	β-Galactosidase
9-(3,4-Dioxopentyl)hypoxanthine (reacts with arginine)	Purine nucleoside phosphorylase
2-(4-Bromoacetamido)anilino-2-deoxypentitol-1,5-bisphosphate	Ribulose bisphosphate carboxylase

Exo-affinity reagents and the photoactivatable reagents have a particularly useful application in terms of mapping the residues around a binding site. With exo-reagents it is often possible to locate the reactive moiety at different parts of the parent compound (provided that the affinity is not reduced to the extent that the advantage of an affinity reagent is lost). This allows the reactive moiety the chance to modify more than one residue in (or near) the binding site. The advantage is greatest with a photoaffinity label where there is little specificity of the generated carbene or nitrene. Exo-reagents, where the reactive moiety is freely mobile about its point of attachment to the parent compound, have the potential to label any reactive side chain within this sphere of influence. Analysis of the various residues that are labeled indicates which side chains are located in the vicinity of the binding site. If there is a chance that more than one binding site exists, protection experiments are of vital importance. All residues located at or near a single site will be protected *to the same extent* at a particular concentration of the protecting ligand. Where two or more sites with differing affinity exist, some modified residues will be protected to a greater extent than others in such an experiment.

Although we have considered site-specific reagents based on purine nucleotides, the list of reagents that have been synthesized is almost limitless. Some of the more interesting are listed in Table 8-1.

SUICIDE INHIBITORS

A suicide inhibitor is the product of an enzymic reaction on a substrate analog which produces a highly reactive enzyme-bound intermediate that chemically modifies the active site that has produced this intermediate. In general, the enzyme-generated reactive species is an electrophile that reacts with a protein nucleophile in the active site to give an irreversible covalent derivative. The scheme in Fig. 8-15 illustrates that there is always a competition, once the electrophile has been generated, between the inactivation event (k_4) and the dissociation of the electrophile prior to the modification event taking place (k_3).

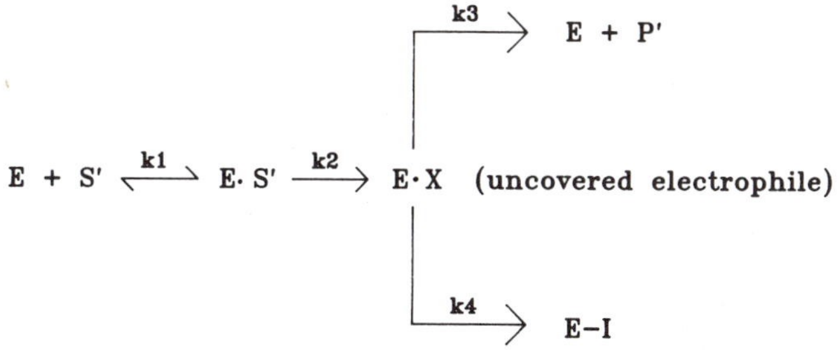

$$E + S' \xrightleftharpoons[]{\mathbf{k1}} E \cdot S' \xrightarrow{\mathbf{k2}} E \cdot X \quad \text{(uncovered electrophile)}$$

with branches:

$$\xrightarrow{\mathbf{k3}} E + P'$$

$$\xrightarrow{\mathbf{k4}} E\text{-}I$$

$$\mathbf{k3/k4} = \textbf{Partition Ratio} = \frac{\textbf{No. of Product Molecules}}{\textbf{Inactivation Event}}$$

Figure 8-15 Scheme of a mechanism of a suicide inhibitor.

The ratio of k_3/k_4 is known as the partition coefficient for inactivation, and for an effective suicide inhibitor should be a low number. This is especially true where the generated active species may diffuse and modify enzyme nucleophiles other than those in the active site, leading to inactivation. Studies of the time course of inactivation allow a distinction to be made between these two cases. In true suicide inhibition, activity will in general be lost in a first-order process, while in the latter case loss of activity will follow an initial lag period. There are many ways that suicide inhibitors can be generated, but we consider only a few examples here.

Figure 8-16 illustrates the process by which the enzyme β-hydroxydecanoyl thiol dehydrase is activated by the acetylenic analog of its substrate. The enzyme catalyzes

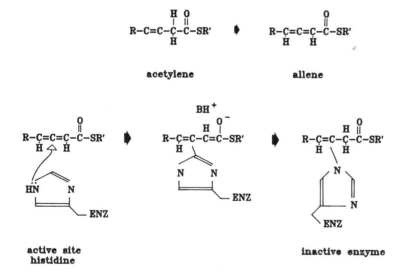

Figure 8-16 Suicide inhibition of β-hydroxydecanoyl thiol ester dehydrase.

TABLE 8-2   Enzymes subject to inhibition by acetylenic substrates

Enzyme	Inhibitor
Aromatase	10-Propargyl-4-ene-3,17-dione
GABA transaminase	
DOPA decarboxylase }	α-Acetylenic DOPA
Prostaglandin synthase	

the isomerization of the unreactive 3-acetylenic moiety to a 2,3-allene conjugated to the thioester group, and it is this allene that reacts with an active-site histidine to give the inactive enzyme. Acetylenic substrate analogs have been widely used as suicide inhibitors in a number of systems, a few of which are shown in Table 8-2.

The serine proteases have been targets for suicide inhibitor design, and two basic approaches are employed. The enzyme can generate an intermediate which is so reactive that it leads to unwanted (by the enzyme) covalent modification.

Incubation of chymotrypsin with the substrate, benzyl chloropyrone, leads to suicide inhibition of chymotrypsin (with a partition ratio of about 20) by generation of the active acyl chloride moiety as a result of acyl-enzyme formation. The acyl chloride generated reacts with an enzyme nucleophile to give the inhibited enzyme, as shown in Fig. 8-17.

In the second strategy, which can be used where covalent intermediates such as the acyl-enzyme intermediate lie on the normal reaction pathway of the enzyme

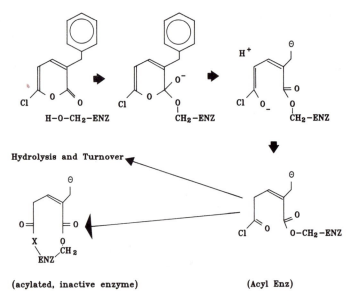

Figure 8-17   Suicide inhibition of chymotrypsin by benzoyl chloropyrone.

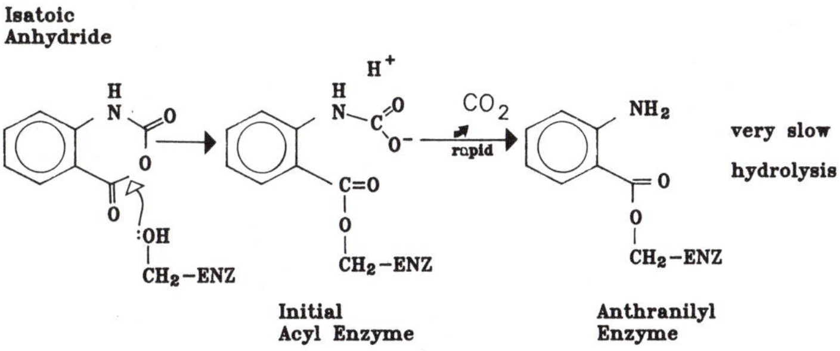

Figure 8-18   Suicide inhibition of chymotrypsin by isatoic anhydride.

mechanism, the substrate is modified so that the acyl-enzyme intermediate cannot break down to give product.

Chymotrypsin acts on isatoic anhydride (Fig. 8-18) to form an enzyme-acyl intermediate that in this case is rapidly hydrolyzed, not at the point of attachment to the enzyme, but at the generated carbamate, resulting in an anthranilyl enzyme intermediate that is hydrolyzed exceedingly slowly.

EXAMPLES OF SITE-SPECIFIC REAGENTS

Modification of Glutamate Dehydrogenase with
5'-p-Fluorosulfonylbenzoyl Adenosine (FSBA)

References: K. V. Saradambal et al., *J. Biol. Chem.*, *256*, 11866–11872 (1981).
J. A. Schmidt and R. F. Colman, *J. Biol. Chem.*, *259*, 14515–14519 (1984).

In a successful chemical modification experiment with a site-specific reagent a number of different experimental approaches must be used to establish that the reagent is indeed site specific prior to the identification of the modified residue or residues. The example of the modification of glutamate dehydrogenase by FSBA is a good one to examine since over a period of years the work has been taken to its logical conclusion: The residues modified by the reagent have been identified.

Kinetics and Quantitation of the Modification. With glutamate dehydrogenase, incubation with FSBA (at pH 8 in buffer containing 10% ethanol to assist solubility of the FSBA) results in a time-dependent *increase* in activity when the enzyme is assayed under conditions giving inhibition by excess NADH. This suggests that the site modified by FSBA is the inhibitory NADH site the enzyme is known to possess. As shown in Fig. 8-19, a semilog plot of the data is linear, allowing calculation of a rate constant of the modification. Protection studies show that an unrelated ligand, GTP, has no effect on the rate constant for modification, but NADH is an effective competitor of the modification.

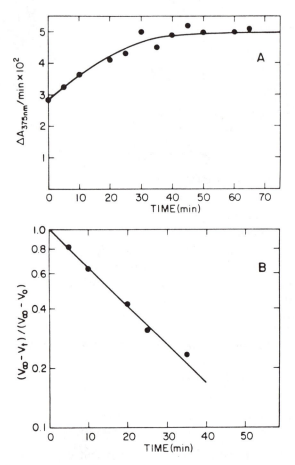

Figure 8-19   Reaction of GDH with FSBA: (A) rate of enzyme reaction of aliquots of enzyme withdrawn from an incubation mix containing FSBA and assayed at time intervals up to 70 minutes; (B) semilog plot of the data from part (A), where V_0 and V_t are the enzyme rates at $t = 0$ and any time t, and V_∞ is the velocity when the modification is complete. [From K. V. Saradambal, A. Bednar, and R. F. Colman, *J. Biol. Chem.*, *256*, 11,866–11,872 (1981). Reprinted with permission of the copyright owner, The American Society of Biological Chemists, Inc., Bethesda, Md.]

The reaction can be quantitated using radioactive FSBA, and Fig. 8-20 shows the percent change in NADH inhibition as a function of the number of moles of FSBA incorporated per subunit. The results clearly indicate that complete loss of NADH inhibition is the result of the modification of significantly *less* than one site per subunit. The possibility that such results could be due to a concentration-dependent aggregation of the protein masking some sites is eliminated by the observation that the stoichiometry of modification is unchanged at two protein concentrations.

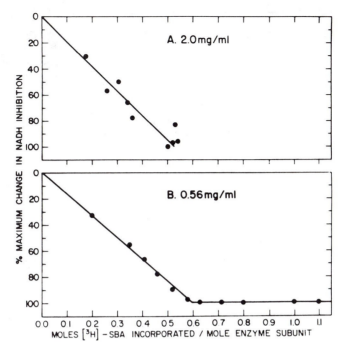

Figure 8-20 Quantitation of FSBA modification. [From K. V. Saradambal, A. Bednar, and R. F. Colman, *J. Biol. Chem.*, *256*, 11,866–11,872 (1981). Reprinted with permission of the copyright owner, The American Society of Biological Chemists, Inc., Bethesda, Md.]

Characterization of the Effects of Modification. Once the stoichiometry of modification is established, the kinetic properties of the modified enzyme are examined. The data in Fig. 8-21 show that the substrate inhibition seen at high NADH concentrations with the native enzyme is absent in the modified enzyme. Although in this particular example it was shown that the modified enzyme is still subject to inhibition by GTP, the various kinetic parameters of the enzyme were not examined.

Identification of Amino Acids Modified. The sulfonyl-fluoride moiety of FSBA is capable of reacting with a number of side chains in proteins, especially serine, cysteine, lysine, and tyrosine. In this example, modification of serine and cysteine is eliminated from further consideration since (1) after modification and acid hydrolysis no pyruvate could be detected (serine modification produces a dehydroalanine residue following incubation with sodium hydroxide which will release pyruvate on acid hydrolysis), and (2) the total cysteine residues in the protein, as determined by DTNB titration, were unchanged after modification.

After acid hydrolysis of protein modifed to varying extents with FSBA, carboxy-benzenesulfonyl (CBS)-lysine and CBS-tyrosine, the expected acid hydrolysis products of modified lysine and tyrosine, are both identified in amino acid analysis.

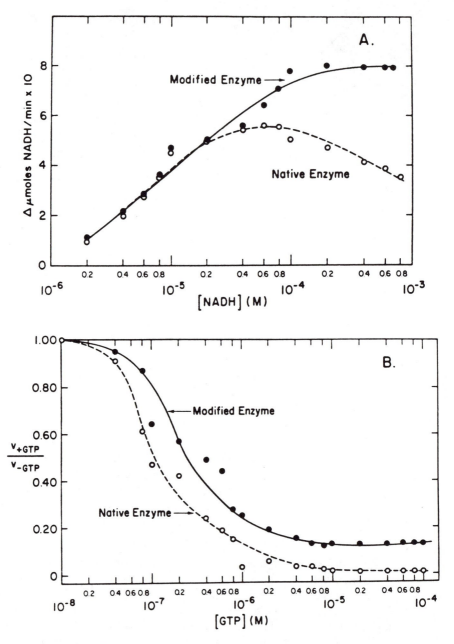

Figure 8-21 Comparison of kinetic properties of native and modified enzyme. [From K. V. Saradambal, A. Bednar, and R. F. Colman, 11,866–11,872 (1981). Reprinted with permission of the copyright owner, The American Society of Biological Chemists, Inc., Bethesda, Md.]

Furthermore, both residues are modified to approximately the same ratio no matter what the total incorporation of FSBA per enzyme molecule.

Location of Modified Residues in the Sequence. The modified protein is subjected to thermolysin proteolysis after denaturation (with urea), reduction, and carboxymethylation. After digestion the modified peptides are initially purified by affinity chromatography using dihydroxyboryl-substituted cellulose, which interacts reversibly with the *cis*-diol moiety of the modified residue. Two peptides are subsequently purified by HPLC and subjected to amino acid analysis. Since the entire amino acid sequence of glutamate dehydrogenase is already known, the amino acid analysis of the two modified peptides is sufficient to uniquely characterize the thermolysin fragments modified. The two modified residues are identified as tyrosine-190 and lysine-420.

β-Lactamase Suicide Inhibitors

References: D. G. Brenner and J. R. Knowles, *Biochemistry, 20,* 3682–3687 (1981). C. J. Easton and J. R. Knowles, *Biochemistry, 21,* 2857–2862 (1982). M. Arusawa and S. Adam, *Biochem. J., 211,* 447–454 (1983).

A variety of compounds that behave as suicide inhibitors of β-lactamase have been described. Sulbactam, whose reaction is shown in Fig. 8-22, results in a double

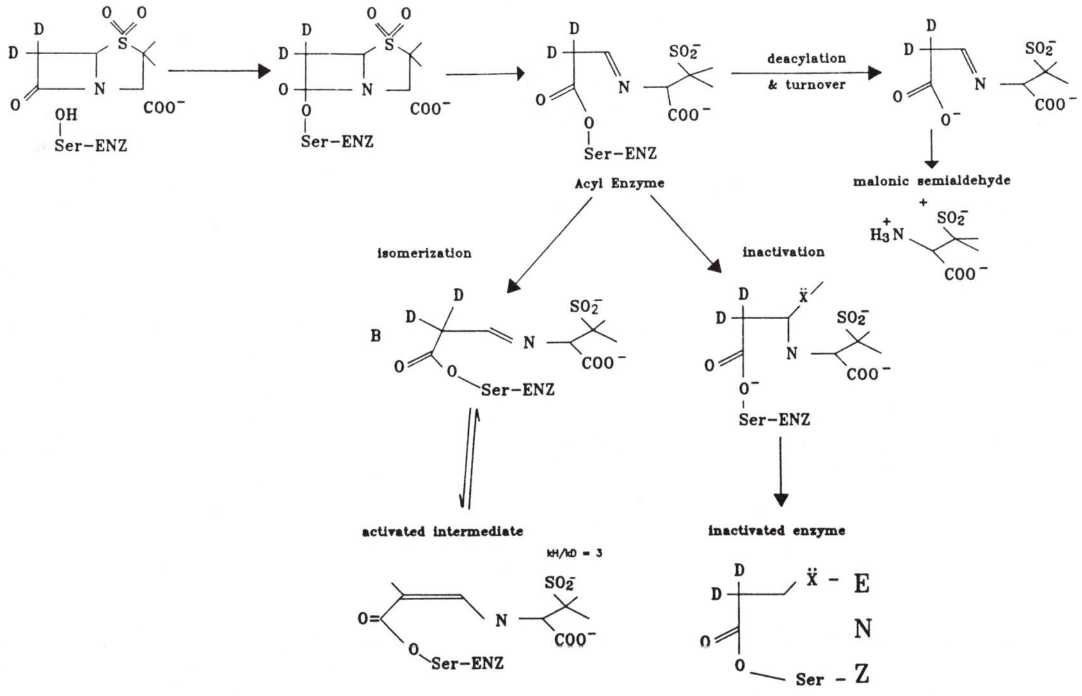

Figure 8-22   6,6,D$_2$-sulbactam as a suicide substrate of β-lactamase.

covalent modification, the acyl enzyme intermediate being stabilized by reaction of a generated electrophile with an enzyme nucleophile.

A somewhat different mechanism of suicide inhibition results from the use of Δ^2-pyroline; two forms of the acyl enzyme intermediate are formed, one of which, the Δ^2-acyl enzyme, is readily hydrolyzed, while the Δ^2-acyl enzyme is hydrolyzed very slowly (Fig. 8-23).

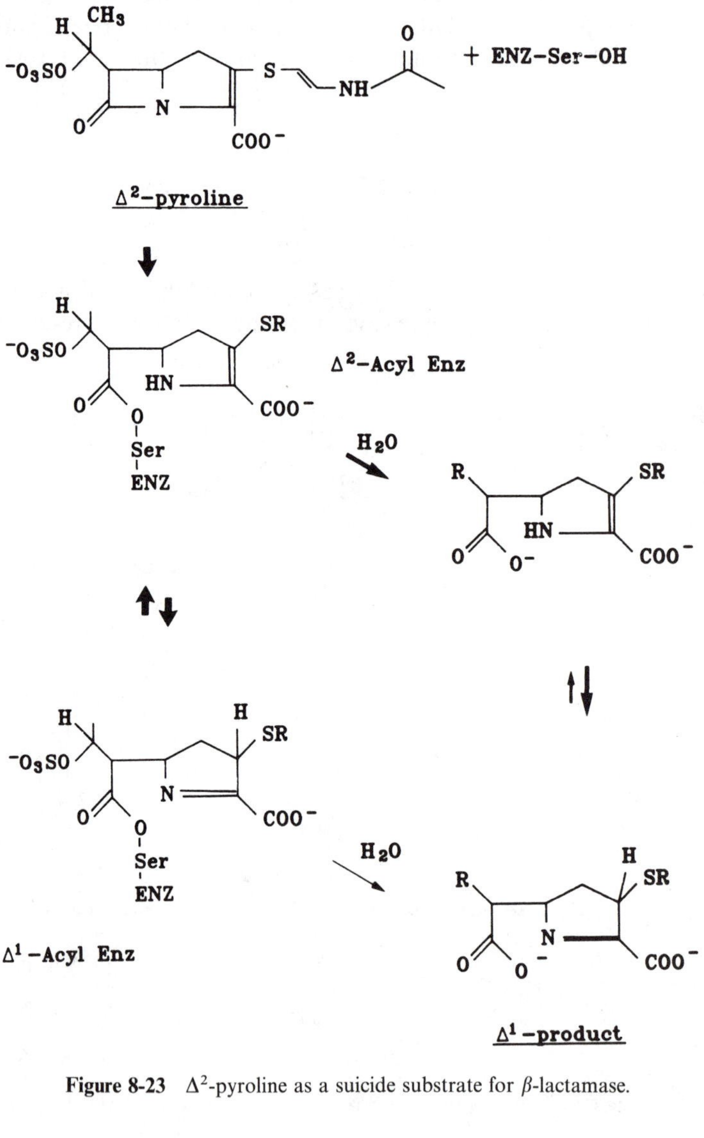

Figure 8-23 Δ^2-pyroline as a suicide substrate for β-lactamase.

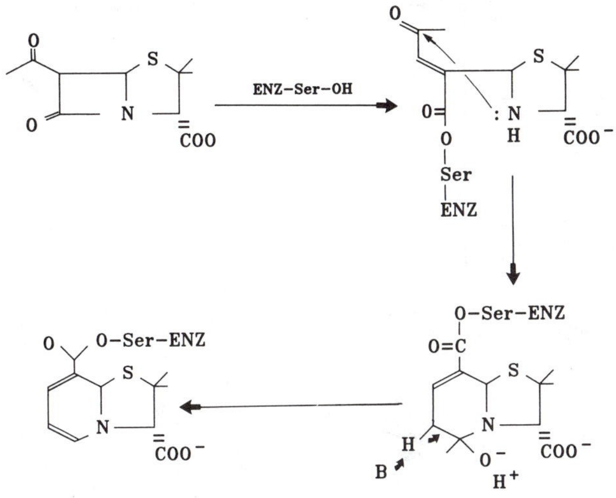

Figure 8-24 AcMPA as a suicide substrate of β-lactamase.

Finally, yet a third type of mechanism is observed, with acetylmethylene peni-cillanic acid (AcMPA) as substrate (Fig. 8-24). Acetylmethylene penicillanic acid leads to intramolecular cyclization of the acyl enzyme intermediate, which gives a product that does not hydrolyze, thus irreversibly inhibiting the enzyme.

Photo-Affinity Labeling of RNA Polymerase with
8-Azido-adenosine 5'-Triphosphate

Reference: A.-Y. Woody et al., *Biochemistry, 23*, 2843–2848 (1984).

Initial rate kinetic studies (Fig. 8-25) show that 8-azido-ATP is a competitive inhibitor of the polymerase, with respect to ATP. When the kinetics with respect to UTP are examined, no inhibition by 8-azido-ATP is observed.

After irreversible labeling and analysis of the subunit distribution of the label, it is found that the β' and σ subunits contained the majority of the label. The presence of ATP, but not UTP, causes a significant decrease in the amount of labeling, as shown in Table 8-3. These results indicate that the β' and σ subunits of this multi-subunit enzyme contain the ATP binding sites, but that the β and α subunits are in close proximity. Labeling of the β and α subunits appears to be specific since it is reduced by the presence of ATP.

Although these examples have been carried to different extents, it is clear that affinity reagents have a tremendous potential to give information about the spatial location of binding sites, and their function. To get the maximum information from such studies requires a careful analysis of the location of modified residues in the primary sequence of the protein, together with documentation of the effects of specific

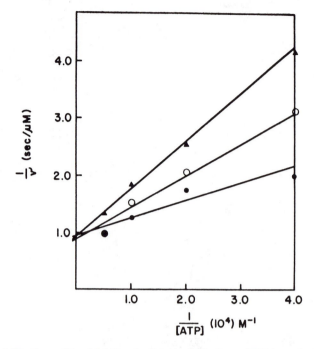

Figure 8-25   Competitive inhibition by 8-azido-ATP. Rate of RNA synthesis was followed in the absence of 8-azido-ATP (●) or in the presence of 25 M_μ (○) or 50 M_μ (▲) 8-azido-ATP. (Reprinted with permission from: A.-Y. M. Woody, C. R. Vader, R. W. Woody, and B. E. Haley, *Biochemistry*, *23*, 2843–2848. Copyright 1984 American Chemical Society, Washington, D. C.)

modifications. The latter point is particularly necessary where conclusions regarding function of the site are to be drawn.

Affinity labeling has also found use in the introduction of specific, conformationally sensitive probes or fluorophores to be used in resonance energy transfer studies to estimate distances between points on a protein molecule.

TABLE 8-3   Effects of ATP and UTP on photoincorporation of 8-azido-ATP in subunits of RNA polymerase

Concentration (μM)		Percent incorporation in absence of ATP or UTP in each subunit			
ATP	UTP	β	β'	σ	α
0	0	100	100	100	100
35	0	75	40	66	57
150	0	20	11	15	11
0	100	100	120	110	
0	200	90	98	94	

9

Secondary Structure of Proteins

INTRODUCTION

Proteins are made up of linear polypeptides that, after synthesis on a ribosome, fold up spontaneously to give a unique and biologically active three-dimensional structure. As was originally proposed by Linderstrom-Lang, three-dimensional structure can be considered at a number of levels. We have already discussed (Chap. 5) the *primary structure* of a protein: its amino acid sequence and various chemical characteristics such as disulfide bonds and covalently attached cofactors, carbohydrates, or other derivatives. In this chapter and in Chaps. 10 and 11 we consider secondary, tertiary (the other two levels originally described by Linderstrom-Lang), and quaternary structure.

Secondary structure may be defined as the local spatial organization of the poly-peptide backbone without consideration of the side-chain conformations. As we will see, however, when considering the prediction of secondary structure from the amino acid sequence of the protein, the nature of the side chains in a particular region of polypeptide chain does influence whether a certain secondary structure is found. The secondary structure is defined by four basic categories: α helix, β strand (often associated into so-called "sheets"), β turn, and random coil.

The *tertiary structure* of a protein is defined as the packing of the foregoing secondary structural elements *within* a polypeptide chain into a three-dimensional structure. Although as just defined, a tertiary structural element should involve a single polypeptide chain, there are instances where an apparent tertiary structural element involves two or more polypeptide chains. As will be seen, the initial stages of the folding of a polypeptide chain are dominated by local secondary structure in the nascent polypeptide.

Quaternary structure is the assembly of tertiary structural elements into an oligomeric form. Such an oligomer can be homologous (i.e., consisting of multiple copies of a single type of polypeptide chain; for example, *glutamate dehydrogenase*, which has six chemically identical polypeptide chains) or heterologous (i.e., consisting of two or more chemically distinct polypeptide chains; for example, *aspartate transcarbamoylase*, which is made up of 12 polypeptide chains, six each of two different types).

In this chapter we are concerned primarily with secondary structure (although, as will become evident, this cannot be done entirely without some consideration of tertiary structure) and the ways in which a protein may fold into its native structure. Within a single polypeptide chain it is possible that several distinct "domains" may occur. In such regions the tertiary structure of one part of the polypeptide appears to be independent of other domains. Such domains are often associated with particular functional regions of the protein. Although a wide variety of experimental evidence is discussed, much of the later part of the chapter focuses on whether or not the secondary structure of a protein can successfully be predicted from its primary sequence, and what insights such an endeavor gives to the basic mechanism of protein folding and to our understanding of the possible mechanisms of conformational changes.

PATHWAY OF PROTEIN FOLDING

The general mechanism of protein folding has been debated for many years, and the arguments fall into two categories. It has been suggested that the formation of a three-dimensional structure of a peptide chain is a totally thermodynamically controlled process. A wide variety of studies involving reversible denaturation support this contention. Studies of proteins such as staphylococcal nuclease have shown not only that complete activity can be recovered after denaturation, but that a number of measurable physical parameters, all of which reflect somewhat different aspects of the native conformation, show similar reversible acid-induced transitions. As we discuss in this chapter and in Chaps. 10 to 12, many physical parameters can be measured that reflect the conformation of a protein. Various spectroscopic properties of amino acid side chains or the polypeptide backbone structure can be followed as a protein undergoes a denaturation process. Staphylococcal nuclease shows a reversible acid-induced denaturation that has been followed by measurements of tryptophan or tyrosine fluorescence, which, as shown in Fig. 9-1, both reflect the same denaturation process.

Also shown in Fig. 9-1 is the correspondence of tyrosine absorbance changes with fluorescence changes. The same conformational transition has also been studied, in circular dichroism (CD) measurements, by following the molar ellipticity at 220 nm, which monitors various aspects of the secondary structure of the protein. Similarly, the overall hydrodynamic properties of the molecule, as measured with viscosity determinations, also show the transition. NMR measurements of the four histidine

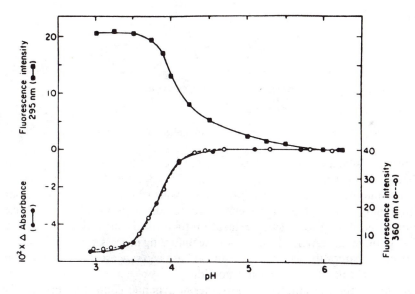

Figure 9-1   pH-induced denaturation of nuclease followed by tyrosine fluorescence
(□), tryptophan fluorescence (○), or tyrosine absorbance (●) measurements. (From
H. F. Epstein, A. N. Schechter, R. F. Chen, and C. B. Anfinsen, *J. Mol. Biol.,* *60,*
499–508. Copyright 1971 Academic Press, Inc., New York.)

resonances in the protein demonstrate the same transition. These various measurements are shown in Fig. 9-2.

The observation that this pH-induced transition is monitored by techniques
reflecting individual amino acid side-chain environments (fluorescence and NMR),
the state of the protein's secondary structure (CD) or the overall conformation of the

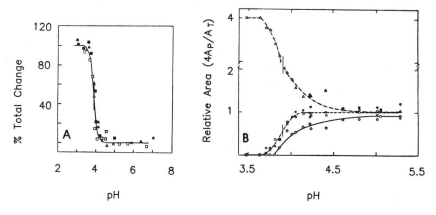

Figure 9-2   pH-induced denaturation of nuclease followed by (A) viscosity (⊔, ■)
or CD measurements, (△, ▲) at 220 nm and (B) NMR measurements of the four
histidines in the molecule. (○–○) H-1; (△–△) H-2; (□–□) H-3; (◇–◇) H-4.

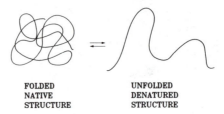

FOLDED   UNFOLDED
NATIVE   DENATURED
STRUCTURE   STRUCTURE

Figure 9-3   Two-state model of protein folding–unfolding.

protein (viscosity measurements) is completely consistent with a simple two-state mechanism of protein folding (represented in Fig. 9-3) under thermodynamic control.

In a number of proteins (e.g., chymotrypsin, trypsin, and ribonuclease) temperature-jump studies have been employed to follow transitions associated with unfolding and have given a single relaxation time consistent with a simple two-state model. In these cases Arrhenius plots for the unfolding process are linear (see Fig. 9-4), but are nonlinear for the refolding process. It has been suggested that water molecules may play a role in refolding but not unfolding.

Although these examples support thermodynamic control of protein folding, much contradictory evidence has been presented that favors the involvement of kinetic (i.e., path-dependent) processes in a variety of other proteins.

A number of proteins cannot be renatured to give active protein: Glutamate dehydrogenase undergoes a two-stage dissociation in guanidine hydrochloride. At intermediate concentrations of guanidine hydrochloride the hexamer dissociates to a trimer with little loss of native structure, as assessed by CD. At higher concentrations the trimer dissociates to a monomer form, with considerable loss of integrity of tertiary structure. The reversibility of these induced transitions has been followed by measurements of the enzyme activity. Activity can be regained only by removal of guanidine hydrochloride when the dissociation and denaturation have proceeded

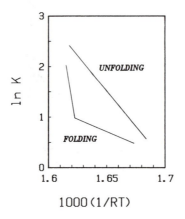

Figure 9-4   Arrhenius plots for protein folding and unfolding.

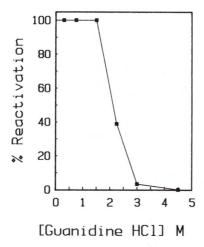

Figure 9-5 Renaturation of glutamate dehydrogenase after denaturation in guanidine hydrochloride.

no further than the trimer stage. When denatured polypeptide chains are formed, enzymatic activity cannot be regained, as shown in Fig. 9-5.

The amount of renaturation possible in this case can be correlated with the quaternary structure (as determined by light scattering) and status of the polypeptide chain conformation (as judged by CD), and this comparison (see Table 9-1) shows that the trimer can be renatured, but once the denaturation has reached the level of individual polypeptide chains, activity cannot be regained.

TABLE 9-1 Renaturation of glutamate dehydrogenase after guanidine hydrochloride denaturation

[Guanidine hydrochloride] (M)	Percent reactivation	Molecular state
4.5	0	100% monomer
3.0	3.4	95% monomer, 5% trimer
2.25	38.9	60% monomer, 40% trimer
1.5	100	100% trimer
0.75	100	85% trimer, 15% hexamer
0.3	100	10% trimer, 90% hexamer

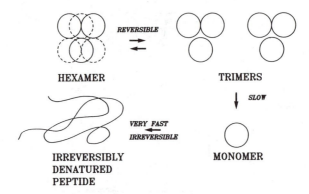

Figure 9-6   Multistage model for the denaturation of glutamate dehydrogenase.

These data are consistent, in the case of glutamate dehydrogenase, with a model (Fig. 9-6) showing a reversible dissociation of the hexamer to trimers and an irreversible dissociation of the trimer to monomers that rapidly lose "native" structure.

In similar types of experiments, enzymes such as muscle aldolase (which has four subunits) have been shown to regain a large portion (in the case of aldolase approximately 80%), but not always all, of their native activity, and that the *rate* of regain of activity is dependent on the presence of a substrate or ligand (fructose diphosphate in the case of aldolase). This fact suggests that the refolding and renaturation process is pathway dependent (i.e., kinetically controlled). In some proteins the presence of a ligand or cofactor is *necessary* for renaturation. Alkaline phosphatase requires the presence of zinc for refolding. With carbonic anhydrase, also a zinc-dependent enzyme, zinc is not required for refolding, but its presence does increase the rate.

There is a certain amount of evidence that protein folding can occur prior to the completion of synthesis. This suggests that *all* of the sequence may not be required to allow enzymatically active structures to form. During biosynthesis of a number of proteins, polysomes with attached partially complete, but immunologically competent, folded protein can be isolated by immunoprecipitation techniques. In some cases these partially complete but folded peptides also exhibit enzymatic activity. If the polypeptide chain folds during synthesis, one would expect to find evidence of *native* conformation—as judged by activity or antigenicity—only if the entire sequence is not required for native folding to occur. If this is true, one would also expect that isolated N-terminal regions of proteins might have some native conformation.

In experiments with staphylococcal nuclease, which has 149 residues, three fragments result from tryptic cleavage: residues 1 to 5, residues 6 to 49, and residues 50–149. When separated they have no enzymatic activity and no detectable helix structure (as judged by CD measurements). When the fragments are mixed, approximately 8% of the expected native activity is regained. The mixture has approximately 10% α helix compared to the 18% estimated for the native protein. This would argue against the expectation of finding some evidence of native structure in the isolated

N-terminal regions of a polypeptide, and thus favors the idea of protein folding being governed by thermodynamic criteria.

Antigenic Detection of Protein Folding

As indicated previously, antibodies to native structure have been used to detect the existence of such native structure in fragments or during synthesis. In general, antibodies can be made to both native or unfolded, denatured structure. It is possible to show that an antibody reacts with the *native* conformation of a peptide by showing that ligands which stabilize the native conformation do not inhibit the interaction of antibody with the protein but in an equilibrium situation where both unfolded and folded protein exist, increase the interaction. Although antibodies have found uses in establishing the existence of native structure, studies that have attempted to use them (to either the native or denatured states of a protein) to follow the kinetics of folding or unfolding have been criticized on the grounds that the antibody will act to stabilize or destabilize one or the other component of an equilibrium between the folded and unfolded protein's form.

Mechanism for Protein Folding

From the preceding discussion it appears that some proteins may fold in accordance with the concept of folding being governed only by thermodynamic criteria, while others fold by kinetically controlled pathways. It may not be coincidence that most of the proteins that can be reversibly denatured successfully are globular proteins of relatively small size, while those that cannot be renatured, or which require the presence of co-ligands for refolding, are larger and often multi-subunit allosteric proteins. The folding process itself cannot be completely random since it would take too long relative to the overall life of the protein. A relatively small protein of, for example, 100 amino acids has approximately 10^{40} possible conformations, depending on its primary structure. Small proteins generally fold on a time scale of seconds. The molecular motions involved in the folding of a polypeptide chain occur on a nanosecond-to-picosecond time scale, suggesting that at most about 10^{11} conformations could be randomly screened during the folding process.

It seems likely that in most instances protein folding is a kinetically governed process and that nucleation events may direct the pathway. The scheme shown in Fig. 9-7 outlines a *suggested* process by which a polypeptide chain acquires its native conformation and is proposed purely as a basis for considering the possible importance of secondary structure in directing protein folding.

In this figure the early events are the *transitory* formation of small regions of formal secondary structure: The lifetimes of these regions vary with their individual stability. As regions of the polypeptide chain with areas of local secondary structure randomly interact with one another, some interactions lead to stabilization of these structures, whereas others lead to destabilization. These interactions, which of course can occur between widely separated regions of the primary sequence, result in new regions of secondary structure being formed and the refining of existing regions. The

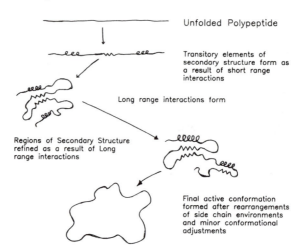

Figure 9-7   Proposed pathway for protein folding.

formation, via such long-range interactions, of a *primal* binding site for a ligand offers a mechanism for the frequently observed ligand requirement for folding or ligand enhancement of the rate of folding. Regions of transitory secondary structure stabilized by such long-range interactions become *nucleation* sites that now direct subsequent folding of the remainder of the molecule. It is probable that a variety of local conformational rearrangements happen after the formation of the principal elements of secondary structure in the protein.

In small proteins this process occurs rapidly and in all likelihood in a highly cooperative manner, giving rise to the simple two-state kinetics often observed. In larger proteins, which may have more than one domain or may contain subunits, a more complex process exists. The initial stages are undoubtedly the same, however, as the various domains independently form local nucleation centers and acquire some type of tertiary structure, interactions between the nascent domains occur and can have further stabilizing or destabilizing effects on elements of secondary structure within the domains. A similar situation holds with subunits: subunit–subunit interactions may have significant effects on the tertiary and secondary structure of the individual subunits. Subunit structures with high energies of interaction would especially be expected to suffer from secondary and tertiary rearrangements as a result of subunit–subunit interactions, which may explain the difficulty in renaturing such proteins.

This suggests that it may not be possible to refold a denatured protein to the same conformation as that obtained during biosynthesis. During biosynthesis local domains may form (these can certainly be detected antigenically) and some formed early in synthesis direct (or at least influence) the folding of other regions. This is a quite different situation from that found in a renaturation experiment, where all the domains in the molecule are simultaneously folding and as a result have somewhat different effects on each other, indicating the possibility that different final conformations are arrived at in the two cases. Some proteins may not refold after denaturation as a result of the fact that they undergo post-translational (and hence post-folding) modification. Such modifications may involve proteolytic processing or covalent derivatization of certain amino acid side chains which could affect folding pathways.

During this discussion we have emphasized that folding occurs at two stages: an early stage involving local regions of secondary structure, which is dominated by the nature of the polypeptide chain in local regions, and a later stage where interactions between local regions of secondary structure occur as the tertiary structure is being built up. The first stage depends on short-range interactions between near-neighbor residues, while the second involves long-range interactions. Each stage is important in the *final secondary structure* of the protein as well as, of course, in the tertiary structure.

FORMAL SECONDARY STRUCTURE

A polypeptide chain consists of a series of amino acids chemically linked by peptide bonds. The peptide bond, illustrated in Fig. 9-8, can be considered as a planar structure. Although usually drawn as in Fig. 9-8, the planarity can be construed as being imposed by the partial double-bond character of the amide group resulting from resonance stabilization. Given this planarity it is evident that the conformation of a peptide can be described by two angles, the ϕ angle and the ψ angle, which are those describing the rotations of two planar peptide bonds on either side of an α-carbon atom in the polypeptide. These angles are indicated in Fig. 9-8. To describe the backbone conformation of a polypeptide chain, all one needs is a description of the

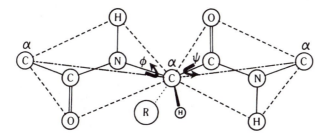

Figure 9-8 Peptide bonds joining three α-carbon atoms. – – –, Planar nature of the rotations about the ϕ and ψ angles; – – –, virtual bonds connecting the α-carbons.

ϕ and ψ angles for each successive α-carbon atom in the chain. Because of steric considerations, certain of these angles are not possible: ϕ and ψ are *interdependent*. This holds only for those angles about a particular α-carbon atom. The ϕ and ψ angles of a given α-carbon are *not* affected by those angles of other α-carbon atoms within the peptide.

The steric effects on ϕ and ψ angles are conveniently represented in a *Ramachandran plot*, which shows energy contours in a plot of ϕ versus ψ. Figure 9-9 shows

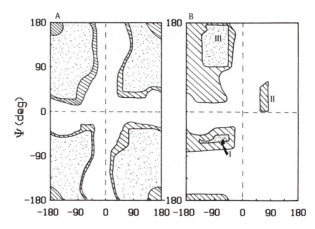

Figure 9-9 Ramachandran plots for (A) glycine and (B) alanine in peptides indicating allowable configurations (hatched areas) and areas of "contact" (stippled areas). Other areas are not accessible for steric reasons.

such plots for glycine or alanine in a peptide. The plot for glycine is essentially symmetrical, whereas alanine, due to unfavorable side-chain contacts, has some additional areas of conformational space restricted. Onto diagrams such as these, energy contours can be superimposed that indicate within the sterically allowable areas which ones are energetically favored. Three types of interactions must be considered in assessing the energy of a particular conformation.

Nonbonded Interactions

These are forces (both attractive and repulsive) between atoms or groups of atoms (e.g., side chains) whose separation distance depends on ϕ and ψ. The energy, E_{xy}, between two such groups, x and y (separated by ϕ_i and ψ_i), is given in

$$E_{xy}(\phi_i, \psi_i) = \frac{A_{xy}}{R_{xy}} - \frac{C_{xy}}{R_{xy}^6} \tag{9-1}$$

where A and C are parameters characteristic of the groups involved and R is the distance of separation. At large values of R there is no interaction, but as R decreases, first an attractive force, and subsequently a repulsive force, operate. The repulsive force becomes significant as the groups penetrate each others atomic radii.

Dipolar Interactions

As might be expected, the amide groups in the peptide bonds have dipole moments, which are oriented approximately parallel to the N—H bond in the direction toward the H. Since the amide dipole is quite large, dipolar interactions play a significant role in backbone conformation. The energy of interaction, E_d, between two dipoles U_p and U_q separated by the vector **r** is given by Eq. (9-2), and, since it arises from the orientation of two adjacent amide groups in the peptide, is governed by ϕ and ψ angles,

$$E_d = \mathscr{E}^{-1}\left[\frac{U_p U_q}{r^3} - \frac{3(U_p \cdot \mathbf{r})(U_q \cdot \mathbf{r})}{r^5}\right] \tag{9-2}$$

where r is the scalar magnitude of **r** and $\mathscr{E}$ is the dielectric constant. The energy E_d is usually computed from Eq. (9-3) using partial charges and summing the charge–charge interactions to give the total electrostatic energy,

$$E_d = \frac{\sum\limits_{xy} Q_x Q_y}{\mathscr{E} R_{xy}} \tag{9-3}$$

where Q is the partial charge and R is the distance of separation. The dielectric constant, $\mathscr{E}$, is much less than the dielectric constant of water since the dipolar interaction passes through the protein. Dipolar interactions longer than nearest-neighboring amide groups are not usually considered since the energy of interaction decreases rapidly with distance.

Intrinsic Torsional Potential

Earlier we discussed the planar nature imposed on the peptide bond by its partial double-bond nature. The inference is that the single bonds to the α-carbons allow free rotation. In reality this is not true, and as a result, ϕ and ψ rotations have an intrinsic rotational hinderence with an associated torsional energy, $E_{tor}(\phi_i, \psi_i)$, given by

$$E_{tor}(\phi_i, \psi_i) = \frac{E^0\phi}{2}(1 + \cos 3\phi) + \frac{E^0\psi}{2}(1 + \cos 3\psi) \tag{9-4}$$

where $E^0\phi$ and $E^0\psi$ are the energy barriers associated with the ϕ and ψ rotations.

The total energy, $E(\phi_i, \psi_i)$, is simply the sum of these contributions and is given by

$$E(\phi_i, \psi_i) = \sum_{xy}\left[E_{xy}(\phi_i, \psi_i) + E_d(\phi_i, \psi_i) + E_{tor}(\phi_i, \psi_i)\right] \tag{9-5}$$

Using Eq. (9-5), energy contour diagrams analogous to the Ramachandran plots can be obtained. Figure 9-10 illustrates such contour maps for glycine and alanine in general detail. As in the Ramachandran plots, glycine is symmetrical, whereas the alanine plot is not. With alanine three low-energy regions are indicated (I → III). Regions I and II correspond to right- and left-handed helices, and from the diagram it is apparent that the right-handed (region I) is favored over the left-handed (region II). Region III is clearly the overall lowest-energy region and corresponds to an extended form of the residue that includes the β-strand conformation. If dipolar interaction energies are omitted from the calculations, region I would have an overall lower energy than III. This emphasizes the importance of dipolar interactions involving "hydrophobic" residues in governing a tendency to be associated with β-strand

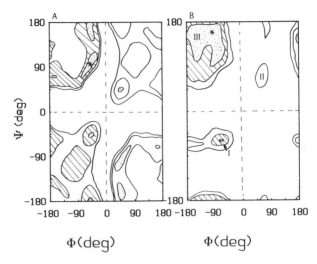

Figure 9-10   General detail of energy contour maps: (A) glycine; (B) alanine. The overall energy minimum is indicated as ∗.

conformations. As the microscopic dielectric constant decreases (in a hydrophobic environment), the energy of dipolar interaction increases.

Although we have considered only glycine and alanine in this discussion, alanine can be considered as representative of most amino acids containing a side chain. From Figs. 9-9 and 9-10 it is clear that glycine is far more flexible than other residues. The one other residue that should be considered is proline. Proline, because of the rigid pyrrolidine ring, has a fixed ϕ angle of about $-60°$, and the conformational energy depends only on the ψ angle. A plot of energy versus ψ has two minima, at $\psi = -55°$ and $\psi = 145°$, which fall into regions I and III in the plot (of alanine) shown in Fig. 9-10. These two allowed conformations correspond to a relatively compact form and an extended form, and are energetically quite similar. The compact form ($\psi = -55°$) is ideal for accommodating a turn or a bend in a chain. Perhaps the most important point regarding proline residues, however, is the effect they have on the conformational energy map of the residue preceding them. With the exception of a glycine, residues preceding proline have energy maps lacking region I, which involves unfavorable steric overlaps involving, in the case of alanine, the methyl side chain and the CH_2 group attached to the imido nitrogen. The effect of this is to deny a residue preceding proline the conformational space associated with a right-handed α helix. Proline itself can adopt this conformation, but it prevents the preceding residue from doing so. As a result, proline may occur at the start of a helix but rarely within a helix.

On a purely mechanical basis it is possible to describe α-helix, β-strand, and β-turn secondary structures in terms of the ϕ and ψ angles of adjacent residues in the peptide. If four or more consecutive residues have ϕ and ψ angles within 40° of $(-60°, -50°)$ the region of peptide is in a right-handed α helix. If three or more residues have ϕ and ψ angles within $40°$ of $(-120°, 110°)$ or $(-140°, 135°)$, the structure is a β strand (either parallel or antiparallel): β turns consist of four consecutive residues where the polypeptide chain folds back on itself by about $180°$, and they can be either right-handed or left-handed.

α Helix

Figure 9-11 shows several representations of α helices. The α helix has approximately 3.6 residues per turn and is stabilized to a large extent by hydrogen bonds formed within the backbone of the chain between amide protons and carbonyl oxygens.

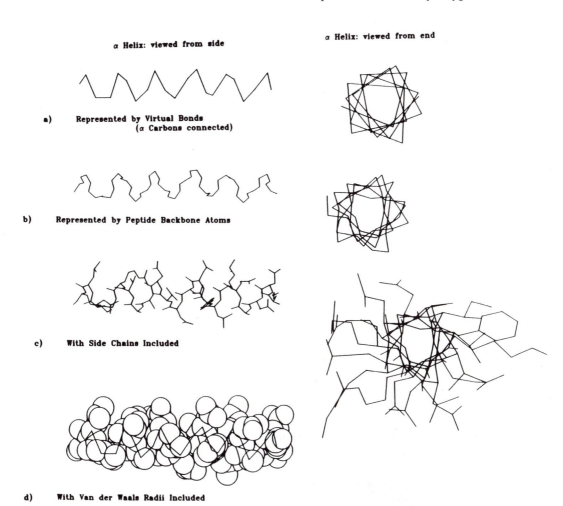

α Helix: viewed from side

α Helix: viewed from end

a) **Represented by Virtual Bonds (α Carbons connected)**

b) **Represented by Peptide Backbone Atoms**

c) **With Side Chains Included**

d) **With Van der Waals Radii Included**

Figure 9-11 An α helix, taken from the crystal structure of lysozyme, is shown as a side view and as an end view.

β Sheet: viewed from side

2 Strands: antiparallel connected
by β Turn

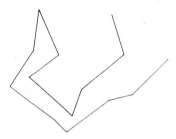

a) Represented by Virtual Bonds

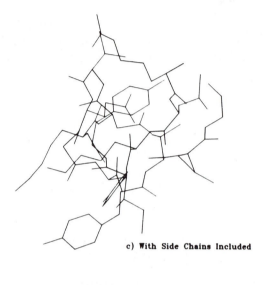

c) With Side Chains Included

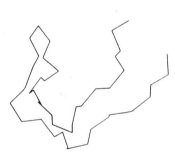

b) Represented by Peptide Backbone Atoms

Figure 9-12 Antiparallel β sheet, connected by a β turn, taken from the crystal structure of lysozyme.

β Strand

The β-strand structure involves an extended conformation of the polypeptide chain seen in edge view in Fig. 9-12. The strands associate to form sheets that can contain strands from quite widely separated regions of the primary sequence.

As with the α helix, stabilization for a β-sheet secondary structure comes largely from hydrogen bonds. In the case of a β sheet, however, the hydrogen bonds are formed between β strands within the sheet. The strands in a β sheet can be parallel or antiparallel, as indicated in Fig. 9-13, and may involve quite widely separated regions of polypeptide.

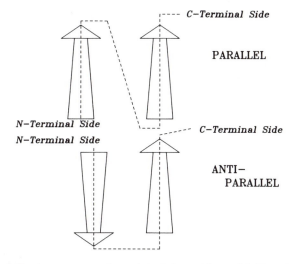

Figure 9-13   Arrangements of strands in β sheets: (A) parallel; (B) antiparallel.

β Turn

Figure 9-14 shows two views of an observed β-turn structure. As with the other formal secondary structures we have considered, much of the stability for this structure comes from the formation of hydrogen bonds between amide protons and carbonyl oxygens. Because of their conformational energies, discussed earlier, proline residues are often found in turn regions, usually in the second or third position.

Each of these structures is taken from the actual secondary structure observed in lysozyme and is shown in several representations. When the α carbons are connected by "virtual" bonds the simplest representation of each structure is obtained. Tracing the peptide-bond backbone also gives a simplified but clear view of each structure. This view might be considered the most appropriate. Inclusion of the side chains tends to obscure the basic element of the secondary structure, particularly with the β turn. Several important features of these secondary structure elements are demonstrated by Figs. 9-11 to 9-14.

The end view of the α helix (Fig. 9-11) clearly illustrates the "sided" nature of a helix: The amino acid side chains form "ridges" and "grooves" along the length of the helix. This helix feature is apparent only by considering the structure with side chains.

The β sheet (Fig. 9-12) shows a clear "twist" to its structure, and from examining the different views of the β turn it is apparent that it is not planar. As might be expected, the nature of the amino acid side chains in these two structural elements affects the twist of the sheet and the orientation of the turn.

ß Turn: viewed from above   ß Turn: viewed from inside

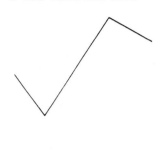

a)   **Represented by Virtual Bonds**

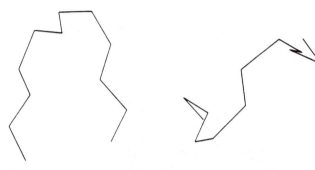

b)   **Represented by Peptide Backbone Atoms**

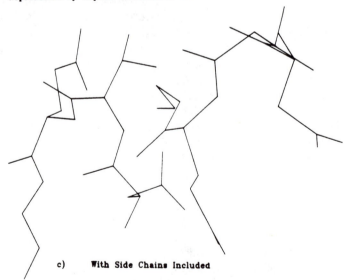

c)   **With Side Chains Included**

Figure 9-14   A β turn, taken from the crystal structure of lysozyme, shown from the top and from inside the turn.

PREDICTION OF SECONDARY STRUCTURE

Because the prediction of secondary structure is based largely on the character of the amino acids that are found in nature to be in certain of the secondary structures, it is informative to consider briefly how procedures for these predictions from primary sequence have been developed and applied. Many of the problems are similar to those the protein itself must encounter during the folding process! Predictive methods are based on the probability that a particular type of amino acid residue is found in a certain type of secondary structure. These data are obtained in one of two ways: In the first, probabilities are obtained by examining the crystal structures of known proteins and counting the number of times particular residues appear in α helices, β strands, or β turns. Alternatively, polymers of single amino acids are used and their secondary structure determined. From the tendency of such polymers and various copolymers to form α helix and β sheet, an assessment of the contribution of individual residues to these structures can be made. In some instances it is useful to keep information on *where* in the type of secondary structure these residues appear most frequently, as this can be helpful in defining starting points and termination points for the type of secondary structure. Table 9-2 gives such information concerning probabilities of residues appearing in α helices, β strands, and β turns.

Several generalities can be drawn from Table 9-2: (1) the charged residues are unfavorable for β-strand formation, and three of them (Asp, His, Arg) are also α-helix

TABLE 9-2 Conformational parameters for amino acid residues appearing in α helix, β strand, and β turn based on frequencies of occurrence in proteins of known structure

Residue	α Helix	β Strand	β Turn
Glu	1.51	0.37	0.74
Met	1.45	1.05	0.6
Ala	1.42	0.83	0.66
Leu	1.21	1.30	0.59
Lys	1.16	0.74	1.01
Phe	1.13	1.38	0.6
Gln	1.11	1.10	0.98
Trp	1.08	1.37	0.96
Ile	1.08	1.60	0.47
Val	1.06	1.70	0.50
Asp	1.01	0.54	1.46
His	1.00	0.87	0.95
Arg	0.98	0.93	0.95
Thr	0.83	1.19	0.96
Ser	0.77	0.75	1.43
Cys	0.70	1.19	1.19
Tyr	0.69	1.47	1.14
Asn	0.67	0.89	1.56
Pro	0.57	0.55	1.52
Gly	0.57	0.75	1.56

PRIMARY SEQUENCE:
α-HELIX PREDICTION:

FOR 3rd Residue: Look at preceding 2 and
following 3 residues. α-Helix prediction
takes into account the nature of neighbor
residues. Classify 3rd residue on 5 point
scale.
REPEAT FOR 4th RESIDUE etc. to give
prediction for each residue in sequence.
ASSIGN regions of α-Helix to stretches
containing at least 4 strong helix residues
out of 6 consecutive α-Helix positive
residues.
TERMINATE α-HELIX when previous condition
not satisfied.

β-STRAND PREDICTION:

Process is similar to that used for α-helix
except that only 2 residues to each side
of predicted residue are used. Stretch of

5 β-Strand favoring residues containing
at least 3 strong β-Strand formers gives
a β-Strand.
TERMINATE β-STRAND when previous
condition not satisfied.

β-TURN PREDICTION:

Examine all Tetrapeptide sequences in
polypeptide sequence. Proline in the
second position of the tetrapeptide is
given additional weighting.
A β-TURN is predicted when a tetrapeptide contains at least 2 strong turn
forming residues.

OVERALL PREDICTION:

A region of STRONG prediction takes
precidence over a region of weaker
prediction. Regions with equal tendency
for α-Helix and β-Strand are indicated
as such.

Figure 9-15 Protocol used for the prediction of secondary structures in proteins.

indifferent; (2) residues that tend to break α helices (Pro, Gly, Asn, Tyr) also tend to be residues with high probability of appearing in β turns, and (3) residues with a strong tendency to be in β strands are rarely found in β turns.

This type of information has been applied to secondary-structure prediction in a variety of ways. One of the more successful is outlined in Fig. 9-15. In this predictive scheme the influence on neighboring residues is taken into account in attempting to assign a propensity of each residue in a peptide to be in an α helix, a β strand, or a β turn. Each type of secondary structure is "predicted" independently and the final "prediction" based on a comparison not only of the "strength" of the prediction but also on the predictions for adjacent residues. For example, it is quite possible for a region of peptide to contain a residue that has a high probability of being in either a β strand or an α helix; if the neighboring residues are predominantly helical, this weights the final choice between β strand and α helix for the prediction. Finally, regions of secondary structure are predicted based on certain "nucleation" rules. For an α helix to be indicated six adjacent helical residues must be present, for a β strand to be indicated five adjacent strand residues must be present, and for a β turn two residues, of a tetrapeptide sequence, must be indicated as strong turn formers. With β-turn predictions, weighting is given to proline in the *second* position in the turn.

Any such predictive scheme is limited by several factors: (1) the data base from which the probability values of the individual amino acids are taken, and (2) the lack of consideration of long-range interactions.

Little can be done about the second point, of course, but the first gives a way of optimizing predictions for members of a class of proteins. Instead of using a random selection of proteins as the data base, a considerable improvement in prediction accuracy (which for general proteins is around 65%) can be achieved if related proteins only are used. In many cases this is possible; for example, by optimizing the parameters using known dehydrogenase secondary structures the accuracy of prediction using schemes such as that shown in Fig. 9-15 is of the order of 92 to 95%, which leads to increased confidence in the prediction of an unknown dehydrogenase.

Although secondary-structure predictions can only be confirmed after the three dimensional structure of the protein is known, it is possible to "test" predictions in terms of the percentage of the protein in α-helix or β-strand structures, where such gross parameters can be experimentally measured.

Perhaps the most useful information that can be obtained from secondary-structure predictions, at least in terms of protein chemistry, is some idea of potential nucleation sites in folding or regions of conformational flexibility. Potential nucleation sites may be regions of the polypeptide chain where strong predictions for a particular secondary structure are made. As with nucleation, prediction is based largely on short-range interactions. When protein sequences are used in prediction schemes, some regions end up having essentially equivalent probability of being helical or strand. Such regions probably represent areas of the protein where long-range interactions contribute the deciding influence with respect to formal secondary structure. Since these long-range interactions are imposed by the tertiary structure of the protein, such ambiguous areas may be regions involved in conformational changes within the protein, because small changes in long-range interactions may cause such an "ambivalent" region to switch from one formal secondary structure to another.

EXPERIMENTAL ASSESSMENT OF SECONDARY STRUCTURE

X-ray crystallography gives a clear picture of the various types of secondary structure present in proteins, although it is not a readily accessible method for determining the amounts and types of formal secondary structure. The only other approach that gives some idea of the amount of secondary structure present in a molecule is circular dichroism (CD). Figure 9-16 shows CD spectra for poly-L-lysine in an α-helical, a β-strand, and a random conformation.

From these spectra it is evident that the longer-wavelength edge of the negative spectrum observed for the helical form is somewhat characteristic of α helix. CD measurements at 222 nm have been used to "quantitate" the amount of helix present in a protein. The approach is largely correlative, using crystal structure α-helix contents to calibrate the dependence of the measured CD at 222 nm on helix content. In an adaptation of this approach, CD spectra between 200 and 245 nm have been used to estimate both α-helix and β-strand content, using computer fitting to equation

$$\mathrm{CD_{meas}} = f_\alpha \mathrm{CD}_\alpha + f_\beta \mathrm{CD}_\beta + f_r \mathrm{CD}_r \tag{9-6}$$

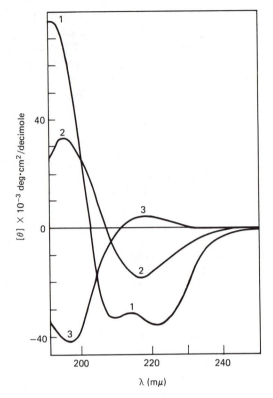

Figure 9-16   CD spectra of poly-L-lysine in various conformations. Curve 1, 100%
helix; curve 2, 100% β; curve 3, 100% random oil. (Reprinted with permission from:
N. Greenfield and G. D. Fasman, *Biochemistry*, *8*, 4108–4116. Copyright 1969
American Chemical Society, Washington, D. C.)

where f_α, f_β, and f_r are the fractions of helix, strand, and random coil, and CD_α, CD_β,
and CD_r are appropriate constants calculated from proteins with known structures.

Application of such measurements to proteins with unknown structure can yield
a useful approximation of the possible amounts of α helix and β strand. Such
measurements with glutamate dehydrogenase give approximately 40% helix and 13%
strand. These can be compared with estimates made by secondary-structure predictions using the approaches described earlier, which give approximately 24% each of
helix and strand structures. The dilemma lies in the fact that both approaches are
subject to unknown errors. The measured CD spectra are probably best used to
indicate protential changes in secondary structure for a particular molecule rather
than to give amounts of particular types of secondary structure.

CD spectral measurements have also been used to follow denaturation profiles
of proteins. Figure 9-17 shows the pH-induced unfolding of tropomyosin followed
by CD at 222 nm as well as by absorbance measurements at 295 nm, which have
essentially similar profiles, consistent with two-state folding–unfolding, as discussed
earlier.

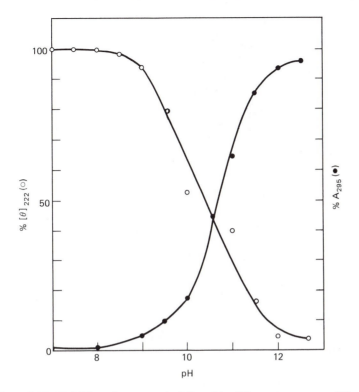

Figure 9-17   Unfolding of tropomyosin followed by CD measurements at 222 nm
(○) and absorbance (●) measurements. B. Nagy, *J. Biol. Chem.*, *252*, 4557–4563.
(Reprinted with permission of the copyright owner, The American Society of Biological
Chemists, Inc., Bethesda, Md.)

On the other hand, in many instances, following α helix (by CD 222 nm measurements) and aromatic residue environments (by CD measurements at 284 nm) gives
quite different profiles. Figure 9-18 shows the urea-induced unfolding of a protein,
where clearly the requirements of a simple two-state model are not met, since measurements at 222 nm (helix) give a quite different concentration dependence from
measurements at 284 nm, where the signal arises from tyrosine side chains.

In summary, CD measurements for the assessment of secondary structure have
found some uses: α-helical secondary structure can be experimentally estimated in
solution with some degree of confidence, and estimates of helical content based on
CD 222-nm measurements are often used to support amounts of protein α helix
estimated by secondary-structure prediction. Because of the underlying spectrum of
β strand, however, such measurements can be regarded as qualitative at best. They
are of considerable use, however, in examining *changes* in secondary protein structure
induced by denaturation (as we have discussed here) or induced by ligand binding
(as we discuss in Chap. 12).

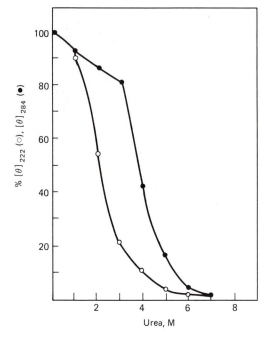

Figure 9-18   Urea-induced protein unfolding followed by α-helix CD at 222 nm and tyrosine side-chain CD at 284 nm. B. Nagy, *J. Biol. Chem., 252,* 4557–4563. (Reprinted with permission of the copyright owner, The American Society of Biological Chemists, Inc., Bethesda, Md.)

In this chapter we have examined various aspects of the secondary structure of proteins. Although much insight into the folding pathway and tertiary-structure assembly is obtained by considering secondary structure, perhaps the most important consideration involves the mobility inherent in protein structure. As developed in detail in Chaps. 10 to 12, the stabilities and mobility of formal secondary structure, dominated by rotation about the dihedral angles of the peptide bonds, are intimately involved in the acquisition of tertiary structure and mechanisms of conformational changes in proteins.

10

Tertiary Structure of Proteins

INTRODUCTION

Many experimental approaches have been developed to examine the tertiary structure of proteins. Of these, x-ray crystallography has been the most powerful and influential in directing the way we think of tertiary structure and (considered in Chap. 12) changes in tertiary structure.

On the basis of the structure of myoglobin obtained from the x-ray crystallographic work of Kendrew, four generalizations were formulated to describe tertiary structure.

1. Proteins are compact structures having very small amounts of internal solvent molecules, which are present internally and presumed to have been trapped during the folding process.

2. Almost all the polar side chains in the protein are at the surface of the molecule, where they can interact with solvent and solute molecules in the bulk solvent. Any exception to this would indicate that a "nonsurface" polar group is involved in some internal function. In the case of myoglobin, for example, a histidine side chain is buried internally but is associated with the heme ring of the molecule.

3. All nonpolar residues, with the possible exceptions of glycine and alanine, are located in the interior of the molecule. Glycine and alanine, because of their "short" side chains, can be located at the surface.

4. All polar groups at the surface of the molecule, whether they are side-chain or main-chain $C{=}O$ and $N{-}H$ groups, have bound water molecules.

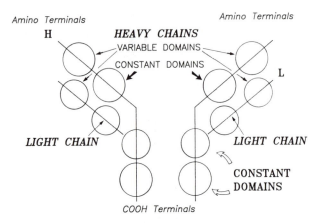

Figure 10-1   Schematic drawing of a typical antibody molecule composed of heavy (H) and light (L) chains.

When one considers that these "generalizations" were made on the basis of the examination of a single protein, it is surprising that numbers 1 and 4 have stood the test of time (and several hundred more crystal structures). The second and third generalizations, although essentially correct, have been modified in the light of further work.

Many proteins with molecular weights of more than about 16,000 Da have been shown to contain "domains" in which the polypeptide obeys the concept of a compact structure, with side chains forming a hydrophobic core and the peptide backbone wrapping around the outside. A typical "domain" containing protein is the IgG molecule, shown in a schematic representation in Fig. 10-1. In this molecule there are several domains, each containing contiguous segments of polypeptide chain, yet each quite separate physically from its adjacent domain. The domains in this instance are so distinct that they can be separated by proteolytic cleavage to give functionally active but separate molecules, each having different domains. The IgG molecule may be regarded as a somewhat anomalous protein, as its domains each contain two chemically separate polypeptide sequences. Although this is by no means unique, most protein domains consist of a single polypeptide chain.

The generalizations regarding the localization of polar and nonpolar side chains in proteins must be reassessed in light of the data shown in Table 10-1. These data were obtained by examination of the x-ray crystallographic structures of chymotrypsin, carboxypeptidase, and thermolysin. Although the observed distributions of the hydrophobic side chains of valine, leucine, isoleucine, and phenylalanine, and the charged side chains of lysine, arginine, aspartate, and glutamate do in general follow Kendrew's generalizations, it is quite apparent that among the "hydrophobic" residues, tyrosine, tryptophan, and proline contain anomalous members, while among the "hydrophilic" residues, serine and threonine also have "black sheep." Of the

TABLE 10-1 Side-chain environments in proteolytic enzymes

Residue	Location		
	Inside	Outside	Surface
Val, Leu, Ile, and Phe	100	43	57
Tyr and Trp	7	20	42
Gly and Ala	52	67	34
Pro	6	13	8
Ser and Thr	33	85	40
Asn and Gln	5	43	35
Lys and Arg	0	44	20
Glu and Asp	8	52	35

charged residues only glutamate and aspartate appear to be occasionally buried. As discussed in earlier chapters, the localization of these carboxyl groups in "nonpolar" regions of the protein leads to perturbed pK values. In some instances these perturbed carboxyl groups (e.g., Glu-35 in lysozyme) seem to have important biological functions, although in other proteins buried carboxyl groups do not appear to have direct biological function.

In this chapter we first consider some of the means available for obtaining tertiary-structural information. From there we examine ways of predicting tertiary structure, some aspects of the dynamic (as opposed to static) nature of protein structure, various aspects of the evolution of tertiary structure, and finally, the processes involved in the acquisition of the tertiary structure of a polypeptide chain.

METHODS OF STUDYING TERTIARY STRUCTURE

A variety of experimental approaches are available to examine different aspects of a protein's tertiary structure. Some of these have already been discussed in other contexts in earlier chapters and are referred to only briefly here. In many cases they give qualitative rather than precise structural information. Although of use in giving a general picture of the tertiary structure, the methods that provide such information are of more use to the protein chemist in providing evidence for, and a means of following, the various conformational changes a protein may undergo.

Solvent Perturbation

The spectral properties of a protein (we restrict the present discussion to UV absorbance and fluorescence) reflect the various environments of the chromophores or fluorophores that contribute to the overall absorbance or fluorescence. For example, a protein with several tryptophan residues usually shows a red shift in its

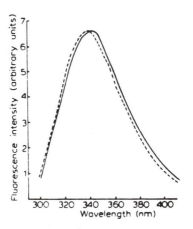

Figure 10-2 Effects of different excitation wavelengths on the emission properties of glyceraldehyde-3-phosphate dehydrogenase. [Reprinted with permission of J. E. Bell and K. Dalziel, *Biochim. Biophys. Acta, 410,* 243–251 (1975).]

fluorescence emission if the exciting wavelength is increased from 280 nm to 290 nm (see Fig. 10-2).

This reflects the fact that the various tryptophan residues are in different environments with resulting different spectral properties. An examination of the different spectral properties of the chromophores or fluorophores thus provides information concerning the various environments (i.e., the tertiary structure of the protein) in which the residues are located. Without detailed information regarding the location of individual residues in the primary sequence, in terms of their contribution to the overall spectral properties, only qualitative information can be obtained: comments such as "most of the tryptophans in the protein appear to be in hydrophobic environments" are made. A detailed analysis of individual residues is possible if their spectral properties can be specifically modified. In bovine α-lactalbumin, for example, specific tryptophan residues can be oxidized by *N*-bromosuccinimide and, from the effects on the protein fluorescence, the fluorescence properties of individual tryptophan residues estimated. Experiments of this type have allowed the environment of Trp-118 in the molecule to be elucidated.

When tryptophan-118 in α-lactalbumin is modified, there is a far greater quenching of protein fluorescence observed using 290- or 300-nm excitation than using lower wavelengths (Table 10-2). This indicates that Trp-118 has a red-shifted absorbance spectrum and contributes more to the overall fluorescence upon excitation at 290 to 300 nm than the other three tryptophan residues in the protein, and therefore that Trp-118 is in an exposed environment. This is also reflected in the *blue* shift of the excitation spectrum of α-lactalbumin when Trp-118 is modified (Fig. 10-3C). As is also indicated in Fig. 10-3 and Table 10-2, modification of a second tryptophan residue (Trp-26) results in minimal changes in fluorescence, thus suggesting that Trp-26 has a much lower quantum yield than Trp-118.

TABLE 10-2 Quantum yields and quenching of fluorescence in α-lactalbumin modified by N-bromosuccinimide

Excitation wavelength (nm)	Quantum yield			Percent quenching	
	Native α-lactalbumin	α-Lactalbumin with 0.95 Trp modified	α-Lactalbumin with 1.6 Trp modified	α-Lactalbumin with 0.95 Trp modified	α-Lactalbumin with 1.6 Trp modified
270	0.033	0.0279	0.0265	15	20
280	0.033	0.0292	0.0281	12	15
290	0.033	0.0245	0.0239	26	28
300	0.033	0.0196	0.0192	41	42

In a similar vein, small molecules that perturb the environment of surface chromophores can be used to estimate the "exposure" of such residues to solvent. If perturbants of differing radii are used, a picture of the surface topography of the protein can be obtained. This method depends on resolving the difference spectrum obtained in the presence of the perturbant into its constituent tyrosine and tryptophan components. Figure 10-4 shows DMSO-induced difference spectra of model compounds.

The protein difference spectrum is then compared to difference spectra obtained using the same perturbant and model compounds. The following equations are used:

$$\Delta_{\lambda_1}(\text{protein}) = a\,\Delta_{\lambda_1}(\text{Trp}) + b\,\Delta_{\lambda_1}(\text{Tyr}) \tag{10-1}$$

$$\Delta_{\lambda_2}(\text{protein}) = a\,\Delta_{\lambda_2}(\text{Trp}) + b\,\Delta_{\lambda_2}(\text{Tyr}) \tag{10-2}$$

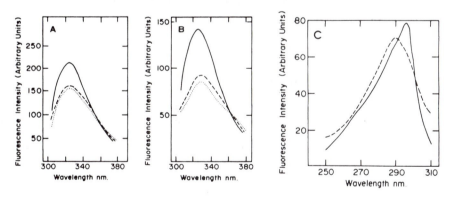

Figure 10-3 Fluorescence emission spectra: pH 7.0, 0.1 M NaCl, 10 M α-lactalbumin. (A) Excitation at 280 nm. (B) excitation at 300 nm; ——, Native α-lactalbumin; – – –, with 0.95 tryptophan residues modified by N-bromosuccinimide; · · ·, with 1.6 tryptophan residues modified by N-bromosuccinimide. (C) Fluorescence excitation spectra: emission monitored at 370 nm; ——, native α-lactalbumin; – – –, α-lactalbumin with 0.95 tryptophan residues modified. [From J. E. Bell, J. Castellino, I. P. Trayer, and R. L. Hill, *J. Biol. Chem.*, *250*, 7579–7585 (1975). Reprinted with permission of the copyright owner, The American Society of Biological Chemists, Inc., Bethesda Md.]

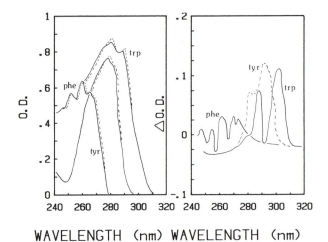

WAVELENGTH (nm) WAVELENGTH (nm)

Figure 10-4 (A) Absorption spectra of the *N*-acetyl ethyl esters of tryptophan, tyrosine, and phenylalanine in water (——) and in 20% dimethoxysulfoxide (DMSO) (–––); (B) DMSO-induced difference spectra for the *N*-acetyl ethyl esters of tryptophan, tyrosine, and phenylalanine.

where *a* and *b* are the apparent number of exposed tryptophan and tyrosine residues in the protein and the $\Delta_{\varepsilon\lambda}$ are the molar absorptivity differences of the protein and the free tryptophan and tyrosine model compounds at wavelength λ, the difference spectra maxima. Chromophore exposure is estimated by first neglecting the tyrosine contribution to the long-wavelength peak to obtain an approximate value of *a*.

Rearranging Eq. (10-1) gives

$$a \simeq \frac{\Delta_{\lambda_1}(\text{protein})}{\Delta_{\lambda_1}(\text{Trp})} \tag{10-3}$$

Thus an approximate value for *b* is obtained by rearranging Eq. (10-2).

$$b \simeq \frac{\Delta_{\lambda_2}(\text{protein}) - a\,\Delta_{\lambda_2}(\text{Trp})}{\Delta_{\lambda_2}(\text{Tyr})} \tag{10-4}$$

The values of *a* and *b* may then be refined by using Eqs. (10-1) and (10-2) reiteratively.

Limited Proteolysis

Susceptibility to proteolysis has often been used to study conformational changes. Altered susceptibility of a protein in the presence of a ligand suggests that the protein has undergone a conformational change that alters the susceptibility of proteolytically cleavable bonds. In such experiments decreased proteolysis could simply reflect physical protection of a susceptible bond from the added protease. Usually, these studies are used only when *increased* proteolysis results from the conformational change.

Because of the availability of a wide variety of proteases with defined specificities and the increased sensitivity of the detection of protein fragments (either by HPLC methods or by gel electrophoresis methods with such sensitive detection as that afforded by silver staining), it has become possible to use limited proteolysis to obtain defined information concerning the environment of various peptide bonds in a protein and changes in their environment induced by ligands or protein–protein interactions. This requires the *limited* proteolysis of the protein being studied such that only a single peptide bond is cleaved per protein molecule, ensuring that only peptide bonds that are exposed in the native protein are studied. Two experimental situations can be envisaged: (1) the primary sequence of the protein is known and information concerning the exposure of various peptide bonds to solvent is sought, and (2) the primary sequence may or may not be known, but information about conformational changes is sought.

The second situation is simply a matter of examining the peptide maps of the *partially* cleaved protein obtained in the presence and absence of the conformational perturbant and looking for differences, and will not be dealt with in more detail.

In the first situation, however, there is the potential for obtaining quite detailed information on the exposure of various bonds to solvent (i.e., the added protease). Two approaches can be employed.

1. After limited proteolysis, the peptide map is obtained. From each cleavage event (only one per individual polypeptide chain) two fragments result; thus for each available cleavage site there are two fragments. Isolation and identification of the various fragments, in comparison with the primary sequence of the protein, allows identification of each susceptible bond.

2. The second varient requires a chemical modification of the original protein prior to proteolysis. If the protein can be reversibly denatured, it may be possible to N-terminal label the peptide with, for example, radioactive cyanate. If the protein does not undergo such reversible denaturation, a *specific* chemical modification (e.g., use of a radioactive site-specific reagent or a radioactive general modification reagent to label a single reactive sulfhydryl or lysine residue with a radioactive derivative) suffices. The modification is used as a *reference point* after proteolysis to assist in the identification of fragments. Fragments are detected after separation by their radioactivity, that is, only one of each pair of fragments is detected. If N-terminal labeling has been possible, the size of the fragment may be sufficient to indicate which susceptible bond has been cleaved. If an internal label has been used, the detected fragments fall into one of two classes, as outlined in Fig. 10-5. After limited cleavage and separation of labeled fragments, the C-terminal and the N-terminal residues of the fragments are determined to indicate which of the two classes each fragment falls into. Once this has been established, the identification of the cleavage sites is the same as with N-terminal-labeled proteins.

In summary, limited proteolysis readily gives information about conformational changes and in conjunction with the primary sequence can give precise information

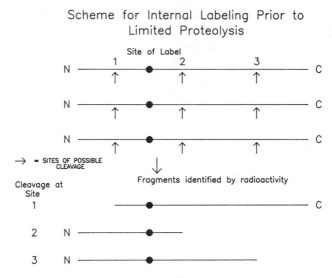

Figure 10-5 Scheme for internal labeling prior to limited proteolysis.

regarding exposed peptide linkages in the native structure and whether or not particular regions are involved in conformational changes.

X-ray Crystallography

Clearly, x-ray crystallography is a most powerful tool in determining the tertiary structure of a protein. Later in this chapter and in Chaps. 11 and 12 we consider in some detail the *type* of information that it gives about tertiary structure. At this point, however, we examine some of the problems, experimental approaches, and questions that are encountered when x-ray crystallography of a protein is attempted. There is no discussion of the technical and theoretical aspects.

Let us begin with a quote from Karplus: "The x-ray structure of a protein is an average structure of many molecules obtained over a long period of time. As such, the probability that a single molecule has the structure, as determined, is essentially zero". We use this quotation not to discredit x-ray crystallography but to give focus to the various questions a protein chemist or enzymologist must consider when examining a protein's crystal structure.

The major question that must be addressed is whether or not the crystal structure resembles that of the protein in solution. Several comments must be made. The conditions necessary to crystallize proteins often are very nonphysiological (such as high salt concentrations, the presence of organic solvents, or high/low pH) and usually require long periods of time (often, 1 to 2 weeks, which may limit the type of protein that can successfully be crystallized, as it must be stable for these time periods). Related proteins, for example α- and γ-chymotrypsin, have essentially identical amino acid sequences but may crystallize quite differently. The tertiary structures obtained for these chymotrypsins are, however, with the exception of a few surface residues, essentially identical, suggesting that the crystallization conditions do not affect the final elucidated structure.

It is often possible to examine the *activity* of the protein while it is in its crystal lattice. Clearly, there are many factors that would be expected to affect the activity of a protein in a crystal lattice, such as diffusion of substrates or products; these can be taken into account. Ribonuclease, carboxypeptidase, chymotrypsin, papain, and alcohol dehydrogenase have all been shown to retain activity in the crystal form. When attempts to correct for diffusional effects are made, it appears that at least with chymotrypsin and papain, the crystal forms have their expected activity. With alcohol dehydrogenase and carboxypeptidase, however, approximately a 200-fold decrease in activity in the crystal form is estimated, which may be attributed to an overall decrease in protein motility, even though there is no major tertiary structural differences between the solution and the crystal structures.

This brings up the question of resolution. In small-molecule crystal structures resolution to about 0.01 Å is possible. This level is *not* achievable in proteins: There is an inherent limitation arising from the fact that protein molecules do not align as precisely in crystal structures as do small molecules. This seems to be due in part to the purity of the proteins and in part to the solvent content of the crystals. In crystals with high solvent content, and consequently less intermolecular contact, the crystal gives diffraction patterns showing a greater level of random disorder, suggesting that unrestrained surface residues may have multiple conformations.

Finally, we must consider the effects of the *heavy atom derivatives* that are used to solve the crystal structure. This is usually achieved in one of three ways.

Trial-and-Error Approach. Crystallization either occurs from a solution containing a heavy atom salt, usually platinum chloride, or the crystal, once formed, is soaked in a solution containing a heavy atom salt, with the hope that a specific derivative will be formed. Platinum chloride seems to bind to exposed methionine or histidine residues in many proteins, and this approach is often successful.

Specific Chemical Modifications. Amino acid side chains in the protein can often be specifically modified with a chemical reagent containing a heavy atom. Mercury-based reagents (e.g., *p*-hydroxymercuribenzoate or *S*-mercuric, *N*-dansyl cysteine) are often employed to modify sulfhydryl groups and introduce a heavy atom. Reagents containing iodine are often useful, and the iodination of tyrosine residues or of amino groups (with iodo derivatives of phenylisothiocyanate) has been done.

Heavy Metal Substitution. In many cases a naturally occurring metal ion in a metalloprotein can be removed and replaced with a heavy metal. The zinc in carbonic anhydrase can be removed and the zinc-free enzyme dialyzed against mercuric acetate, which results in Hg replacing Zn. Calcium ions in proteins can often be replaced with barium, as in the case of staphylococcal nuclease. Various proteins, which on the basis of activity have no metal requirement, may in fact have specific metal binding sites (often for calcium or magnesium) that can be substituted for by a heavy metal ion.

Heavy metal derivatives of proteins may, of course, cause conformational alterations that can be reflected in the determined crystal structure. Activity measurements in solution in the presence of whatever chemical modification or substituted metal are used can give an indication of whether the heavy metal ion produces such altered states.

Chemical Methods

In Chap. 7 a wide variety of chemical modification reagents were discussed, and general mention of their use in assessing availability of particular amino acid side chains for modification was made. In addition, several instances of chemical cross-linking were considered briefly in the context of tertiary structure determination.

In general, the fact that a particular amino acid side chain can be chemically modified by a water-soluble modification reagent implies that the residue is solvent accessible (i.e., exposed at the surface of the protein). Information concerning residues in the primary sequence that are exposed in the tertiary structure can thus be obtained. In conjunction with secondary- and tertiary-structure prediction, chemical modification data can give meaningful insight into the tertiary structure. It is important that the modification is to a single residue, so that tertiary-structural changes as a result do not lead to additional modification events, which may be misinterpreted in terms of exposure in the original tertiary structure. Table 10-3 details some of the chemically modifiable residues in bovine glutamate dehydrogenase, together with their predicted secondary and tertiary structures. In each case, where defined secondary structure is thought to be present, the tertiary-structural assessment indicates a hydrophilic (i.e., solvent exposed) character (the ? in Table 10-3 indicates that the modified residue ap-

TABLE 10-3 Analysis of chemical modification in relationship to predicted secondary structure

Residue	Predicted secondary structure	Possible amphiphilic nature
Lysine-27	α Helix	Hydrophilic
Lysine-105	α Helix	Hydrophilic
Cysteine-115	β Sheet	Hydrophilic
Lysine-126	Random	—
Lysine-154	α Helix	Hydrophilic
Methionine-169	Random	—
Tyrosine-190	β Sheet	?
Tyrosine-262	Random	—
Lysine-269	Random	—
Cysteine-319	β sheet (N terminal)	?
Lysine-333	Random	—
Lysine-358	α Helix	Hydrophilic
Tyrosine-407	β Sheet (N terminal)	?
Lysine-420	α Helix	?

pears at the hydrophilic end of a possible amphiphilic structure). This type of analysis gives confidence in assigning modifiable residues to hydrophilic environments.

Chemical cross-linking studies of the type described in Chap. 7 can also, with the use of cross-linking reagents of differing lengths and specificities, give much valuable information, either of a direct nature or in conjunction with prediction studies, regarding tertiary structure.

Solution Methods

In Chap. 4 it was indicated that various approaches for the determination of molecular weight depend on assumptions concerning the *shape* of the protein whose molecular weight is being evaluated. The converse of this is, of course, that if the molecular weight of the protein is known, such methods give information on the overall shape of the protein. Although such approaches as sedimentation, gel filtration, and light scattering do not yield precise information concerning the surface topography of a protein, they can give valuable data on the overall shape, usually in terms of an axial ratio. Such methods are often sensitive enough to detect differences in shape under different conditions, and as such can be used to follow conformational changes. These approaches are discussed in more detail in Chap. 12.

A potentially very useful way of obtaining tertiary-structural information (at the level of the separation between points in a protein molecule) is fluorescence resonance energy transfer (see Figs. 10-6 and 10-7). If several *specific* sites can be used for such measurements, distances between specific points on a protein can be estimated. Such

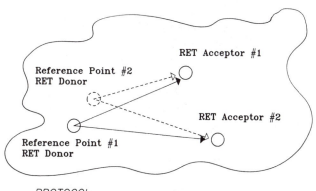

PROTOCOL
1. Measure efficiency of RE Transfer between Donor #1 and each RET acceptor: allows calculation of "Distance" between Donor #1 & Acceptors.
2. If possible use a second reference Donor site to calculate additional "Distances" to Acceptors.

Figure 10-6 General resonance energy transfer measurements used to obtain information about the relative location of various fluorescence donors and acceptors that might be introduced into a protein.

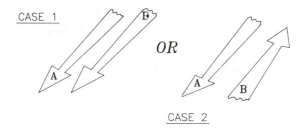

EXPERIMENTAL PROTOCOL
1. Add RET Donor @ point A
2. Add RET Acceptor @ point B
3. Measure distance between A & B
RESULTS distinguish between case 1 and case 2

Figure 10-7   Use of fluorescence energy transfer measurements to resolve specific questions with regard to the orientation of elements of tertiary structure.

distances can, for example, be used to estimate the separation of specific binding sites on a protein or the distance of a specific site from a metal-ion binding site. In addition, they (as with chemical cross-linking data) can be useful in the prediction of tertiary structure by restricting the number of possible arrangements of secondary structure that have to be considered.

PREDICTION OF TERTIARY STRUCTURE

The prediction of the tertiary structure of a protein from its amino acid sequence alone, although in theory possible, is probably a practical impossibility. Some workers have attempted such feats based on thermodynamic calculations but have, in large measure, been frustrated by the magnitude of the task. (Such attempts have, however, given much insight into the dynamic nature of protein tertiary structure.) We concern ourselves here with the prediction of what we shall term *tertiary structure packing elements*. The basis for this approach comes from the model of protein folding described in Chap. 9. Any region of *secondary* structure within the tertiary structure of the protein is packed according to the general principles espoused by Kendrew and summed up in the phrase "hydrophobics in, hydrophilics out." A particular α helix or β sheet is oriented in the tertiary structure of the molecule such that it obeys this simple rule. If some idea of the amphiphilic nature of particular elements of secondary structure can be obtained, some idea is also obtained of how such secondary structure may be oriented in the tertiary structure of the protein.

With β sheets there is a distinct "sidedness" of the secondary structure. Shown in Fig. 10-8 is a representation of the predicted β-sheet regions of glutamate dehydrogenase. Alternating residues in each region of secondary structure lie on opposite

Residues

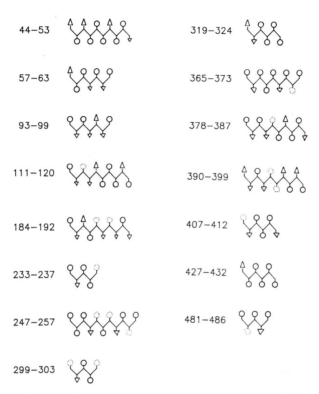

Figure 10-8 Depiction of β sheets predicted for glutamate dehydrogenase.

sides of the sheet, and depending on the nature of the side chain, the sheet may acquire a hydrophobic or a hydrophilic side.

In Fig. 10-8 hydrophobic residues are represented as circles and hydrophilic residues as triangles. Obviously, many of these β-sheet regions have clearly hydrophobic or hydrophilic sides. In other instances, a predicted sheet region has one-half of the sheet (in terms of the linear sequence) hydrophobic. In such a case it is be inferred that the sheet might extend from the inside of the tertiary structure toward the solvent-exposed exterior. β Sheets with a hydrophobic side are expected to have the hydrophobic side facing inward, with the more exposed, hydrophilic side oriented toward the surface of the tertiary structure. Sheet regions with a completely hydrophobic nature are internally located.

α-Helical regions of secondary structure can be assessed in terms of their regions of hydrophobicity by means of "Edmundson wheels"; several are shown in Fig. 10-9 for predicted helical regions of glutamate dehydrogenase. These wheels represent a

LONG HELICAL REGIONS

RESIDUES 138-155   RESIDUES 453-470

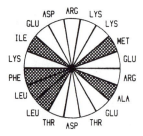   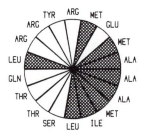

POTENTIAL LONG HELICES

RESIDUES 10-27   RESIDUES 325-342

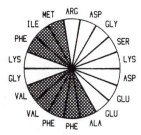   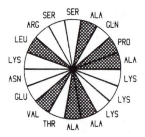

Figure 10-9   Edmundson wheels for several regions of α helix predicted for glutamate dehydrogenase.

projection, onto a flat surface, of an α helix, looking down on the axis of the helix. The construction of such a projection is shown in Fig. 10-10, where the number of each segment of the "wheel" represents the position of the residue in the linear sequence.

In the final representations, illustrated in Fig. 10-9, the hydrophobic side chains are shown as shaded segments and the hydrophilic side chains as open segments. Not only do these representations allow an assessment of stabilizing hydrophobic ridges along a predicted helix, but they also show the amphiphilic nature of the helices. As with the β-sheet tertiary-structure packing elements, α helices can thus be oriented with regard to whether or not they are likely to face internally or externally within the tertiary structure of the protein.

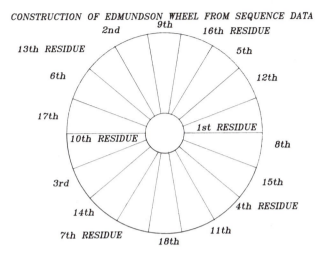

CONSTRUCTION OF EDMUNDSON WHEEL FROM SEQUENCE DATA

Figure 10-10 Scheme for Edmundson wheels: prediction from sequence.

An analysis of such tertiary-structural packing elements in conjunction with the results of chemical modification, chemical cross-linking, fluorescence resonance energy transfer, and active (or binding)-site labeling studies can give considerable insight into the overall tertiary structure of a protein.

A more precise conception can be obtained by further consideration of the nature of the interaction between packing elements in the formal tertiary structure. Strands of β sheet can interact to give a parallel or an antiparallel sheet. From an analysis of β-sheet regions of proteins whose crystal structure has been elucidated, several generalizations can be made.

1. In antiparallel sheet structures, branched side chains are found next to un-branched side chains to give better packing.
2. In parallel sheet structures, like-type side chains are found next to one another, either branched or unbranched.

Similarly, in proteins where α helices must pack with one another, it is found that the various helices have "rows" and "ridges" and that they pack by their intercalation.

Although we have not discussed β turns in any detail in the context of tertiary structure, it is found, again by examination of known crystal structures, that β-turn regions usually lie at the surfaces of proteins and that the side chains of the residues involved usually point in the same direction.

In summary, although we stated at the start of this section that the prediction of the tertiary structure of a protein based only on its amino acid sequence remains a practical impossibility, it does seem that a judicious application of some of the principles of tertiary-structural packing elements and their interactions, together with chemical and biophysical experimental results, does allow a tertiary-structural model

SEQUENCE

Secondary Structure Prediction

SHEET I   HELIX I   SHEET II   HELIX II

eg:

Lys–I   Lys–II   Lys–III

CHEMICAL CROSS–LINKING

    Find Lys–I cross–linked to Lys–III

This places restrictions on possible conformations

Consider two possibilities:

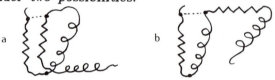

    a   b

EXAMINE PACKING ELEMENTS:

    Look for hydrophobic surfaces

For Example, possible interactions might include:

    SHEET I – SHEET II

    HELIX I – SHEET II

    HELIX I – HELIX II

IF RESULTS SHOW :

    1. That sheets I & II both have strong hydrophobic sides– consistent with "a"

    2. If helix I and sheet I both have hydrophobic sides, could favor "b".

INTRODUCE R.E.T COUPLE:

    If acceptor – donor pair can be introduced at known residues in helix I & II, then distance measurements could favor "a" or "b".

FURTHER CROSS–LINKING & CHEMICAL MODIFICATION:

    Use photo–activatable reagent specific for lysine   – look for other cross–links. General Chem.Mod gives exposed residues.

FINAL TESTS:

    Once a predicted structure is obtained it can be tested using site directed muta- genesis to alter key residues in secondary structure and examine effects on function & structure: correct model gives prediction

Figure 10-11   Scheme for building a tertiary structure model based on prediction and experimental evidence.

of a protein to be formulated. A simplified scheme for how such an endeavor might proceed is given in Fig. 10-11.

As indicated, a variety of biochemical and biophysical experimental methods can be used to obtain a relatively low resolution picture of the tertiary structure. Although such pictures cannot compare to the detailed information obtained from, for example x-ray crystallography, they can contribute significantly to the state of knowledge about the tertiary structure of a given protein.

It is not unusual to have available a primary structure from which secondary-structure predictions can be made. The limitations of such predictions were discussed in Chap. 9 with the principal criticism involving the lack of any knowledge of tertiary structure and the effects that might have on secondary structure. It is necessary to examine environmental effects of experimentally determined facets of tertiary structure on the predicted secondary structure: For example, does the interaction of a helix with a particular region of sheet lead to increased or decreased stability of either element?

Although such combinatorial approaches to examining the secondary and tertiary structure of a protein are low resolution, they can give insight into the types of structures associated with the binding sites and surfaces of a protein and lead to an increased understanding of its tertiary structure. More important, it is possible that such approaches permit the design of experiments to further test postulated relationships between regions of the protein.

DYNAMIC ASPECTS OF TERTIARY STRUCTURE

General Experimental Evidence for the Dynamic Nature of Proteins

For many years there was discussion as to whether proteins were best regarded as essentially rigid structures (suggested by early ideas of enzyme mechanisms and given some credence by x-ray crystallographic structures of proteins) or as "floppy bodies" with large amounts of conformational mobility. In terms of enzyme mechanisms it was pointed out that an enzyme needed conformational flexibility so that it could lower the free-energy level of a transition state for an enzyme–substrate complex via a set of conformational fluctuations; furthermore, transient fluctuations of structure were inevitable even at thermodynamic equilibrium. Direct evidence of such conformational fluctuations; on a nanosecond time scale were provided by the fluorescence quenching experiments of Lakowicz and Weber. They demonstrated that all regions of a protein molecule were accessible to collisional quenchers such as oxygen, even though the crystal structure showed no possible mechanism for such quenching: the protein *must* undergo rapid structural fluctuations, allowing direct, collisional quenching to occur.

We can consider the *time scale* of various events involved with protein structure and function, as shown in Table 10-4. The majority of these processes occur at the surface of a protein molecule and involve relaxation of bound water molecules. A

TABLE 10-4   Time scale of events

Determinant	Time (s)
Protein surface	
Bound-water relaxation	10^{-9}
Side-chain rotational correlation	10^{-10}
Ionization of side chains	10^{-7}–10^{-9}
Protein conformations	
Local motion	10^{-8}–10^{-9}
Isomerization	10^{-2}–10^{-7}
Folding–unfolding	10^{2} –1
Enzyme substrate reactions	
Encounter rate	Diffusion control
Lifetime of transition state	10^{-10}
Metal-ion coordination events	10^{-6}–10^{-9}
ES conformational changes	10^{-2}–10^{-4}
Covalent ES intermediate lifetimes	10^{-2}–10^{-4}

variety of different experimental approaches have been used to give some insight into the dynamic nature of protein tertiary structures, and here we briefly consider several types.

Cryogenic Experiments. If at room temperature a protein is in a dynamic state of flux between a number of energy equivalent conformations (i.e., has a dynamic structure), it should be possible to trap various structures by lowering the temperature. This has been examined by studying the kinetics of ligand binding to hemoglobin at low temperature. It is found that nonexponential rebinding of ligand occurs after photo-dissociation, indicating that there exists a continuous spectrum of activation energies due to the existence of many conformational states in which the thermal energy, at these low temperatures, is less than the kinetic energy barrier between the states. At room temperatures the thermal energy is such that these many states are rapidly interconverted and the rebinding of ligand occurs as a simple, single exponential process.

Hydrogen-Exchange Kinetics. In a protein there exists many protons which exchange if exposed to tritiated water. In an unstructured polypeptide the NH protons of the peptide bond exchange very rapidly. In a protein in its native structure these same protons exchange at a much slower rate than in the unfolded form, but they do still exchange. Although many other side-chain protons in a native protein can exchange with solvent protons, their rate is too fast to be measured. In some cases, buried side-chain protons exchange at rates slow enough for measurement.

As with the fluorescence quenching experiments referred to earlier, an examination of the x-ray crystallographic structure of a protein does *not* reveal a mechanism by which such exchanges can take place with internal NH protons. In effect, hydrogen exchange samples the solvent accessibility to interconverting forms of a protein.

There are a number of ways that this can be followed experimentally. For the NH group of the peptide bonds the simplest is to follow the infrared amide II band, which differs for NH and NT (or ND). Where the chemical shift of a group whose proton exchanges is known, NMR spectroscopy gives information about exchange rates. This approach, however, can be applied only when the resonance is shifted away from the main protein signal. In a crystal lattice the positions of H and D (or T) can be determined by neutron diffraction. For such experiments proteins are crystallized from aqueous solution, and deuterated solvent is perfused into the crystal. Neutron diffraction studies give information about *which* protons can be exchanged, although it must be recognized that the information obtained is relevant to exchanges that can take place *within the constraints* of the crystal lattice.

Several nonspectroscopic methods of following hydrogen exchange are available. In general, they depend on the rapid separation of solvent from protein and direct estimation of tritium in either solvent or protein. Such separations can be achieved by freeze-drying to obtain the solvent and/or protein, or by rapid dialysis or gel filtration.

As indicated earlier, hydrogen-exchange kinetics reflect the conformation and accessibility of exchangeable protons to solvent (the most slowly exchanging protons are often those which are hydrogen bonded). There are two basic mechanisms by which such dynamic exchange can take place: local unfolding and exchange from a folded conformation. It should be noted that in most proteins there is a core of protons that exchange at rates too slow to be measured when temperatures are reduced below those where continual thermal unfolding occurs. This suggests that some exchanges do not occur from the folded conformation but require the dynamic structure of the protein. It is also generally observed that any effect causing an increased stability of the protein to denaturation results in *decreased* exchange rates.

Two mechanisms by which solvent can have access to buried regions of the protein allow exchanges to take place. These are summarized in Fig. 10-12. In the first, there exists a conformation of the protein in which certain regions are totally inaccessible, and exchange of protons in that region cannot occur unless there is a reversible, localized cooperative unfolding of an internally hydrogen-bonded segment. In the second mechanism there may exist a number of smaller-amplitude motions, any of which individually would not allow exchange to take place, but which together lead to a finite probability that solvent can progressively penetrate into the exchange site.

Evidence for *local unfolding* as the mechanism of such exchanges has come from studies of oxygen binding to hemoglobin. When oxygen binds there is an *increase* in the exchange rates of some, but not all, of the protons that could be exchanged in deoxyhemoglobin, thus allowing preferential labeling of protons that exchange slowly in deoxyhemoglobin but rapidly in oxyhemoglobin. This differentially labeled species allows the transfer kinetics of these protons from the rapidly to the slowly exchanging type of proton to be followed. It is found that the transfer is approximately first order, indicating that the process results from a perturbation of a local unfolding equilibrium as a result of oxygen binding.

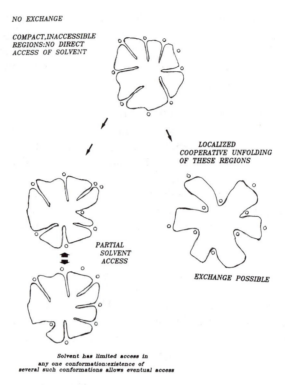

Figure 10-12   Possible mechanisms for the exchange of protons from regions of a protein that appear inaccessible.

Hydrogen exchange is a very sensitive probe for induced conformational changes. An increase in exchange rate reflects an *increased* internal accessibility and mobility and a *decreased* conformational stability. In lysozyme, the activation energy for exchange increases after binding of an inhibitor that decreases the exchange rate. It is also observed that binding of the inhibitor stabilizes the protein against thermal unfolding. In general, decreased exchange rates have correlated well with increased stability to temperature or urea denaturation, while increased exchange rates have correlated with increased susceptibility to proteolytic cleavage.

Models for the Dynamic Nature of Polypeptide Structure

The period between 1965 and 1975 can be described as the decade of the rigid molecule. Replicas of double-helical DNA and various protein molecules dominated many textbooks and much of the thinking. Results from high-resolution x-ray crystallography strongly influenced protein characterization. The intricate and detailed drawings of proteins that developed led to the image of a protein having each of its atoms fixed in position. When x-ray data made visible cases of conformational change, such as ligand or substrate binding, the conclusions proposed were abrupt transitions between otherwise static structures.

Today the static picture of protein structure is replaced by the view of proteins as delicately balanced dynamic systems. It is now realized that the atomic positions determined by x-ray diffraction studies represent the average equilibrium geometry of the molecule; these atoms exhibit fluidlike motions of sizable amplitudes around their average positions. The growing importance of protein dynamics has led to more accurate interpretations of protein function not possibly described by a static viewpoint. Some examples are the following:

1. The functional interactions of flexible ligands with their binding sites, which often require conformational adjustments in both ligand and binding protein.
2. Structural changes in binding proteins, which regulate activity through induced-fit and allosteric effects.
3. The chemical transformations of substrates by enzyme, which involve atomic displacements in the enzyme–substrate complexes. The mechanisms and rates of such transformations are sensitive to dynamic properties of these systems.
4. Electron transfer processes, which depend strongly on vibronic coupling and fluctuations that alter the distances between the donor and acceptor.
5. Various structural transitions that sometimes alter points of contact between proteins and other molecules which affect the activities of (a) contractile proteins such as myosin, (b) other enzymes, and (c) antibodies, as well as proteins involved in (d) membrane transport and (e) genetic regulation. Overall, it appears that understanding the functions of proteins requires an investigation of the dynamics of structural fluctuations and their relation to activity and conformational change.

In this section we discuss in detail two methods used to study specific motions that occur in proteins. The first presents a simplified model of a globular protein and the second presents a simplified model of a polypeptide chain, more specifically an α helix.

Globular proteins exhibit a variety of motions that can be classified according to their amplitude, energy, or time scale (Tables 10-4 to 10-6). Such structural flexibility implies that the secondary bonds are strong enough to fit different conforma-

TABLE 10-5 Classification of internal motions of globular proteins

Scales of motion (300 K)
  Amplitude: 0.01–100 Å
  Energy: 0.1–100 kcal
  Time: 10^{-15}–10^3 s

Types of motion
  Local: atom fluctuations, side-chain oscillations, loop and "arm" displacements
  Rigid body: helices, domains, subunits
  Large scale: opening fluctuation, folding and unfolding
  Collective: elastic-body modes, coupled atom fluctuations, soliton and other nonlinear
    motional contributions

TABLE 10-6   Thermal motions in proteins

Frequency range (s^{-1})	Type of motion	Proteins studied in detail
10^{10}–10^{12}	Surface group rotations (CH$_3$, CH$_2$NH$_3$, etc.), wobble of N- and C-terminal residues, α—CH oscillations	BPTI, MCBP, lysozyme, RNase, myoglobin
5×10^8–5×10^{10}	Interior side-chain oscillations (aliphatic and aromatic)	BPTI, MCBP, lysozyme
10^7–10^9	Rotational diffusion (may be anisotropic)	All
10^7–10^8	Backbone warp	BPTI, MCBP, lysozyme
10^4–10^8	180° aromatic ring flip	BPTI, lysozyme, MCBP
?–10^4	Opening of secondary structure (^{1}H—^{2}H exchange)	BPTI, RNase
?–10^6	Conformational changes and translocations	

tions but weak enough to allow configurational and conformational transitions in the presence of slight perturbations. In other words, the protein exhibits flexibility in its normal functional state, being not entirely rigid but "breathing" about its equilibrium. A simplified procedure was developed to study this dynamic aspect: It treats the globular protein as a spheroidal structure, sets up dynamic differential equations, and solves them to determine the nature of the spatial dimensions and time parts of the local fluctuations.

Model Protein Systems

The simplified model for globular proteins is confined to the fluctuations that affect the relative distances between parts of the protein such that the topological connections are maintained while the parts deviate from their original positions and then return to their original positions from the new ones. In this way, the globular protein is defined as an isotropic body with definite shape and elastic properties. Such a system fits classical mechanical equations of motion whose solutions establish the variations of the fluctuations quantitatively.

Theory. The differential equation of motion for a mass $\rho\, dx\, dy\, dz$ of an elastic-deformable isotropic body located at the point (x, y, z) along the radial direction is given by

$$(\lambda + \mu)\frac{\delta\Delta}{\delta r} + \mu\frac{\delta^2 U_r}{\delta r^2} + \rho R = \rho\frac{\delta^2 U_r}{\delta t^2} \qquad (10\text{-}5)$$

where λ and μ are defined as Lamé's constants. They are related to Poisson's ratio and Young's modulus by Eqs. (10-6) and (10-7).

$$\text{Poisson's ratio} = \frac{\text{lateral strain}}{\text{axial strain}} \qquad (10\text{-}6)$$

Strain is defined as change in dimension and stress is the force per area that causes this strain.

$$\text{Young's modulus (of elasticity)} = \frac{\text{stress}}{\text{strain}} = \frac{\text{force/area}}{\Delta \text{ in dimension in}}$$
$$x, y, \text{ and } z \text{ planes}$$

(10-7)

Thus these constants describe both the strength and relative position of protein fluctuation. In Eq. (10-5), Δ is the sum of the longitudinal strain components along the three directions, ρ the density of the body, R the body force, and U_r the displacement along the radial direction.

Next, the protein is limited to a prolate spheroidal shape, which means that its polar axis is longer than its equatorial diameter, and Eq. (10-5) transforms to

$$(\lambda + 2\mu) \sinh \xi \cosh \frac{\delta^2 U\xi}{\delta\xi^2} + (\lambda + 2\mu) \cosh^2 \xi \frac{\delta U\xi}{\delta\xi} + [\lambda \sinh \xi \cosh \xi$$

$$-(\lambda + 2\mu) \coth \xi \cosh^2 \xi - 2(\lambda + \mu) \tanh \xi \sinh^2 \xi] U\xi + \rho R = \frac{\rho \delta^2 U\xi}{\delta t^2} \quad (10\text{-}8)$$

The Cartesian coordinates (see Fig. 10-13) are converted to prolate spheroidal coordinates by the relations given in

$$x = a \sinh \xi \sin n \cos \theta \qquad (10\text{-}9a)$$

$$y = a \sinh \xi \sin n \sin \theta \qquad (10\text{-}9b)$$

$$z = a \cosh \xi \cos n \qquad (10\text{-}9c)$$

where a is the radius or U_r, the distance from the center of the protein, ξ is the deformation parameter, n the angle between the radius and the x-z plane, and θ the

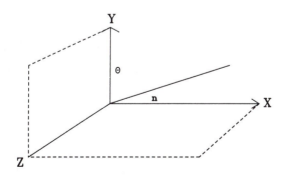

Figure 10-13   Relationship of Cartesian and prolate spheroidal coordinates.

angle between the radius and the y-z plane. The defined conditions are $a = 1$ and $n = 90°$. The sinh and cosh functions are given by

$$\cosh = \frac{e^x + e^{-x}}{2} \qquad (10\text{-}10)$$

$$\sinh = \frac{e^x - e^{-x}}{2} \qquad (10\text{-}11)$$

The sinh and cosh functions are hyperbolic analogies for the circular trigonometric identities, sin and cos; sin x and cos x define point (x, y) on a circle, while sinh x and cosh x define point (x, y) on a hyperbola. The solution, $u\xi$, for Eq. (10-8) is given in

$$u\xi = A \cos(\omega t + \varepsilon_0)u(\xi) \qquad (10\text{-}12)$$

This solution is best defined in two parts: one, the time part, specified by a cosine term, the other, the space part, specified by a function $u\xi$, which depends on ξ. Substituting Eq. (10-12) into Eq. (10-8) and simplifying leads to the differential equation

$$\frac{d^2u}{d\xi^2} + \coth \xi \frac{du}{d} + [B - \coth^2 \xi - C \tanh^2 \xi + D \text{ sech } \xi \text{ cosech } \xi]u = 0 \quad (10\text{-}13)$$

where

$$B = \frac{\lambda}{\lambda + 2\mu}$$

$$C = \frac{2(\mu + \lambda)}{\lambda + 2\mu}$$

$$D = \frac{\rho\omega^2}{\lambda + 2\mu}$$

Equation (10-13) defines the radial distance from the center of the protein and the slope of the fluctuation. To solve Eq. (10-13) it is split into two equations,

$$\frac{du}{d\xi}(=p) = F_1(\xi, u, p) \qquad (10\text{-}14)$$

$$\frac{dp}{d\xi} - p \coth \xi - (B - \coth^2 \xi - C \tanh^2 \xi + D \text{ sech } \xi \text{ cosech } \xi)u = F_2(\xi, u, p)$$

$$(10\text{-}15)$$

Equation (10-14) indicates that the differential coefficient of displacement of the elementary mass, with respect to its position from the centroid of the protein, is equal to the slope of the space–displacement relation. Equation (10-15) defines the differential coefficient of the slope of the space-displacement relation of an element in the protein with respect to its radial position.

A set of initial values for the variables ξ, u, and p defines the initial values of the functions F_1 and F_2. The initial values of ξ and u characterize the fluctuational condition of the centroid of the protein, whereas the initial value of p characterizes the fluctuational condition of the outermost residue on the surface of the protein. After considering values of ξ at regular intervals of 0.2 Å, from 0 to the maximum radius distance, r, values of F_1 and F_2 are computed. They are used to calculate values of u and p for the different ξ values up to the maximum r. X-ray data of the protein's atomic coordinates define the centroids of the amino acid residues and the protein centroid. The maximum radius distance is defined as the distance of the farthest residue centroid from the protein centroid. Under ordinary temperatures, the molecules that neighbor the protein centroid fluctuate much less than the molecules on the surface of the protein. To define the fluctuation boundary conditions, it is assumed that the protein centroid experiences no fluctuation; at $\xi = 0$, $u = 0$, and the surface experiences fluctuations near 10% of the maximum r value in the protein; at $\xi = r$, $u = r/10$, as suggested previously.

To determine the initial p value to use, a displacement value, u, close to 0 is assumed, since the ideal situation of 0 displacement does not occur, and possible p values are tried, starting at a value of 0.1 and varied by increments of 0.1. These u and p values are changed and iterated through the equation until the assumed u value and subsequent initial p value are obtained. This initial p value becomes a characteristic parameter of the protein.

Application to Model Proteins. The described limiting conditions and dynamic equations were applied to two proteins, bovine pancreatic trypsin inhibitor (BPTI) and tuna ferrocytochrome c (CYTC). Their solutions provided information on the motion of specific amino acid residues relative to their location from the centroids of the two proteins. Once the specific p values for the two proteins were obtained, the displacement, or fluctuations, were determined at points along the length of the radius vector. From these values, a space plane of the plot was constructed: ξ, position from the centroid versus u, displacement.

Note that the slope of these lines, $du/d\xi$, is defined in Eq. (10-14) as p, which depends on the specific size and amino acid distribution of the protein. Both curves in Fig. 10-14 show the displacement values increasing slowly until the mid r value, and then increasing rapidly. The displacement varies in a nonlinear manner from the centroid to the surface of the proteins.

To determine the specific fluctuations of the amino acids in the two proteins, the magnitudes of the position vectors for the individual residues, obtained from the crystal structures of BPTI and CYTC, are paired with the actual displacements of the residues from the curves, and fluctuation plots are constructed that connect residue sequence number and respective displacement.

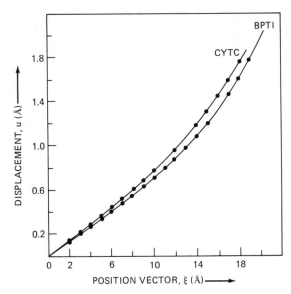

Figure 10-14   Space-plane plots for BPTI and CYTC. Points on a curve indicate the extent of displacement (vertical axis) of elements of the protein at their respective radial positions (horizontal axis) from the centroid of the respective protein. (Reprinted with permission from: P. K. Ponnuswamy and R. Bhaskaran, *Int. J. Pep. Protein Res.*, **24**, 168–179. Copyright 1984 Munksgaard International Publishers Ltd., Copenhagen, Denmark.)

Fluctuations in BPTI.   See Fig. 10-15A.

1. The amino and carboxyl ends of the molecule experience much fluctuation due to their freedom of motion.

2. The large peaks observed at and around residues Leu-6, Arg-17, Arg-39, Glu-49, Arg-53, Gly-57, and Ala-58 reveal their highly exposed nature.

3. Generally, larger displacements are expected for prolines and glycines, which intrinsically appear in bends and loops on the surface. However, smaller displacements are observed for Pro-8 and Pro-9, due to their association with the β-sheet structural residues 29 to 36. These prolines are stabilized by hydrogen bonding and other intra- and interchain interactions.

4. A loop region, residues 25 to 28, connects two β-sheet regions, residues 18 to 24 and 29 to 36, and displays much higher displacement peaks than they do.

5. The β-sheet segment, residues 43 to 46, shows the smallest displacement because it is buried.

6. A comparison of the fluctuations of the α-helical segments, residues 2 to 7 and 47 to 56, and the displacements of the β-sheet segments reveals that the β sheets fluctuate less than the α helices.

7. The low displacement troughs noted for residues Cys-5, Arg-20, Phe-22, Phe-33, Tyr-35, Asn-44, Cys-51, and Arg-53 indicate their buried positioning.

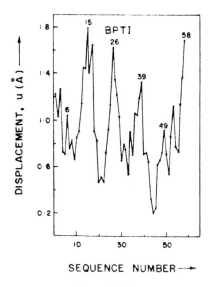

Figure 10-15A

8. The cysteine residues of the disulfide linkages between pairs 5 to 55, 14 to 38, and 30 to 51 fluctuate slightly.

9. The polar residues have an average displacement of 1.006 Å. Nearly 60% of the polar residues fluctuate at or beyond this average, which indicates their exposure to an aqueous environment.

10. The nonpolar residues have an average displacement of 0.736 Å. About 70% of these residues fluctuate at or less than their average, which indicates their buriedness.

Fluctuations in CYTC. See Fig. 10-15B.

1. The five turn segments, residues 21 to 24, 32 to 35, 35 to 38, 43 to 46, and 75 to 78, show high displacement peaks, indicating their exposure.

2. The exposed residues, Asp-2, Lys-5, Lys-25, Lys-39, Lys-73, and Lys-87, display high displacement peaks.

3. The residues Cys-14, Cys-17, His-18, and Met-80 have low displacements. The decreased fluctuation may reflect their role as linkage points of the heme group to the polypeptide chain.

4. As for BPTI, both the amino- and the carboxyl-end residues exhibit higher fluctuations.

5. About 70% of the polar residues have an average displacement of 1.30 Å, indicating their exposure, while about 80% of the nonpolar residues have an average displacement of 0.790 Å, indicating their buriedness.

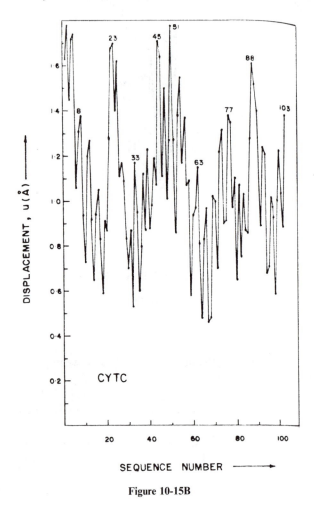

Figure 10-15B

Figure 10-15 Residue-displacement plots: (A) BPTI; (B) CYTC. (Reprinted with permission from: P. K. Ponnuswamy and R. Bhaskaran, *Int. J. Pep. Protein Res., 24,* 168–179. Copyright 1984 Munksgaard International Publishers Ltd., Copenhagen, Denmark.)

After analyzing the proteins BPTI and CYTC, the average displacement for each amino acid (see Table 10-7) was computed to characterize the behavior of the individual residues. A study of Table 10-7 leads to the following conclusions.

1. The two polar residues, Asp and Ser, have the largest amplitudes of fluctuation.
2. The two nonpolar residues, Phe and Leu, have the lowest amplitudes of fluctuations.

TABLE 10-7   Average displacements of amino acid residues as found in
BPTI and CYTC

Nonpolar residue	Displacement (Å)	Polar residue	Displacement (Å)
Phe	0.605	His	0.849
Leu	0.679	Asn	0.861
Tyr	0.798	Thr	0.907
Cys	0.844	Arg	1.023
Met	0.871	Gln	1.109
Trp	0.876	Glu	1.126
Ile	0.940	Lys	1.233
Val	1.043	Ser	1.262
Ala	1.159	Asp	1.315
		Special Type	
Pro	0.0963	Gly	1.224

3. The varied polar residues, Arg, Gln, Lys, Glu, and Asn, all display similar fluctuations. Of the nonpolar residues, Trp, Met, Cys, Ile, and Tyr have similar amplitudes of fluctuation.

Note that these conclusions are based exclusively on the results from two small proteins, BPTI and CYTC.

It is apparent that the amplitude of displacement depends on the residue's spatial position; similarly, certain residues characterize defined positions. Table 10-8 and the curves in Fig. 10-14 emphasize the nonlinearity of displacement with respect to spatial position. This displacement can be further analyzed with a curvilinear equation of the form

$$u = 1.309(\text{Sp})^{0.571} \qquad (10\text{-}16)$$

where u is the displacement and Sp the spatial position.

TABLE 10-8   Average displacement of residues in shells of 3-Å thickness in the
protein CYTC

Outer radius of the shell (Å)	Number of atoms in the shell	Average displacement (Å)
6	12	0.456
9	118	0.619
12	263	0.813
15	294	1.107
18	101	1.483
21	12	1.954

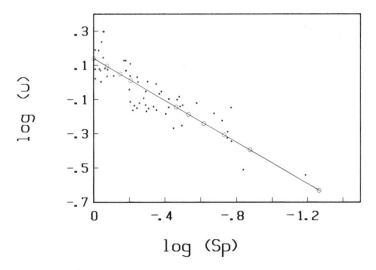

Figure 10-16   Plot of log (u) versus log (Sp) for BPTI. (Reprinted with permission from: P. K. Ponnuswamy and R. Bhaskaran, *Int. J. Pep. Protein Res.*, *24*, 168–179. Copyright 1984 Munksgaard International Publishers Ltd., Copenhagen, Denmark.)

Figure 10-16 shows a plot of log (u) versus log (Sp) for the values obtained from BPTI and a computer-fitted least-squares line: $u = 0.1117 + 0.571$ log (Sp) best describes this line. For a residue whose spatial position coincides with the protein centroid, this yields a displacement of 0.095 Å, whereas for a residue positioned on the surface, where the spatial position is unity, the computed displacement value equals 1.31 Å. The boundary condition value for such a surface displacement equals 1.94 Å; however, the average displacement value of the residues with the largest spatial position (Sp = 0.99) equals 1.25 Å, which agrees closely with the computed value of 1.31 Å. Few points deviate from the least-squares fitted line.

The solvent accessibility for a residue correlates well with its spatial position; therefore, the magnitude of displacement of a residue depends on its solvent exposure level. A correlation analysis was made between the water-accessible surface areas (A_s) of the amino acid residues in BPTI and their displacements (u). This was plotted in Fig. 10-17 as displacement versus accessible surface area. A least-squares line was fitted to the data, and the line was defined by $u = 0.0128A_s + 0.4748$. This line plot implies that the larger the contact area of a residue with the solvent medium, the larger the fluctuation. A closer analysis revealed about a 30% possibility of deviation from the linear relationship. For example, in BPTI, the polar residues, Lys-41, Arg-42, Lys-46, and Asp-53, show lower displacements despite their surface positions and highly accessible surface areas. One explanation may be asymmetry in the shape of the protein. These deviations may also occur at kinks and turns in the protein. The model proves more valid for proteins of nearly spherical or spheroidal shapes. In addition, it does not consider the surface atom's hydrogen bonding or any other types of interactions with the solvent medium.

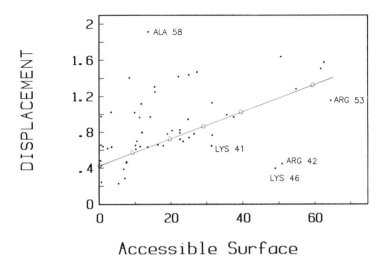

Figure 10-17   Plot of displacement versus accessible surface area. (Reprinted with permission from: P. K. Ponnuswamy and R. Bhaskaran, *Int. J. Pep. Protein Res., 24,* 168–179. Copyright 1984 Munksgaard International Publishers Ltd., Copenhagen, Denmark.)

Peptide Models: The Helix–Coil Transition

The helix–coil transition has several roles: Frequently, it nucleates the folding of globular proteins and has also been implicated in the promotion of certain hormone–receptor interactions. Despite its importance, it is difficult to evaluate the dynamics of this transition, as the time scale limits analysis. Whereas most dynamic simulation methods for polypeptide chains utilize a time span of 10^{-12} to 10^{-9} second, the helix–coil transition requires a range of 10^{-9} to 10^{-6} second. To overcome this, the following model for the polypeptide chain, in the aqueous solution, was proposed. The simplified character of the chain permits the calculation of overall peptide energies and the forces acting on individual residues.

Theory. In the simplified model, each amino acid residue is depicted by a soft sphere with a volume comparable to that of the corresponding amino acid. A consideration of a variety of x-ray structures suggests that with an α helix this approximation is adequate. Each sphere is defined as a single interaction center, R_i, which is connected to other interaction centers by virtual bonds, as shown in Fig. 10-18.

The centroids of the interaction centers are placed at the C^{β} positions of the detailed structures. These centers are linked by virtual bonds and corresponding virtual dihedral angles, which replace the ϕ and ψ dihedral angles about the C^{α}. The virtual dihedral angle Φ_i is defined in terms of the residue centers $i, i + 1, i + 2, i + 3$; $\Phi_i = 0$ for the eclipsed conformation. Each virtual dihedral angle has an associated torsional potential that represents the average sum of nonbonded interactions between the atoms of near-neighbor residues. Reference to the detailed model of the standard

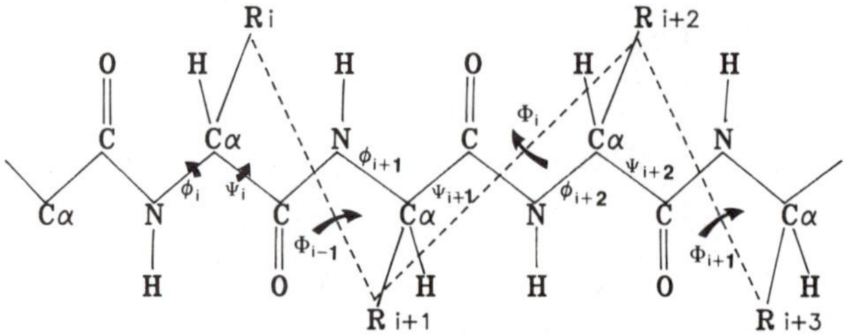

Figure 10-18 Geometry of the simplified polypeptide model compared with that of the atomic model. The dashed lines are virtual bonds connecting the interaction centers R_i.

α-helix geometry, presented by Pauling et al., leads to the following conditions for the simplified chain: b, equilibrium bond length = 5.14 Å, θ_0, equilibrium bond angle = 87.2°, and Φ_α, characteristic α-helix dihedral angle = 38.3°. Levitt determined a 30° width for α-helix region; thus the range $25° < \Phi < 55°$ defines the α-helix boundary for the simplified chain.

In the model, torsional motions about the virtual bonds depict large-scale conformational changes. The energy function representing the chain defines a balance between interactions that stabilize either the helical conformation or the coil conformations. This energy function is obtained by averaging inter-residue interactions over all possible local atomic configurations within each residue; thus the model incorporates the separate time scales for local and overall chain motions. The reduced number of degrees of freedom allows rapid calculation of the energy and forces, which in turn permits significantly longer dynamic simulations than a more detailed model. Although it does not accurately represent detailed atomic interactions, such as side-chain hydrogen bonding, the model does describe overall structure flexibility directed by simple packing effects, such as steric and hydrophobic interactions, within the polypeptide–solvent system. The lengthened time scale of these fluctuations causes rapid averaging of localized motions, such as side-chain internal rotations. The dynamic simulation applied to the model simulates the overall chain motions for periods of several hundred nanoseconds. While the method appropriately simulates local unfolding and folding of proteins and their secondary structural elements, such as helix–coil transitions, it could not accurately simulate folding of an entire protein, which requires a longer time scale and more detailed atomic interactions.

To study dynamics theoretically, it is necessary to understand the potential-energy surface, the energy of the system as a function of the atomic coordinates. This potential energy is used to determine the relative stabilities of different possible structures of the system. The forces that act on the atoms of the system are derived from the first derivative of the energy potential with respect to the atom positions. These forces are then used to calculate dynamic properties of the system by solving equations of motion that determine the change in atomic positions over time. Most of the motions occurring at ordinary temperatures leave bond lengths and bond angles of polypeptide chains near their equilibrium values, which remain relatively constant throughout the protein; thus the standard dimensions of the α helix proposed by Pauling in 1951 were used to determine the dimensions in the simplified model. It is the contacts among nonbonded atoms that are significant in the potential energy of the higher structures.

The following energy function approximates the energy potential, reflecting the average *over the polypeptide degrees* of freedom omitted in the simplified model and *over degrees of freedom* of the solvent molecules. The function is a sum of six distinct interactions:

$$E = E_{\text{bond}} + E_\theta + E_\Phi + E_{\text{sol}} + E_{\text{ev}} + E_\alpha \tag{10-17}$$

E_{bond} and E_θ represent the bond and bond-angle interactions, respectively;

$$E_{\text{bond}} = \sum k_b (b - b_0)^2 \tag{10-18}$$

and is the summation over all virtual bonds; and

$$E_\theta = \sum k_\theta (\theta - \theta_0)^2 \tag{10-19}$$

and is the summation over all virtual bond angles.

k_b and k_θ are force constants and, as indicated earlier, $b_0 = 5.14$ Å and $\theta_0 = 87.2°$. E_{bond} and E_θ are harmonic functions, which implies that they represent the vibrational energies of oscillating, springlike systems.

E_Φ represents the energy potential of near-neighbor atomic interactions, averaged over all dihedral angles consistent with a given Φ value. Recall that the simplified polypeptide model has only one backbone rotational degree of freedom per residue (Φ_i), while the detailed atomic model has 2, ϕ, and ψ. The position of the interaction center, R_{i+3}, is essentially determined by ψ_{i+1} and ϕ_{i+2} through their sum, $\psi_{i+1} + \phi_{i+2}$; therefore, Φ_i depends on ψ_{i+1} and ϕ_{i+2}. Variations in the difference $\psi_{i+1} - \phi_{i+2}$ correspond to rotations of the plane of the amide group between R_{i+1} and R_{i+2}; such rotations minimally affect the overall chain direction. Thus this energy potential or mean force is averaged over amide plane orientations as well as sidechain and solvent molecule configurations. It is assumed that the local reorientations of the amide planes are rapid compared to the variations in Φ, which reflect the slow changes in the overall shape of the chain.

E_{sol} represents interactions between residues separated by three or more virtual bonds. More specifically, E_{sol} represents the contribution to the energy potential of net attractive effects, from van der Waals interactions, excluded volume effects, and solvent interactions, and is given by

$$E_{\text{sol}} = \sum \sigma g(r_{ij}), \qquad i > j \tag{10-20}$$

where r_{ij} is the distance between interaction centers i and j, and $g(r)$ is a sigmoidal function which varies from $g(0) = 1$ to $g(r) = 0$ for r_{ij} 9 Å (Fig. 10-19). This summation is taken over centers separated by three or more bonds. It represents the free energy of attractive interactions that occur upon transfer of a residue from coil or aqueous surroundings, to helix or more hydrophobic surroundings.

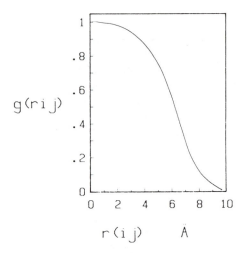

Figure 10-19 Dependence of $g(r_{ij})$ on r_{ij}.

E_{ev} represents the excluded-volume term. This free-energy contribution, which is a purely repulsive potential, opposes the change in attractive interactions introduced by E_{sol}. E_{sol} implies that once the coil residue relinquishes its freedom of mobility and commits itself to helix structure, it gains hydrophobic stabilization energy. However, its presence among the other residues eliminates a crevice of "free space" and depletes this latent stabilization source. So the E_{ev} term is needed to prevent the equation from overemphasizing the stabilization energy and thereby to prevent the excessive stability of "globules" (i.e., overlapped nonbonded residues).

E_{α} represents helix-stabilization energy and is given by

$$E_{\alpha} = \sum A_{\Phi} f(\Phi) \tag{10-21}$$

It is summed over all the virtual dihedral angles. The function $f(\Phi)$ is bell shaped; it reaches its maximum value of unity at $\Phi = 40°$; it diminishes to 0 if $\Phi > 55°$ or $\Phi < 25°$, the conditions when Φ is outside the α-helical range. The coefficient A_{Φ} for each dihedral angle depends on adjacent dihedral angles, as demonstrated in Fig. 10-20.

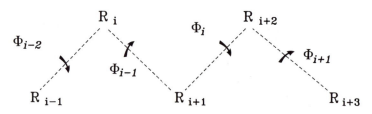

Figure 10-20 Schematic representation of virtual dihedral angles.

$A_{\Phi_i} = -6.0$ kcal/mol if Φ_{i-1}, Φ_{i-2}, and Φ_{i+1} are in the α-helix range. This value reflects the difficulty of nucleating a coil sequence in the interior of an α helix. $A_{\Phi_i} = -1.4$ kcal/mol if Φ_{i-1}, Φ_{i-2}, and Φ_1 are in the α-helix range, but Φ_{i+1} is not. Thus the α-helix stabilization term contributes up to -1.4 kcal/mol upon the addition of a residue to the helix. This term includes free-energy changes associated with the formation of a backbone hydrogen bond and the freezing of amide-plane rotational motions. Various calorimetric and pH titration studies of helix–coil transitions suggest that hydrogen-bond formation provides an estimated -1.5 kcal/mol of free-energy change, the larger of the contributions. The entropy reduction upon freezing the amide-plane orientation is less than 2 eu, which corresponds to a free-energy increase at 25°C of less than 0.6 kcal/mol. Therefore, the stabilization energy used appears to be of a reasonable magnitude. $A_{\Phi_i} = 0$ if I_i is not in the helix interior or at the helix-coil interface.

Stochastic Dynamics. The method of stochastic dynamics is used to study the dynamics of the helix–coil transition. First, the system is divided into two parts: One part serves as a heat bath for the other part whose dynamics are to be examined; this situation could be a protein in a solvent or a portion of protein within the surrounding protein. Second, this method assumes that the displacement of the dynamic part is analogous to molecular diffusion in a liquid or solid. Third, the energy potential is determined. This potential-energy function corresponds to the free energy of displacement of the elements being studied due to surrounding bath atoms; in our simplified model, these bath atoms are both solvent molecules and other chain residue atoms, while the dynamic element is a specific residue. The motion of the chain is largely determined by the time variation of its nonbonded interactions with the neighboring atoms. Finally, forces are derived from the energy potential which speed or slow the motion of the chain in a given direction. The chain motion is described by a set of Langevin equations of motion. The forces are plugged into these equations and the solutions define a new chain position. For example, the Langevin equation for a particle in one dimension is given by

$$m\frac{d^2x}{dt^2} = F(x) - f\frac{dx}{dt} + R(t) \tag{10-22}$$

where m is the mass of the particle, x the position of the particle, t the time, $m(d^2x/dt^2)$ the acceleration of the particle, and $F(x)$ represents the force on the particle derived from the energy potential. The remaining terms represent effective bath forces: $f(dx/dt)$ is the average frictional force caused by the motion of the particle relative to its surroundings, f the friction coefficient, and $R(t)$ represents any remaining randomly fluctuating forces not categorized by the mean force: electrostatic, dipole influences. For local denaturations such as helix–coil transition, the acceleration term is neglected in comparison to the others. This motion has no inertial character and imitates Brownian motions.

To study the dynamics of the helix–coil transition, the internal Brownian motion of a 15-residue chain in solvent surroundings is computer simulated. The system of

Brownian particles, placed in their initial configuration, is simulated by space trajectories. Each particle is moved along a trajectory calculated from the appropriate
chain diffusion equation. This equation includes diffusional, hydrodynamic, and
interparticle interactions as well as a weighted sum of all other forces acting on the
Brownian particles. The trajectories are composed of successive displacements taken
over a short time (Δt). These displacement values are derived from the appropriate
equation of motion. To minimize the number of time steps required to calculate a
trajectory, Δt is chosen as large as possible provided that the forces on the particles remain nearly constant during each step. Each successive configuration of the
chain is selected from a probability distribution, which is the short-time solution of
the chain diffusion equation, with the previous configuration as an initial condition.
In this particular study, hydrodynamic interactions among the residues are neglected.
The chain is initially in an all-helical configuration. The first 11 residues are held fixed
in space, while the last four residues (R_{15}, R_{14}, R_{13}, and R_{12}) are allowed to move.
The virtual dihedral angles Φ_{12}, Φ_{11}, Φ_{10}, and Φ_9 correspond to the mobile residues, respectively.

Figure 10-21 shows the components of the potential energy for residue 12 as a
function of Φ_9. Results for the other mobile residues are nearly identical. Note the

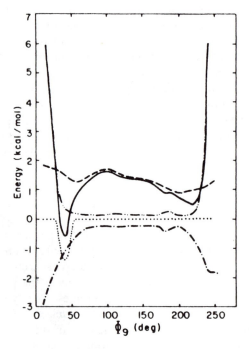

Figure 10-21   Potential of mean force and its components for a residue at the helix–
coil interface for E(——), E_Φ(– – –), E_{sol}(– · –), E_{ev}(– ·· –), and E(· · ·). (Reprinted with
permission from: J. A. McCammon and S. H. Northrup, *Biopolymers*, *19*, 2033–2045.
Copyright 1980 John Wiley & Sons, Inc., New York.)

TABLE 10-9 Maximum relaxation times, $1/\tau_{max}$, and the rate constants, k_f, of the helix growth process at various temperatures[a] for poly(α-L-glutamic acid)

T (°C)	$1/\tau_{max}(10^5 \text{ s}^{-1})$	$k_f (10^7 \text{ s}^{-1})$
15	2.9 ± 0.2	2.4 ± 0.2
25	3.1 ± 0.2	2.6 ± 0.2
35	3.5 ± 0.2	2.9 ± 0.2
45	4.1 ± 0.3	3.4 ± 0.2

[a] In the estimations of k_f, a σ value of 3×10^{-3} was used.

free-energy barrier of approximately 1 kcal/mol per residue for the helix–coil transition. Other studies support the validity of this calculated value. Researchers used the electric field pulse (EFP) apparatus, which applies an electric field density and detects changes in electric conductivity, to follow the lifetime of the helix–coil transition. Generally, lifetime is defined as the time required for an atom or group of atoms in a high-energy state to return to a more stable, lower-energy state. In the case of the helix–coil transition, the lifetime defines the time required for a residue to leave its coil state and enter the helix state, leading to helix growth. This lifetime is expressed by

$$\frac{1}{\tau} = k_f[(s' - 1)^2 + 4\sigma] \tag{10-23}$$

where k is the forward rate constant for helix growth, s' the equilibrium constant for helix growth, and σ the nucleation parameter. The temperature dependence of these parameters is shown in Table 10-9.

The rate constant k_f is related to the activation enthalpy ΔH and entropy ΔS by

$$k_f = \frac{kT}{h} e^{-(\Delta H° - T\Delta S°)/RT} \tag{10-24}$$

The calculated values are

$$\Delta H = 1.5 \pm 0.4 \text{ kcal/mol}$$

$$\Delta S = -20 \pm 2 \text{ kcal/mol}$$

These values support the activation energy calculated from the energy potential.

Three independent simulations were performed, each for a total length of 12 ns. The time histories of the last three dihedral angles, Φ_{12}, Φ_{11}, and Φ_{10}, are shown in Fig. 10-22; the time history for Φ_9, which showed relatively little mobility, is omitted from the figure for clarity. Points are plotted at intervals of 120 ps.

Since the helix–coil transition occurs at the end of a single helical section, Φ_i is either in the α-helical range (when Φ_{i-1}, Φ_{i-2}, . . . , Φ_1, and Φ_{i+1} are in the α-helical range) or at the helix–coil interface (when Φ_{i-1}, Φ_{i-2}, . . . , Φ_1 are in the α-helix range

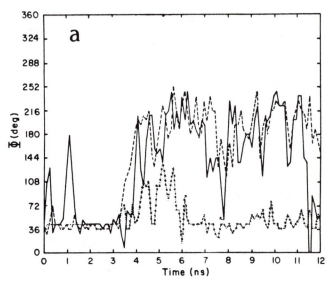

Figure 10-22A

but Φ_{i+1} is not). Nucleation at a second site along the helix chain would be rare on the time scale of the given simulations. Significant helix unwinding occurs in each simulation; thus the mobility of the residues markedly increases at the end of the chain. When two or more residues leave the helix they do so sequentially, although this is not always apparent from Fig. 10-22 due to the 120-ps intervals. After un-

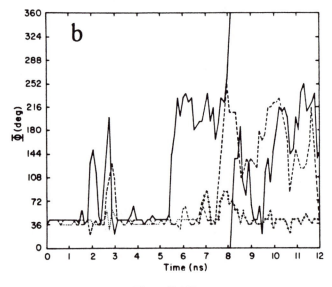

Figure 10-22B

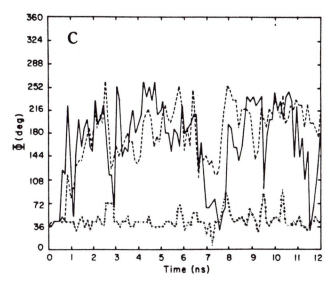

Figure 10-22C

Figure 10-22 (A) Dihedral angle histories during the first helix-unwinding simulation: Φ_{12}(——), Φ_{11}(– – –), and Φ_{10}($\cdot \cdot \cdot$). (B) Dihedral angle histories during the second helix-unwinding simulation: Φ_{12}(——), Φ_{11}(– – –), and Φ_{10}($\cdot \cdot \cdot$). (C) Dihedral angle histories during the third helix-unwinding simulation: Φ_{12}(——), Φ_{11}(– – –), and Φ_{10}($\cdot \cdot \cdot$). (Reprinted with permission from: J. A. McCammon and S. H. Northrup, *Biopolymers*, *19*, 2033–2045. Copyright 1980 John Wiley & Sons, Inc., New York.)

winding, Φ_{12} and Φ_{11} tend to remain out of the helix range; however, Φ_{12} drifts to smaller values while Φ_{11} returns to the helix near the end of the second simulation. In conclusion, the final study results suggest:

1. R_{15} unwinds from the helix readily and exhibits large fluctuations after R_{14} has also unwound.

2. R_{14} unwinds less readily than R_{15} and exhibits smaller fluctuations after unwinding.

3. R_{13} exhibits occasional transient departures from the helix, even after the last two residues are unwound. R_{13} seems to be in limbo between the two states.

Since the helix–coil transition free-energy barriers are similar for all the residues, their reduced mobility in the chain interior indicates that these residues have smaller effective diffusion constants. Unwinding of an interior residue requires simultaneous displacement of residues in the coil, so that larger frictional forces are involved. In addition, the coil region does not move as a rigid body, but its torsional motions are correlated to minimize dissipative effects. For example, examine the large displacements of Φ_{10} during the first simulation. With Φ_{11} and Φ_{12} in extended formations, the positive correlation of $\Delta\Phi_{10}$ and $\Delta\Phi_{11}$ tends to minimize the displacements of residues 14 and 15 when residue 13 moves, while the negative correlation of $\Delta\Phi_{10}$ and $\Delta\Phi_{12}$ also tends to minimize the displacement of residue 15 when residue 13 moves. Thus the movements seem to complement each other and appear controlled even though they are governed by diffusional force.

EVOLUTION OF TERTIARY STRUCTURE

The conservation of amino acid residues involved in a particular catalytic function is a well-documented phenomenon (e.g., the serine proteases). Similarly, homologies between sequences of similar proteins from various sources have been used to establish evolutionary relationships, as seen in the cytochrome c system. Comparison of sequences of superoxide dismutase from bacterial and animal sources has also given further insight into the evolution of mitochondria in animal systems. Such homology in the primary structure of proteins is not entirely unexpected.

However, where no particular catalytic function needs to be conserved, but only a given secondary structure such as a region of helix or pleated sheet, many more types of amino acid replacement are possible, and the conformation of a particular region of polypeptide chain may be conserved even though the sequences are not. For instance, as is indicated in Tables 10-10 and 10-11, a variety of amino acids can be interchanged in a helical region without disrupting the helical structure. Similarly, if a hydrophobic region on a protein needs to be conserved for, say, interaction with another macromolecule or a hydrophobic ligand, a number of amino acid residues can provide essentially similar hydrophobic properties. Yet in terms of specific catalytic functions, the possibilities of replacement are far fewer—threonine and serine may be interchanged in some instances, but the properties of histidine are unique.

In this section we discuss the structural homology within the class of proteins that bind adenine nucleotides—in particular the dehydrogenases; and we examine this structural homology (or in some cases the lack of particular elements of structural homology) in terms of the known enzymology of the systems. Most of the work discussed is x-ray crystallographic, but some sequence work is also included.

TABLE 10-10   Effects of residues on helix structures

Residues that stabilize helix structures: Alanine, leucine, phenylalanine, tyrosine, tryptophan, cysteine, methionine, histidine, asparagine, glutamine, valine

Residues that destabilize helix structures: Serine, isoleucine, threonine, glutamate, aspartate, lysine, arginine, glycine

Residues that break helix structures: Proline, hydroxyproline

---

TABLE 10-11   Roles of residues in protein and enzyme structure and function

Amino acid residue	Possible roles in protein and enzyme structure and function
Arginyl	Hydrophilic; electrostatic interactions
Lysyl	Hydrophilic; electrostatic interactions; attachment of prosthetic group or cofactor in amide bond; interacting to form a Schiff's base; ligand to metal ion
Histidyl	Hydrophilic or hydrophobic (depending on ionization); electrostatic interactions; proton transfer; ligand to metal ion; hydrogen bonding; acceptor in transfer reactions
Glutamyl, aspartyl	Hydrophilic; electrostatic interactions; proton transfer; ligand to metal ion; covalent linking in ester or amide through γ-carboxyl
Glutaminyl	Hydrophilic; hydrogen bonding
Asparaginyl	Hydrophilic; hydrogen bonding
Seryl	Hydrogen bonding; nucleophile; covalent linkage of OH in esters
Threonyl	Hydrogen bonding; nucleophile; covalent linkage of OH in esters; hydrophobic
Glycyl	Lock of side chain permits flexibility of folding and cross-hydrogen bonding
Alanyl, valyl lecucyl, isoleucyl, phenylalanyl	Hydrophobic interactions; determinants of steric and conformation specificity, e.g., numerous alanyl residues favor α-helix formation while numerous valyl or isoleucyl in sequence tend to inhibit formation of such features
Tyrosyl	Hydrophobic; hydrogen bonding; proton transfer; electrostatic interactions; at high pH; ligand to metal ions
Tryptophanyl	Hydrophobic; hydrogen bonding
Cysteinyl	Nucleophile; acyl acceptor; hydrogen bonding; ligand to metal ions
Cystyl	Cross-linking through disulfide bonds
Methionyl	Hydrophobic; hydrogen bonding to S (?); ligand to metal ions
Prolyl	Interrupts α and β structures allowing irregular conformation; hydrophobic

Ways of Assessing Structural Homology

Structural homology between two proteins (or peptides) is assessed by estimating the distances between equivalent backbone atoms of the molecule from their atomic coordinates. We will not discuss in any detail the mathematics of this procedure, but just qualitatively describe a few relevant aspects. Obviously, when two protein structures with closely homologous sequences or the same protein in two conformations are being compared, it is a fairly simple matter to decide which are the equivalent backbone atoms.

When there is no prior knowledge of atomic equivalence it is necessary to establish any structural homology purely on the basis of the tertiary structures. In practice this is usually done by taking a small region of, say, three to four peptides from one of the molecules and comparing it systematically with similar-sized blocks from the second molecule. The mean distance between supposedly corresponding atoms and the scatter from the mean is calculated. If a given group of atoms from molecule II has a scatter of less than, for example, 1 A from molecule I, the two blocks can be equivalenced. If the scatter is too high, the block of atoms for comparison is slid one residue down the chain on molecule I, and the procedure is repeated.

In the cases of the dehydrogenases that we will talk about, lactate dehydrogenase (LDH), malate dehydrogenase (MDH), horse liver alcohol dehydrogenase (ADH), and muscle glyceraldehyde-3-phosphate dehydrogenase (GA-3P-DH), the C atoms that were used in the previous type of procedure were selected from the common hydrogen-bonded scheme within the parallel pleated-sheet region of each of these enzymes. It has been found that when the hydrogen bonds of two different parallel pleated-sheet regions are aligned in an equivalent manner with the polarity of the strands being the same, the corresponding βC atoms are on the same side of the sheet. As a result, the alignment of residues within parallel pleated sheets is relatively simple once the positions of the βC atoms and the hydrogen-bonding scheme have been determined.

Dehydrogenases

Although there are very marked sequence correlations between dehydrogenases of similar function from different species—for instance, yeast and muscle GA-3P-DH, yeast and horse liver ADH, and bovine and chicken GDH—there are no really significant sequence similarities between dehydrogenases with different functions—for example, L-ADH and GDH. However, the crystallographic structures of this class of enzymes have revealed a very interesting pattern of structure–function correlations.

The three-dimensional structure of dogfish LDH was the first of the dehydrogenase structures to be obtained at high enough resolution to follow the chain in its entirety. In the original structure 331 amino acids were used, although it is now known that only 329 residues are present. The outstanding feature of the LDH structure was the region of the molecule that bound the pyridine nucleotide coenzyme, which consisted of residues 24 to 162 and appeared to be arranged in a series of β-pleated sheets connected by either helical regions or loops of polypeptide chain. In LDH there are six parallel strands of pleated sheet (labeled $\beta A \rightarrow \beta F$ in Fig. 10-23, which is a schematic representation of the coenzyme binding site). The βA strand consists of five residues, 24 to 28; the B strand, four residues, 49 to 52; the C strand, five residues, 92 to 96; the E strand, four residues, 135 to 138; and the F strand, also four residues, 159 to 162.

Subsequent to the structure of LDH, those of liver ADH (at 2.9 Å resolution) and lobster GA-3P-DH (at 3.0 Å resolution), as well as the structure of MDH at 3.0-Å resolution, were reported. In each the coenzyme binding domain was found to consist of a series of β-pleated sheets, as was the case with LDH.

Table 10-12 gives the sequence location and the number of residues in each of those six pleated-sheet regions in these three enzymes. Although the coenzyme binding domain is in the first half of the molecule in LDH and GA-3P-DH, it is in the second half in L-ADH. Also shown are the root-mean-square deviations between equivalent atoms, which gives an estimate of the structural homology between the compared regions. All three possible comparisons are shown in Table 10-12: ADH with LDH, ADH with GA-3P-DH, and LDH with GA-3P-DH. The distances between the 26 equivalent αC atoms used in the pleated-sheet regions are no greater than those found

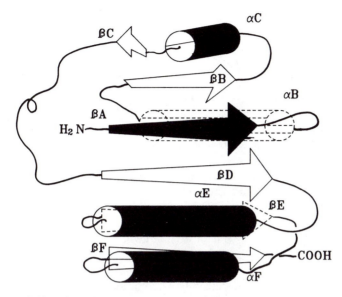

Figure 10-23 Schematic representation of the structure of the coenzyme binding domain in dehydrogenases.

in a comparison of identical structures such as subtilisin BPN and subtilisin novo. In fact, the errors in the measurements of atomic coordinates on which these distance estimates are based are about the same magnitude. The angle of twist of the pleated sheet (about 100°) is constant for each of the dehydrogenases, whereas pleated sheets have been found to have twists varying from almost 0° for concanavalin A to about 220° for carbonic anhydrase.

Despite this marked structural homology there is very little, if any, significant homology in the sequences of these three proteins in this nucleotide binding region. Over the entire domain, comprising approximately 120 residues, there are only five

TABLE 10-12 Correspondence between equivalent αC atoms used in the alignment of the nucleotide binding regions of different dehydrogenases

Structural element	Residue numbers			Root-mean-square deviation between equivalent atoms (Å)		
	ADH 1	LDH 2	GA-3P-DH 3	1–2	1–3	2–3
βA	195–199 (5)	24–28 (5)	3–7 (5)	0.97	0.94	1.05
βB	219–222 (4)	49–52 (4)	28–34 (4)	1.33	1.54	0.89
βC	238–241 (4)	78–81 (4)	71–74 (4)	1.10	2.61	2.04
βD	263–267 (5)	92–96 (5)	90–94 (5)	1.12	1.32	1.30
βE	289–292 (4)	135–138 (4)	116–119 (4)	0.87	1.85	2.04
βF	313–316 (4)	159–162 (4)	142–145 (4)	1.34	1.17	1.21

residues common to all three sequences. In particular, three of these residues, Gly-28, Gly-33, and Asp-53 (in the LDH sequence), lie in the central region of the nucleotide binding domain and have particular functions in coenzyme binding. The outer regions of the domain, in particular α-helix IF and the loop between βC and βD, have significantly different structures in the three dehydrogenases; it appears that for correct coenzyme binding there are no stringent requirements for the outer regions of the domain.

These three invariant residues just mentioned each have a particular functional significance. If residue 28 (in LDH) has a β-carbon atom, this position would overlap with the C_2 position in the ribose of the coenzyme, preventing the adenosine part of the coenzyme from binding in the correct orientation. There appears to be an absolute requirement for a glycine residue in this position.

One of the carboxyl oxygens of Asp-53 forms a hydrogen bond to the oxygen of the adenine ribose, which is apparently important in coenzyme binding. It might be noted that in NADP the extra phosphate group is attached to the oxygen and the hydrogen bond cannot then form. This interaction may well be important in determining the specificity for NAD of these dehydrogenases.

The remaining invariant glycine, Gly-33, although not involved in the coenzyme binding area, does seem to be important for structural reasons. A β-carbon atom in this position would point from helix αB toward the pleated-sheet and would thus interfere with the main chain of the sheet.

So far we have discussed the structural similarities between various dehydrogenases. There are, of course, differences and we can attempt to relate these differences to some aspects of their enzymology.

In LDH (and MDH), a 20-residue chain connects the D and E strands of the β-pleated sheet. This loop undergoes a marked change when the coenzyme, or various adenine-containing derivatives, are bound. In L-ADH only three residues are used to join these strands. Similarly, in GA-3P-DH this loop region is absent. It has been suggested that in LDH (or MDH) the main trigger for the conformational change is an interaction between the adenosine phosphate and an arginine residue in this loop region. Since in L-ADH and GA-3P-DH this region is missing, it was suggested that conformational changes induced by coenzyme binding in these proteins may involve the nicotinamide moiety. Very recent studies with rabbit muscle GA-3P-DH have demonstrated that the conformational change in this enzyme associated with the negative cooperativity in coenzyme binding does indeed require the nicotinamide moiety. Unlike the conformational changes observed in LDH or MDH, ADPR does *not* induce the conformational change. Since cooperative phenomena have been reported in GA-3P-DH as well as ADH, it may be that the *loss* of this loop region between the βD and the βE regions of the pleated sheet is related to the evolution of cooperative phenomena in the dehydrogenases.

Related Proteins

So far we have discussed the dehydrogenases whose crystal structures have been determined. There are, however, a number of other proteins that bind adenine nucleotides whose crystal structures have also been determined.

The crystal structure of flavodoxin, a protein that binds flavin mononucleotide, indicates that the nucleotide binding site is similar in structure to that used to bind a nucleotide in LDH or the other dehydrogenases. This structural homology can be seen in diagonal plots (Fig. 10-24) representing distances between α-carbon atoms as a function of residue number for flavodoxin and LDH. The contour plots represent given αC–αC distances and uniquely reflect the structure of the protein backbone. Clearly, there is quite a marked similarity in the conformation of the FMN binding site in flavodoxin and either half of the dinucleotide binding domain of LDH. Also, from these diagonal plots it can be seen that the remainder of the LDH molecule bears *no* structural resemblance to the coenzyme binding domain.

Recently, similar nucleotide binding structures have been recognized in phosphoglycerate kinase and adenylate kinase. A structure very similar to the nucleotide binding domain of flavodoxin has also recently been identified in rhodanese, a protein whose biological function has not yet been completely established, but which is known to bind FAD and FMN as well as NAD.

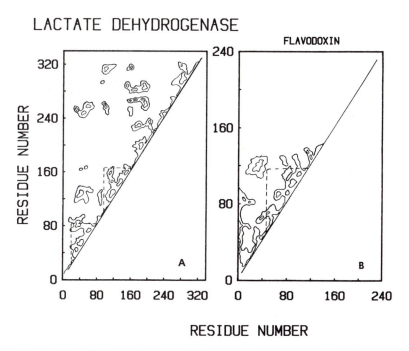

Figure 10-24 (A) Diagonal plot for LDH representing distances between C atoms. Only major contours are shown. Different parts of the structure have been identified along the diagonal. The first two structural domains between which there is less contact are marked in a way that emphasizes the comparison between them. (B) Diagonal plot for flavin. The marked domain in the central part of the polypeptide chain has a structural and functional similarity to each of the first two domains of LDH.

It also appears possible that a structure similar to the nucleotide binding region of the dehydrogenases is present to a lesser extent in some other proteins. For instance, there is a region in subtilisin, residues 121 to 181, that in some ways resembles the mononucleotide binding domain of LDH, residues 22 to 81, which contains βA, βB, and βC regions of the coenzyme domain. Note that in this region the hydrophobic pocket which binds the aromatic side chain of the subtilisin substrate, between Gly-127 and Gly-154, is in a similar position to the hydrophobic pocket that binds adenine in LDH, containing Val-27 and Val-54.

Evolutionary Significance

The retention of an essentially identical fold, despite the wide variety of sequences, demonstrates that there is a stability of tertiary structure even in cases where there is no apparent conservation of primary structure. That these observations are a chance occurrence is unlikely, as such a large number of amino acid residues are involved. Also, it seems unlikely that these large structural similarities are the result of convergent evolution from different precursor molecules, especially as alternative structures of nucleotide binding proteins are known to exist: for instance, ribonuclease or staphylococcal nuclease. Thus it seems that this class of nucleotide binding proteins represents an example of divergent evolution.

If this is the case, a relative estimate of the time elapsed since divergence from a common ancestor can be made using as a measure of evolutionary distance the

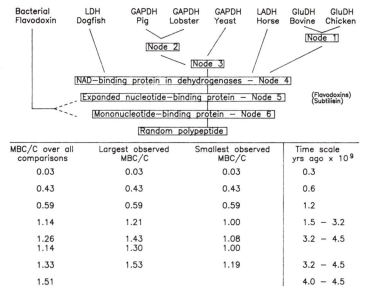

MBC/C over all comparisons	Largest observed MBC/C	Smallest observed MBC/C	Time scale yrs ago x 10^9
0.03	0.03	0.03	0.3
0.43	0.43	0.43	0.6
0.59	0.59	0.59	1.2
1.14	1.21	1.00	1.5 − 3.2
1.26	1.43	1.08	3.2 − 4.5
1.14	1.30	1.00	
1.33	1.53	1.19	3.2 − 4.5
1.51			4.0 − 4.5

Figure 10-25 Evolutionary tree with observed minimum base changes per codon (MBC/C) and possible time scale derived primarily from fossil data.

minimum base change per codon. In the cases of the nucleotide binding proteins where the homology is based on a structural similarity in a given domain, Rossmann and co-workers have obtained an evolutionary tree for that region of each of the considered proteins that is involved in nucleotide binding. This is given in Fig. 10-25.

One point which might be made is that if one considers pig and yeast glyceraldehyde-3-phosphate dehydrogenase, the observed minimum base change per codon in the nucleotide binding domain is 0.59, while for the remainder of the molecules the value is only 0.35. Presumably the differing functions of the two domains within a single polypeptide chain imposes different rates of evolution.

One might also speculate that the dehydrogenases and related nucleotide binding proteins evolved by a process of gene fusion; that is, there is one gene for the nucleotide binding domain common to all dehydrogenases and related proteins that has fused with one gene for a substrate binding domain different for each of these proteins. This is a particularly attractive hypothesis when one remembers that in the case of phosphoglycerate kinase, the substrate binding and nucleotide binding domains are clearly separated in space as well as in sequence.

ACQUISITION OF TERTIARY STRUCTURE

The discussion thus far has been aimed at giving a picture of the tertiary structure of a protein as being in a dynamic state. From this we can rationalize the differing mechanisms of protein folding examined in Chap. 10. With small proteins the dynamic nature of the final tertiary structure permits the folding process to be essentially pathway independent and the "final" structure can be reached rapidly. With proteins, where there is evidence for pathway-dependent folding, the dynamic nature of the process, as well as the final conformation, offers a mechanism by which a substrate, for example, can stabilize an intermediate and "direct" the folding process.

Protein unfolding–folding is often considered in terms of a two-stage process for small proteins that can be renatured after denaturation. This implies that there are no stable intermediates in the process. As more sophisticated approaches for detecting such intermediates become available it is apparent that even in small proteins such as bovine pancreatic trypsin inhibitor (BPTI) or ribonuclease (RNase), various intermediates can be identified. In BPTI, the fully reduced protein appears completely unfolded: When refolding occurs it is found that of 15 possible species containing only a single disulfide, only two are significantly populated, and they are in rapid equilibrium with one another. The major species involves the Cys-30–Cys-51 disulfide bond found in the native protein. Disulfide bonds also play a significant role in stabilizing some protein structures. In RNase, the correct disulfide bonds are formed when the denatured protein is renatured. However, when four residues are removed from the C-terminal end, the correct disulfide bonds are not reformed.

In many proteins when renaturation is attempted it is found that regain of the native structure is a multiphasic process. In cytochrome c, refolding after guanidine hydrochloride–induced denaturation can be followed by monitoring tryptophan fluorescence (which in the native protein is quenched by heme interaction), absorbance at 287 nm (from the five tyrosines in the molecule), or by absorbance at 695 nm (which monitors the heme ligation in the native protein). Over short time scales these three monitors of conformation reflect similar processes. However, at longer time scales more processes are observed with the absorbance measurements at 287 nm than with the tryptophan fluorescence, suggesting that slower processes are occurring, affecting tyrosine environments in the protein. Proline peptide bonds can undergo slow cis-trans isomerizations that appear to be important for these slower processes in cytochrome c refolding. This cis-trans equilibrium constant is near 1 and is affected by the local environment in terms of the primary sequence. Proline isomerization has a characteristic activation energy around 20 kcal/mol, and its pH dependence is governed by the presence of ionizable groups near the peptide bond. Such isomerizations may well be responsible for the slow "refinements" of the folded structure of a protein observed in many instances.

In this chapter and Chap. 9 we examined aspects of secondary and tertiary structure, emphasizing the dynamic nature of protein structure together with the importance of long-range interactions in governing final form. These long-range interactions play a pivotal role in the folding pathway as well as influencing local secondary structure. In addition, as will be discussed in Chaps. 11 and 12, they are intimately involved in the assembly of quaternary structure in oligomeric proteins and in the mechanism of conformational changes in allosteric proteins.

11

Quaternary Structure

INTRODUCTION

The subunit nature of many proteins was first demonstrated by the ultracentrifugation experiments of Svedberg. In studies of the molecular weights of a variety of proteins, he observed that in some cases the apparent molecular weight changed, depending on the nature or concentration of the solvent or other physical conditions that did *not* lead to polypeptide cleavage. This led to the suggestion that some proteins may be made up of aggregates of smaller proteins. Noticing that these smaller "subunits" tended to group into defined categories led to the hypothesis that all proteins were in fact assembled from a limited number of molecular weight class subunits. In the light of later knowledge this is obviously erroneous: The generalization that large proteins are built up from smaller entities does, however, remain valid. (It might be noted that in many instances there is evidence that the original suggestion of a limited number of subunits retains some appeal at the level of tertiary-structure domains rather than at the level of subunits. For example, the adenine binding domain which appears to be common to the dehydrogenases and other classes of proteins that bind adenine nucleotides may well have evolved from a common ancestral gene coding for a nucleotide binding site which subsequently became incorporated into the wide variety of proteins that make up this group).

In 1958, Bernal first adopted the designation of quaternary structure to describe a macromolecular system built up from noncovalently linked subunits. A scheme was proposed by which such subunit assemblies could be described consisting of the following designations.

Stoichiometry: A description of the types and number of each sort of subunit involved in the quaternary structure. Proteins can be classified as homo-oligomers, made up of a single type of polypeptide chain, or hetero-oligomers, made up of two or more types.

Architecture: A description of the geometric arrangements of the subunits and of the types of symmetry found in the quaternary structure.

Assembly: A description of the energetics of subunit interactions within a quaternary structure and of the nature of the interface contacts between the subunits.

Inter-subunit Communications: A detailing of the ways that conformational changes within one subunit of an oligomer affect the conformations of other subunits within the same oligomer.

Functional Aspects of Subunit Proteins: A natural corollary to the last point. What is the need for subunits in oligomeric proteins, and how do inter-subunit communications affect the biological functioning of the entire oligomer? An extension of the latter point is the question of multienzyme complexes, where individual "subunits" in an oligomer are involved in quite separate (though often related in some metabolic pathway) chemical processes.

STOICHIOMETRY

The determination of the types and number of subunits within an oligomer might seem to be quite straightforward. Intellectually, of course, it is; one establishes the number of chemically distinct polypeptide chains and how many of each are present. Practically, however, such an endeavor may be difficult. Initially, one must define the oligomer; in many instances this presents the most problems. Mammalian glutamate dehydrogenase is now regarded as a hexameric enzyme of molecular weight 336,000 Da. For many years the molecular weight could not be accurately determined because the hexamer unit undergoes a concentration-dependent polymerization, and there were questions as to whether the *active* form of the enzyme was a hexamer or some higher oligomer. The issue was finally resolved with the advent of reacting enzyme centrifugation, which showed that the minimum unit capable of catalyzing the oxidative deamination of glutamate was the hexamer. Subsequent studies demonstrated that the concentration-dependent polymerization of hexamers did not affect the enzymatic activity. Perhaps the simplest definition of quaternary structure is the size of the minimum unit capable of catalyzing the reaction supported by the protein. The demonstration that a number of proteins contain regulatory subunits which themselves have no catalytic activity but are an integral part of the native oligomer suggests that this definition should be modified to "the size of the minimum unit having the biological properties of the protein." Even with this definition, care must be taken that regulatory subunits are not lost during purification and thus go unrecognized.

A second problem arises from the possibilities of proteolytic degradation of some subunits during isolation, generating what might appear to be multiple-subunit types in an oligomer which in reality consists of only a single type of polypeptide chain.

This is illustrated by the example of bovine galactosyltransferase, a protein that is conveniently purified by a variety of affinity chromatography procedures, giving a number of bands on SDS-PAGE with molecular weights ranging from 42,000 to 50,000. These bands copurify through all types of purification procedures and all are active, suggesting an oligomeric enzyme with dissimilar but related subunits. When care is taken to add protease inhibitors at the early steps in the purification, and to maintain their presence throughout, a single higher-molecular-weight band is isolated, indicating that rather than consisting of dissimilar subunits, this galactosyltransferase contains a single polypeptide chain. Subsequently, it was shown that the native enzyme is a monomer.

A variety of experimental techniques may be used in establishing the stoichiometry of an oligomer. In many instances these involve molecular weight determinations under native and denaturing conditions, with simple inference of the stoichiometry. In some cases the determination of the native molecular weight may present problems other than those just examined for glutamate dehydrogenase. Proteins such as yeast hexokinase or glyceraldehyde-3-phosphate dehydrogenase have oligomers that are dissociated by their substrates or regulatory ligands. More accurately, one should say that the oligomers exist in equilibrium with the subunits and, for example, glucose in the case of hexokinase, or ATP in the case of glyceraldehyde-3-phosphate dehydrogenase, displace the equilibrium toward the subunits. In other systems—for example, cytidine triphosphate synthetase (ATP or UTP) or homoserine dehydrogenase/aspartokinase (threonine, isoleucine, methionine, and divalent metal ions)—substrates or regulatory ligands displace the equilibrium toward the oligomeric form. In either instance the determination of the molecular weight under "native" (usually meaning nondenaturing) conditions can give results indicative of multiple species.

In bovine glutamate dehydrogenase an even more complex situation holds. The protein undergoes a concentration-dependent polymerization that can be affected towards either the hexamer form (the "monomer" in the concentration-dependent polymerization) or higher-molecular-weight forms. The polymerization can be displaced in either direction by various substrates or regulators. The adenine nucleotides ADP and ATP enhance polymerization, while the coenzyme NADPH or the inhibitor GTP increase dissociation. In a number of instances an oligomeric form of an enzyme can be made to dissociate into its constituent subunits by application of Le Châtelier's principle: Dilution leads to depolymerization. Examples of this type of behavior include enolase, glyceraldehyde-3-phosphate dehydrogenase, and tryptophanase, where in each case dilution yields increased amounts of monomers.

A number of chemical approaches have been successfully used to investigate subunit composition. These can be divided into two groups: (1) chemical modifications to *increase* dissociation of subunits, and (2) chemical cross-linking reactions to covalently "lock" an oligomer into its maximal molecular weight.

Various proteins—for example, glyceraldehyde-3-phosphate dehydrogenase and pyruvate carboxylase—dissociate into subunits when their sulfhydryl groups are

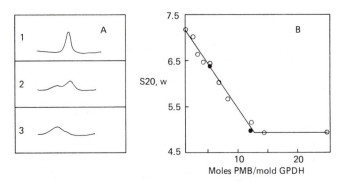

Figure 11-1 (A) Sedimentation velocity patterns of glyceraldehyde-3-phosphate dehydrogenase in the native form (panel 1), with partial modification (panel 2) and with
complete modification (panel 3) by pHMB; (B) dependence of $S_{20,w}$ on moles of
pHMB per mole of protein. (Adapted with permission from: G. D. Smith and H. K.
Schachman, *Biochemistry*, *10*, 4576–4588. Copyright 1971 American Chemical Society, Washington, D. C.)

altered by chemical modification. Rabbit muscle glyceraldehyde-3-phosphate dehydrogenase, which sediments as a tetramer (Fig. 11-1A, panel 1), is dissociated by
modification with *p*-hydroxymercuribenzoate (*p*-HMB). At intermediate levels of
modification a mixture of monomer and tetramer is seen (panel 2), while at higher
levels the tetramer is completely dissociated to the monomer (panel 3). As shown in
Fig. 11-1B, there is a linear dependence of the change in the sedimentation coefficient
on the modification reagent concentration. In addition, the presence of coenzyme
does not affect this induced dissociation of the tetramer.

Although the actual basis for such effects has not been established, they presumably involve conformational changes induced in the proteins by the modification,
which leads to decreased interactions between subunits and hence increased dissociation into the constituent subunits. More easily explained is the increased dissociation
produced by acylation of reactive lysine residues in a variety of subunit-containing
proteins by dicarboxylic anhydrides such as succinic anhydride, maleic anhydride,
or tetrahydrophthalic anhydride, all of which result in the introduction of negative
charges in place of the formal positive charges of the lysine side chains. This presumably leads to increased electrostatic repulsion between the subunits, which, depending
on the stability of subunit interactions, may lead to increased dissociation. Simple
determination of molecular weight before and after such modifications may, however,
be misleading since it is possible that despite modification of each polypeptide chain
in the oligomer by such reagents, dissociation is only prompted to some intermediate
stage rather than to the level of individual subunits. As will be discussed later, such
information can be invaluable in studying the architecture of an oligomer.

Chemical cross-linking studies, using the types of reagents described in Chap. 7,
can be most useful in establishing subunit composition. In principle one can get cross-
links occurring (1) between polypeptide chains within an oligomer, (2) within a single

polypeptide chain in the oligomer, and (3) between polypeptide chains in different oligomers. Clearly, this latter situation is not desirable and can be controlled against by making use of Le Châtelier's principle to overcome oligomer polymerization. In addition to the already mentioned effects of cross-linking, monofunctional reaction may be obtained which, although it may alter enzymatic parameters, does not effect the overall molecular weight distribution. In essence, such studies make use of the molecular weight of the highest cross-linked aggregate to indicate that of the oligomer. In conjunction with the polypeptide-chain molecular weight the stoichiometry is determined; with homopolymers unequivocal results can be obtained. With hetero-polymers, however, it is possible that confusion can arise if the dissimilar subunits have molecular weights not well separated from one another. In such cases chemical cross-linking with cleavable cross-linking reagents, followed by diagonal mapping using SDS-PAGE with cleavage of the cross-links between the dimensions, may be of use.

Perhaps the most useful approach for establishing subunit composition is *hybridization*. The essence is to mix variants of the same protein under conditions where interchange of subunits can occur and then to determine the number of hybrid forms obtained. In a protein containing S subunits it can readily be shown that one obtains $S + 1$ hybrid forms of the oligomer. Although the situation is slightly more complex when the oligomer contains dissimilar subunits, the number of hybrid forms (N) is given by

$$N = \frac{[m + (S - 1)]!}{S!\,(m - 1)} \tag{11-1}$$

where S is the total number of subunits in the oligomer and m is the number of subunit *types* in the oligomer.

A number of important experimental criteria are involved in this approach. The hybrids formed must be experimentally distinguishable. Usually, charge differences are used and the hybrids separated by electrophoresis (either native or isoelectric focusing) or ion-exchange chromatography. With lactate dehydrogenase, two homogeneous isoenzymic forms are available which differ by charge. When the isoenzymes are hybridized (Fig. 11-2), five bands are obtained, indicating that each isoenzyme is

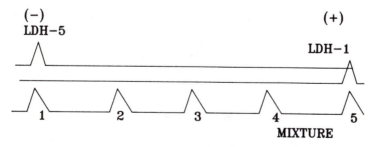

Figure 11-2   Gel scan of LDH-1 and LDH-5 isoenzymes and the hybrids obtained from an equal mixture of LDH-1 and LDH-5 after dissociation and recombination.

tetrameric. Such hybridization may also use genetic or species differences of proteins having different charges.

Yeast and rabbit muscle glyceraldehyde-3-phosphate dehydrogenase do hybridize, giving five hybrid bands constituent with a tetrameric structure. Hemoglobin variants have been hybridized and the hybrids analyzed. In this instance, however, the results suggest that the tetramer dissociates only into dimers taking part in the hybridization. Furthermore, they indicate that the dimers were $\alpha\beta$ dimers rather than $\alpha\alpha + \beta\beta$. Where genetic, species, or isoenzymic forms of the protein are not available, it is possible to use this approach by generating molecular isomers through chemical modification. The modification procedure should be selected to change the charge of the polypeptide chains in one form, allowing easy separation of the hybrids. This method has been used with glutamate dehydrogenase. Tetrahydrophthalic anhydride modification leads to incorporation of negative charge on the subunits of the enzyme. When such a modified enzyme and unmodified native enzyme are mixed in the presence of low concentrations of guanidine hydrochloride removed by dilution, hybrids containing a native trimer and a negatively charged trimer are obtained. This indicates that the hexamer dissociates into two trimers. The hybrid forms in this example were separated by ion-exchange chromatography on DEAE-Sephacel.

The elution profile (Fig. 11-3) achieved with a salt gradient has three peaks, corresponding to the native hexamer, the hetero-hexamer (consisting of one native trimer and one acylated trimer) and the fully modified hexamer. This pattern is consistent with the predictions for a "dimer"-type system where one expects to obtain $S + 1$ hybrid forms. The trimer does not dissociate into individual polypeptide chains under these conditions. Implicit in the previous discussion is the fact that for this approach to be applied, the subunits in the oligomers must be reversibly dissociated and reassociated. This often requires some sort of reversible denaturation. The two

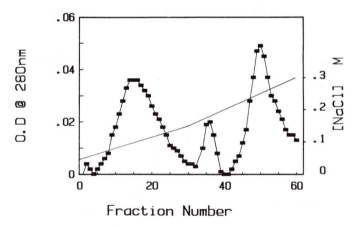

Figure 11-3   DEAE chromatography of GDH hybrids resulting from chemical modification.

parent forms of the protein are mixed, denatured, and then renatured prior to separation of the hybrids (the fact that it is possible to *reversibly* denature the hexamer of glutamate dehydrogenase only to the trimer stage and not to the monomer stage explains the previous observation that the hexamer behaves as a dimer in such experiments; this is an important limitation). Two other restrictions must be placed on this approach. First, it is essential that the association–dissociation equilibrium is *slow* relative to the time it takes to separate the hybrids, and second, it is important that complete equilibration among the variant forms occurs prior to separation.

Finally, before considering the geometric arrangement of subunits in oligomers, some comment must be made on the range of subunit compositions found in oligomeric proteins. Table 11-1 indicates some typical oligomeric enzymes, both homo- and hetero-oligomers, and their established subunit composition. Most proteins that contain subunits have two, four, or six, and almost all oligomeric proteins contain an *even* number. Excluded are multienzyme complexes, which must be dealt with separately. It would seem that there may be several exceptions to this generalization. Rat liver adenylate kinase appears to consist of three polypeptide chains of 23,000 Da

TABLE 11-1   Subunit composition of some oligomeric proteins

Protein	Mol. wt. (kDa)	Number of subunits	Subunit mol. wt. (kDa)
Homopolymers			
Azoferredoxin	55	2	27.5
Prealbumin	62	4	15.5
Malate DH	66.3	2	37.5
Glycerol-1-PO$_4$	78	2	40
Alcohol DH	80	2	41
D-Amino acid oxidase	100	2	50
Glycerol-6-PO$_4$DH	130	2	63
Phosphofructokinase	145	4	35
Glycerol kinase	217	4	55
Glutamate DH	336	6	56
Apoferritin	460	24	18.5
Thyroglobulin	669	2	335
Heteropolymers			
Lactose synthase	64	2	50, 14
Histidine decarboxylase	190	10	9 (5), 30 (5)
Aspartate *trans*-carbamylase	310	12	17 (6), 34 (6)
F$_1$ ATPase	380	9	55 (3), 50 (3), 31 (1), 19.5 (1), 15 (1)

each; 2-keto-3-deoxy-6-phosphogluconate kinase also appears to contain three poly-peptide chains [24 kilodaltons (kDa) each], and histidine decarboxylase, while it does contain a total of 10 polypeptide chains, seems to contain five subunits of 9000 Da and five subunits of 30,000 Da, giving an oligomer of approximately 190,000 Da. The latter case is by no means the largest oligomeric protein known: Apoferritin has 24 subunits of 18,500 Da, each while erythrocruorin contains 162 subunits of 18,500 Da each.

The chemical identity of subunits within the oligomer may need to be established. The observation of two types having quite distinct properties leads to the conclusion that the oligomer contains dissimilar subunits. The converse situation—the obser-vation of a single molecular weight of subunits—is not so easily interpreted. It is possible that two chemically dissimilar subunits have similar molecular weights. Sep-aration by isoelectric focusing as well as by SDS-PAGE can resolve this problem. Identification of terminal amino acid residues may also distinguish identical subunits from dissimilar subunits. However, it is clearly possible that these approaches may still indicate a single type of subunit when in fact, dissimilar ones are present. Amino acid analysis and tryptic mapping represent the most declarative solution to this problem. As shown in Fig. 11-4, if the total number (N) of lysine + arginine for a protein sample is known, $N + 1$ peptides are obtained on a tryptic map, assuming

TRYPTIC MAPPING TO CHECK PURITY

1. Amino acid analysis to get total
   lys+arg per mole of protein
2. Total Tryptic Digest & 2D Map
3. Count Peptides on Map
4. If n+1 then protein is pure
   n=(lys + arg)

IF MIXTURE OF PROTEINS

Protein 1

N ～ lys ── arg ── arg ── lys ── lys ── arg ── lys ── arg ─ C

Protein 2

N ──── lys ──── arg ──── arg ──── lys ── C

Mixture gives an average (lys+arg) of 6 per mole

Would expect to get 7 peptides if pure

In reality will find up to 14 Peptides

Figure 11-4 Scheme of tryptic mapping of a mixture of proteins.

that *all* the molecules in the sample are identical. If two or more different types of polypeptide chain are present, greater than $N + 1$ peptides on the map are obtained.

ARCHITECTURE

An oligomer consisting of a definite number of subunits could have any one of a finite number of spatial arrangements. With some oligomeric proteins it is possible that the molecular architecture can be observed directly using x-ray crystallography or electron microscopy. In other cases such approaches may not be possible and alternative techniques must be used to infer an architecture.

Clearly, the number of possible arrangements a molecule can have increases with the number of subunits. However, they can be limited by two assumptions which appear to hold in many cases (it is important to note, however, that there are exceptions to these assumptions and that they can have significant implications for some considerations of the functional aspects of oligomeric proteins; these are discussed as appropriate).

Assumption 1: *All subunits are in equivalent, or pseudo-equivalent environments.* Strictly speaking, this applies only to homo-oligomers and not to oligomers containing dissimilar subunits. In practice, hetero-oligomers can frequently be considered as consisting of an oligomer of $\alpha\beta$ building blocks, where α represents one type of subunit and β represents a different type. Each $\alpha\beta$ block would be in an equivalent or pseudo-equivalent environment to all other $\alpha\beta$ blocks in the hetero-oligomer.

Assumption 2: *The bonding potentials of the subunits must be saturated.* As a direct consequence of the first assumption, one can disregard linear arrays of subunits or variants of the idea of linear arrays. The second assumption limits the categories of subunit architecture that one must consider to closed sets for homo-oligomers or hetero-oligomers. In the absence of this second assumption, aggregates containing multiple oligomers could be obtained. Such is the case in sickle-cell hemoglobin, where the substitution of a valine for a glutamate at the B_6 position creates an additional bonding potential on the β subunits in the tetramer that is not satisfied by the normal subunit–subunit interactions; the result is an uncontrolled polymerization of the tetramer and resultant precipitation of the variant hemoglobin. Another example, found normally, is the case of glutamate dehydrogenase, a hexamer, but one that undergoes a concentration-dependent polymerization:

$$n(\text{GDH hexamer}) \longleftrightarrow (\text{GDH hexamer})_n \qquad (11\text{-}2)$$

Although this is very well documented and is affected by a variety of ligands and substrates for the enzyme, it appears to have no enzymatic function. Because the polymerization results in long linear arrays of hexamers it seems that the hexamer has some unsaturated bonding potential at its apexes, which results in its ability to undergo this polymerization.

The result of both assumptions is that the architecture of the oligomers is restricted to those that have the subunits arranged in a regular fashion about a central point: *Only point symmetry is possible.* Point groups contain rotation axes and inversion axes; however, for polypeptide chains, which contain *L*-amino acids, inversions are not allowed (these would lead to enantiomorphic differences) and only *rotation axes* passing through the point need to be considered. There are three types of point symmetry that we consider: cyclic symmetry, dihedral symmetry, and cubic symmetry.

Cyclic Symmetry

In a system showing cyclic symmetry (which is the simplest type) there is an *n*-fold rotational axis, where *n* = the number of subunits in the oligomer. As illustrated by Fig. 11-5, an *n*-fold rotational axis indicates that the molecule must be rotated about a point by $360/n°$ to effect a transposition into an equivalent environment for the subunits.

For a *dimer* (where *n* = 2) there is a single two-fold rotation axis and the molecule is said to show C_2 symmetry. A rotation by $360/2 = 180°$ transposes the molecule into itself again. With a trimer (where *n* = 3) there is a single three-fold axis (the

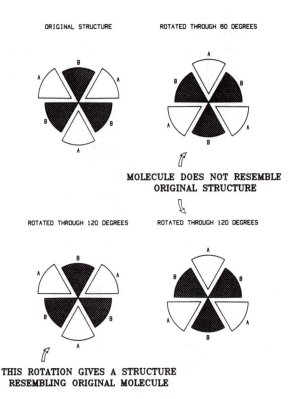

Figure 11-5   Point-symmetry operations.

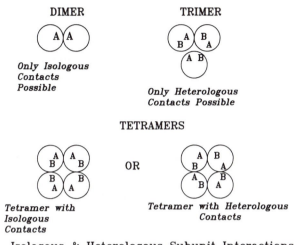

Figure 11-6   Isologous and heterologous subunit interactions for (A) a dimer, (B) a trimer, and (C) a tetramer.

molecule has C_3 symmetry) and a rotation about a point of $360/3 = 120°$ results in an equivalent molecule. C_2 and C_3 symmetries are the only geometries possible for dimers and trimers, respectively, that result in identical environments for the subunits.

In considering cyclic symmetry for dimers or trimers (Fig. 11-6) an important point is apparent. In a dimer the contact sites between the subunits are identical in both subunits—the subunits are said to have *isologous bonding*. In the trimer, however, such a situation is clearly not possible—the subunit contact regions between any two subunits are not identical and the intersubunit bonding is *heterologous*: Any oligomer containing an odd number of subunits must have heterologous bonding. In a system with an even number of subunits (>2), either isologous or heterologous bonding can exist (to maintain the condition of identity in subunit environments, however, a single oligomer cannot have isologous *and* heterologous contacts). Consider the case of a tetramer—in cyclic symmetry there are two ways of orienting the subunits that fulfill both assumptions: one has two types of isologous bonding and the other heterologous bonding.

Dihedral Symmetry

If a molecule has n twofold axes at right angles to a single n-fold axis, it is said to exhibit dihedral symmetry. In such cases the number of subunits must be equal to $2n$. As a result, only oligomers with an even number of subunits can have dihedral symmetry.

The simplest and most common form of dihedral symmetry is *tetrahedral*, where the oligomer contains four subunits and is said to have D_2 symmetry. In tetrahedral

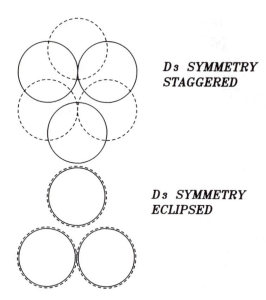

**D₃ SYMMETRY
STAGGERED**

**D₃ SYMMETRY
ECLIPSED**

Figure 11-7   D-3 symmetries of a hexamer.

symmetry the subunits have isologous bonding. For a simple molecule such as methane (CH_4) one gets cubic symmetry with four threefold axes and three twofold axes. However, with proteins containing only L-amino acids (which as a result are inherently asymmetric), the threefold axes cannot exist, and as a result one gets three twofold axes.

With six subunits (i.e., $n = 3$ for dihedral symmetry) there are two possible non-cyclical architectures, both of which are D_3 symmetry. These are represented in Fig. 11-7 and are the *trigonal prism*, which has an eclipsed orientation of the two apparent trimers that make up the hexamer, and an *octahedron*, which has a staggered orientation of the two apparent trimers. In addition one can also get cyclic symmetry in the form of a hexagon, which has C_6 symmetry.

Cubic Symmetry

In addition to quaternary structures with cyclic or dihedral symmetry, oligomers may have cubic symmetry. However, for an oligomer to exhibit this, it must possess $12n$ subunits; as a result, not many have been described showing cubic symmetry. It is an important symmetry, however, in macromolecular assemblies such as the spherical viruses.

Before considering some experimental approaches to establishing the symmetry of an oligomer there are several comments to be made about subunit arrangements and the types of bonding between subunits.

Earlier it was indicated that very few, if any, oligomeric proteins had odd numbers of subunits, yet we have considered symmetries for trimers. This was done because in a number of instances both homo- and hetero-oligomers overall have an even number of subunits but have substructures containing elements with odd numbers of subunits. Aspartate transcarbamylase serves as an example of a hetero-oligomer. It contains a total of 12 subunits, but the architecture of the oligomer indicates that the six catalytic subunits are arranged in two trimers separated from one another by the regulatory subunits. Glutamate dehydrogenase, a hexamer of identical polypeptide chains, appears to have D3 symmetry but consists of two trimers—a dimer of trimers. These considerations are important when examining subunit–subunit interactions since the architecture of the oligomer indicates something about the nature of the contacts between subunits.

If one considers the bonding between subunits in an oligomer, there is no thermodynamic reason to a priori prefer isologous or heterologous bonding. The driving force for the association of subunits into an oligomer *must* come from increased stability due to the formation of inter-subunit bonds. As a result, one might expect the spatial architectures giving the largest numbers of inter-subunit contacts to be the most stable. This would result in a tetrahedral architecture being the most stable form for a tetramer and an octahedral architecture being optimal for a hexamer. Although these are common architectures (thus supporting the idea), there are exceptions. This results from the fact that the energy of interaction in the oligomer contains an intensity factor as well as a quantity factor. Finally, it must be noted that in cyclic symmetry the inter-subunit contacts are *all identical*, whereas in dihedral or cubic symmetry one can get several different bonding regions within a subunit.

EXPERIMENTAL APPROACHES

As discussed previously, direct establishment of molecular architecture by means of x-ray crystallography or electron microscopy gives an unequivocal answer, but it is often not possible or is laborious. A wide variety of other experimental approaches have been used in various systems to give information about molecular architecture.

One of the most used comes from assumption 2. In certain symmetries one finds different types of subunit–subunit interactions within the oligomer. In a tetramer with D_2 symmetry there is one type of dimer bond and two types of tetramer bonds. As a result of the intensity factor discussed, one might expect that a relatively stable dimer could be produced in such a situation. Thus an idea of the tetramer symmetry can be established if the dissociation stages of the oligomer can be identified. This approach has been used in a number of cases, such as hemoglobin, glyceraldehyde-3-phosphate dehydrogenase, aspartate transcarbamylase, and glutamate dehydrogenase. In the former two cases stable dimers have been identified, while in the latter trimer intermediates are found, indicating D_2 or D_3 symmetry respectively. In addition to assisting in the assignment of the molecular architecture, such studies have inherent implications for the strength of different contact regions between subunits in oligomers. This is discussed in more detail later in this chapter.

Chemical cross-linking studies may also be useful in distinguishing between different molecular architectures. In studies with glutamate dehydrogenase the subunits can be covalently cross-linked with bifunctional reagents such as dimethylpimelimidate. If the cross-linked subunits within the oligomer can be experimentally quantitated by, for example, SDS-PAGE, it is possible from computer modeling studies to distinguish between C_6 and D_3 symmetries. Equations giving the probabilities of the distribution of the monomers, dimers, trimers, tetramers, pentamers, and hexamers that from random cross-linking of the subunits in D_3 or C_6 symmetries can be

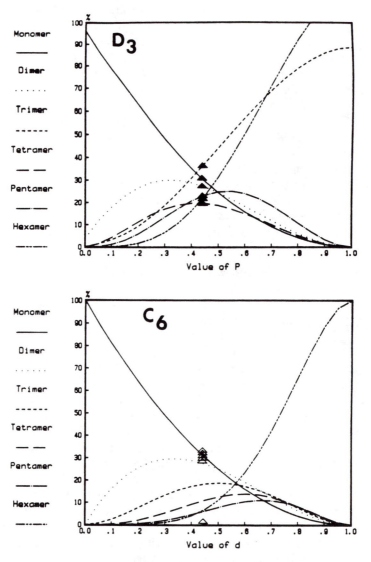

Figure 11-8 Comparison of experimentally observed distribution of cross-linked species of glutamate dehydrogenase with predicted patterns for C-6 or D-3 symmetries.

derived. With D_3 symmetry two parameters, p, the probability of intra-trimer modification, and r, the probability of inter-trimer modification, are required. With C_6 symmetry only the probability of modification between adjacent subunits, d, is required. Figure 11-8 shows computed profiles for each model together with experimental data that clearly indicate that the molecule has D_3 rather than C_6 symmetry. It should be pointed out that distinction between D_3 trigonal prism and D_3 octahedron is not possible by such an approach.

A variation has been used with avidin, a tetrameric molecule of molecular weight 68 kDa which binds biotin. A series of bis-biotin derivatives synthesized with differing chain lengths between the biotin rings were examined for their ability to aggregate avidin. It was established that with a $(CH_2)_n$ chain length of $n = 1$ between the biotin moieties, cross-linking between two tetramers occurs and the aggregated material can be observed by electron microscopy. The polymers are linear chains, with a width as expected for a single avidin tetramer, and have *no* branch points. If the subunits in avidin are in equivalent environments, only C_4 or D_2 symmetries need to be considered. With C_4 symmetry one expects the possibility of branch points and also that the width of the chain is larger than observed. The results therefore indicate that the avidin oligomer has D_2 symmetry.

ASSEMBLY OF SUBUNITS

The assembly of subunits into a quaternary structure can occur in a variety of ways: In homo-oligomers the process may be a gradual, stepwise transition, with one subunit being sequentially added to the growing "core" until the final oligomer is reached. Alternatively, stable, intermediate forms may be formed which then associate to give the final form. With hetero-oligomers this latter process seems inherently more probable. The formation of relatively stable sub-structures, either in the case of homo- or hetero-oligomers, can themselves be considered as structures that must be assembled from subunits, and as initially stated the process presumably consists of the stepwise addition of subunits to a preexistent core. In such schemes, at each aggregation stage the succeeding subunit may be attached to the core with the same intrinsic free-energy change as its predecessor, with a greater intrinsic free-energy change or with a lesser intrinsic free-energy change. These situations can be correlated with an assembly process showing *no* interactions between subunits in self-assembly, cooperative interactions, and antagonistic interactions, respectively. In these cases there is a smooth transition between initial and final states, an extremely sharp transition, or a particularly gradual transition. Figure 11-9 is a schematic energy diagram showing the intrinsic free-energy change for each subunit coming into the oligomer for a tetramer in each of these cases. The intrinsic free-energy change per subunit is of course always negative; otherwise, the oligomeric form would not be favored. The experimental determination of these free energies requires that the equilibrium constants between the various intermediate stages in the assembly process be determined. Any method that can quantitate the concentrations of, for example, monomer, dimer,

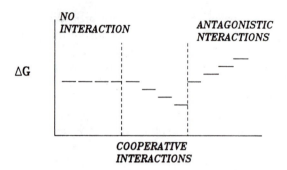

Figure 11-9   Schematic energy diagram for subunit assembly, showing the stepwise free-energy changes for the formation of a tetramer.

trimer, and tetramer, for a tetrameric protein, as a function of the protein concentration, could be used to determine these equilibrium constants.

Such associations are best followed using techniques such as sedimentation or gel filtration of the protein under native conditions, or by using an approach such as light scattering or fluorescence polarization measurements. The first three approaches, essentially based on molecular weight determinations, were discussed in Chap. 4. The final approach is outlined, for the example of the association of α-lactalbumin with galactosyl transferase to form the hetero-dimer lactose synthase, in Fig. 11-10.

If we consider in more detail the case of a tetrameric protein, two extremes become apparent.

1. There may exist, in equilibrium with monomer and tetramer, significant amounts of intermediate forms; for example, in the case of lactate dehydrogenase a significant concentration of dimers is observed.

2. Only monomers and tetramers may exist in significant concentrations. An example is *hemerythrin*, which although it has the final oligomeric size of an octamer, seems to be assembled from two tetramers that appear to exist in equilibrium only with monomers.

These two extremes represent cases of either noncooperative (or antagonistic) interactions between subunits and extremely cooperative interactions during assembly.

A *cooperativity parameter* α can be defined by

$$\alpha = \frac{K_{i+1}}{K_i} \tag{11-3}$$

where K_i is the equilibrium constant for the addition of monomeric A to a molecule containing A_{i-1} subunits.

$$A + A_{i-1} = A_i \qquad \text{with } K_i = \frac{[A_i]}{[A] \cdot [A_{i-1}]} \tag{11-4}$$

*Experimental Design to Show Protein–Protein Interaction
Using Fluorescence Polarization Measurements*

*System: Galactosyltransferase   Mol.Wt 50,000 daltons
           αLactalbumin : Mol.Wt 14,000 daltons*

    1. Fluorescently label smaller protein (eg with FITC or DnsCl)
                                 ----> F–αLactalbumin
          a) Remove Free label(dialysis or gel filtration)
          b) Check Biological activity

    2. Measure Fluorescence Polarization of F–αLactalbumin in the
       absence of Galactosyltransferase

    3. Titrate F–αLactalbumin with Galactosyltransferase: Monitor
       Polarization

    4. Expected Results
          a)If Interaction   b)If no Interaction

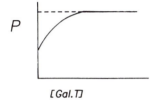 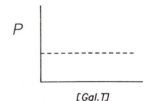

Figure 11-10   Use of fluorescence polarization measurements to study protein–protein interactions in the lactose synthase system.

Such an equation can be written for each step in the assembly process, and from the expression for α [Eq. (11-3)] we get $K_3 = \alpha K_2$, $K_4 = \alpha K_3 = \alpha^2 K_2, \dots, K_i = \alpha K_{i-1} = \alpha^{i-2} K_2$. From Eq. (11-4) it is apparent that

$$A_i = K_2^{i-2}\alpha^{\tau}A^i \tag{11-5}$$

where $\tau = 1 + 2 + \cdots + (i - 2)$; thus $\tau = [(i-1)(i-2)]/2$.

The experimental single association constant for a monomer–i mer equilibrium is given by

$$K_i = \frac{A_i}{A^i} \tag{11-6}$$

From Eqs. (11-5) and (11-6) it is found that

$$K_i = K_2^{i-1}\alpha^{\tau} \tag{11-7}$$

If we consider now a hypothetical system, an octamer made up of identical subunits of 10 kDa each, and arbitrarily assign a value for K_2 of 1×10^5, we can compute the concentration dependence of the molecular weight of the system with values for α of 3.0 (indicating cooperativity between subunit self-assembly), 1.0 (each

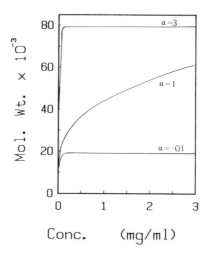

Figure 11-11   Weight-average molecular weight for an associating system, as a function of concentration, with different values of α.

subunit binds with the same affinity as the previous subunit), and 0.1 (antagonistic interactions during self-assembly). Figure 11-11 illustrates the results of such calculations.

With cooperative interactions the concentration dependence shows a rapid rise to the oligomer molecular weight of 80,000, and as expected, with antagonistic interactions the octamer is never formed—even formation of a dimer is unfavorable. Examination of Fig. 11-11 shows that a relatively small increase in the cooperativity parameter $\alpha = 1$ to $\alpha = 3$ results in a large effect on the concentration dependence of the molecular weight. The free energy associated with such a change is given by

$$\Delta(\Delta G^\circ) = \frac{RT \ln K_{i+1}}{K_i}$$

$$= RT \ln \alpha$$

$$= RT \ln 3$$

$$= 0.65 \text{ kcal} \quad \text{at room temperature} \tag{11-8}$$

This quantity, 0.65 kcal, is actually surprisingly small and represents, for example, a small increase in the area of contact between subunits in the oligomer compared to the dimer. Later, we consider where this energy change could come from. However, first we can represent the molecular weight versus concentration profile of Fig. 11-11 in terms not of the value of α, but in terms of the free-energy change associated with α. Such a plot is shown in Fig. 11-12.

As discussed previously and as calculated using Eq. (11-8), the energy changes associated with subunit–subunit interactions during self-assembly required to give

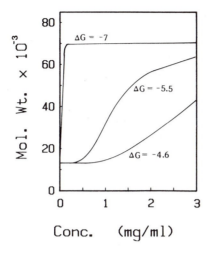

Figure 11-12 Dependence on concentration of weight-average molecular weight for
an associating system with different energies of interaction between the subunits.

cooperative interactions are quite small. It is useful, in the current context, to consider
some of the possible conformational changes that could be associated with subunit
interactions during self-assembly.

   Small conformational changes involving, for example, the change in environment
of a carboxyl group or the change in flexibility of an amide group can easily account
for free-energy changes in the range 1 to 2 kcal. The transfer of a carboxyl group
from an aqueous to a nonpolar environment can be approximated, for the purposes
of considering the free-energy change, by considering the pK of glutamate-35 in
lysozyme. This residue is in a highly nonpolar environment, and as a result the pK
of the side-chain carboxyl is shifted to approximately 6.0, compared to a side-chain
pK of 4.5 in aqueous solution. This change corresponds to a free-energy change of
approximately 2 kcal. The restriction of an amide group from a freely mobile, flexible
conformation to an immobile amide group is basically an entropy-requiring process
for which a $\Delta S°$ of -5 eu has been estimated. At room temperature the size of
this entropy change corresponds to a ΔG of about 1.5 kcal.

   Similarly, one might consider the effects of substituting, for example, an asparagine residue with an aspartate residue at a subunit interface. Such a process is
accompanied by a substantial free-energy change as a result of the ionization of the
carboxyl group. The ΔG can be estimated approximately by considering the free
energy of ionization of a carboxyl, which is about 6 kcal.

   As can be seen by considering these examples, quite large amounts of free energy
(relative to the amount needed to produce highly cooperative subunit self-assembly)
is available via a variety of seemingly inconsequential conformational changes. Large
conformational changes upon subunit–subunit association are not needed to account
for highly cooperative (or antagonistic) interactions.

Other Factors Affecting Subunit Interactions

In some cases the assembly of subunits into an oligomer requires some cofactor that allows assembly to take place. A typical example is provided by aspartate transcarbamoylase. The oligomer contains 6 mol of zinc, which is *extremely* tightly bound; it is not removed from the oligomer by dialysis against the chelating agent 1,10-phenanthroline and is not exchangeable with [65]Zn when the polypeptide subunits of aspartate transcarbamoylase are dissociated. Yet the zinc is *not* required for catalytic activity since the catalytic subunits can be prepared zinc-free, yet still have full activity. When zinc is removed from the isolated regulatory subunits (aspartate transcarbamoylase is a hetero-oligomer) however, they cannot interact with the catalytic subunits to reconstitute the regulatory properties of the holoenzyme. Zinc has no effect on the catalytic trimers but does promote the dimerization of the regulatory subunits. It is apparent that zinc is involved in maintaining quaternary structure, which in this particular case is essential not for catalytic purposes, but for the regulatory properties of the holoenzyme. Zinc apparently enhances the interface interactions between the catalytic and the regulatory subunits and thus allows the association of the two catalytic trimers with the three regulatory dimers in the holoenzyme.

In Chapters 9 to 11 we examined various aspects of the three-dimensional structure of proteins. Although secondary, tertiary, and quaternary structures have been dealt with separately, it is important to realize that each additional level of structure has some impact on the previous one. Thus in considering quaternary structure it must be emphasized that the association of two or more subunits into a quaternary structure *is expected* to have an influence on the tertiary structure of each subunit. It is via this coupling of tertiary and quaternary structure that conformational changes can be transmitted between subunits in an oligomer.

12

Conformational Changes

INTRODUCTION

In Chapters 9 to 11 we examined in great detail the conformation of a protein—from its secondary to its tertiary to its quaternary structure. The forces involved in maintaining the structure and its dynamic nature have been emphasized. In the remainder of this book the emphasis is concerned more with functional properties of proteins and the experimental approaches used to examine them. Throughout Chaps. 12 to 21, the concept that a protein can change its conformation in response to ligand binding or to its environment is implicitly assumed. Proteins may change conformation as ambient conditions such as temperature and pH change, thus affecting the forces that hold the protein in a particular conformation and allowing other conformations to become predominant. When ligands (either small molecule or macromolecule) interact with a protein at a specific binding site, part of the free-energy change of binding involves a free energy for the conformational adjustment of the protein. Such conformational changes include induced-fit conformational changes (which may be necessary to allow a second ligand molecule to bind to a protein or to induce an active conformation of the protein) and conformational changes associated with the regulation of the protein's activity.

As we discuss in Chap. 21, allosteric regulation is an important property of some oligomeric proteins, and a number of models have been proposed to explain allosteric properties. The term "allosteric" indicates that a response in one part of a protein conformation is elicited by a ligand binding at a distant site. In monomeric proteins, or in oligomeric proteins that do not exhibit subunit–subunit communication, such changes involve the secondary and tertiary structures. Where the protein consists of several domains within the tertiary structure, conformational changes may occur within a single domain or may involve interaction between domains. In the context of regulation of oligomeric protein activity, two types of "allosteric" phenomena can be distinguished. Conformational changes may be induced within a subunit in much the same way as can be considered for a monomeric protein; such changes are involved in heterotropic regulation. Many oligomeric proteins display homotropic effects, and this usually implies that ligand binding to one subunit in an oligomer induces a conformational change in a separate subunit. Such conformational changes may involve the secondary, tertiary, and quaternary structures of a protein.

In this context a very important point concerning conformational changes must be raised. As we shall discuss in Chap. 21, the principal burden of "proof" for one allosteric model over another lies in detecting the conformational changes associated with a particular model. As is obvious from the nature of allosteric interactions in oligomeric proteins, this involves detecting conformational changes occurring in subunits other than the one to which the initial ligand is bound. Unfortunately, this is made more difficult by the fact that in an oligomeric protein, anything that affects the conformation of one subunit might be expected to affect the conformation of adjacent ones. Just as when a ligand binds to a monomeric protein there is a free energy of conformational adjustment, so too when a subunit "binds" to another subunit in an oligomer there is an effect from the interaction on the conformation of both subunits. Anything affecting the conformation of one subunit may then affect the subunit–subunit interface and hence the conformation of the adjacent subunit. As a result, we must distinguish between induced conformational changes that might be expected to occur as a natural consequence of ligand binding to one subunit in an oligomer, and those involved in allosteric interactions. In allosteric interactions, the induced conformational change must affect not just an adjacent subunit conformation but specifically the conformation of the appropriate binding site on the adjacent subunit. Rather than simply detecting a conformational change in a specific subunit, we must be able to detect conformational changes at specific points within the subunit.

Since many of the techniques for following conformational changes in proteins involve the methods discussed in Chap. 9 to 11 for examining their structure, in this chapter we discuss a number of examples of the application of these techniques to the problem of following conformational changes. The main emphasis is the type of conformational change that it is possible to follow by the various experimental approaches. For convenience we examine the various methods in three categories: (1) spectroscopic approaches, (2) protein chemistry approaches, and (3) overall shape approaches.

METHODS OF STUDYING CONFORMATIONAL CHANGES

Spectroscopic Approaches

Spectroscopic techniques for following conformational changes in proteins can be grouped as giving information either about overall conformation or specific sites on a protein. Into the first group fall approaches that reflect the overall conformation rather than specific regions of the protein, and involve protein fluorescence, polypeptide backbone circular dichroism (CD), and absorbance changes either in the form of difference spectra or the use of solvent perturbation. Into the second group fall the spectral properties of ligands with specific binding sites, approaches with covalently attached spectral probes, or those that give information on the spectral properties of specific, individual side chains within a protein. Both groups depend on the fact that spectral properties in some way reflect the environment of the spectrally active moiety, and thus changes in that environment affect the characteristic parameters of the signal being monitored. In this section we look at a number of examples that have been used to give information about conformational changes using spectroscopic studies.

Protein Fluorescence. The fluorescence properties of proteins arise from the aromatic residues tryptophan, tyrosine, and phenylalanine. Because of the degeneracy of the protein fluorescence signal it is usually not possible, on the basis of steady-state fluorescence measurements, to determine which particular tryptophan or other residue in a protein is involved in changes in the fluorescence properties, and hence measurements of protein fluorescence changes usually follow conformational changes on a gross scale. It is sometimes possible to resolve steady-state fluorescence measurements by using three-dimensional spectra to give more defined information on the spectral properties of contributing aromatic residues in the protein. In some proteins there is a single tyrosine or tryptophan residue and the fluorescence properties reflect those specific residues.

A particularly sensitive way of using protein fluorescence to detect conformational differences is illustrated by the experiment shown in Fig. 12-1. Here the excitation wavelength dependence of the quenching of protein fluorescence for glyceraldehyde-3-phosphate dehydrogenase was followed as a function of NAD bound per tetramer. If the four subunits of the enzyme behave as identical subunits with no interactions between them, ligand binding to each would be expected to produce the same fluorescence quenching characteristics. As is clear in Fig. 12-1, this is

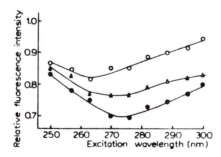

Figure 12-1 Fluorescence quenching spectrum obtained when NAD binds to glyceraldehyde-3-phosphate dehydrogenase. Data show the wavelength dependence of quenching with one molecule of NAD bound per tetramer ($\bigcirc-\bigcirc$), with two molecules bound ($\triangle-\triangle$), and with three molecules bound ($\bullet-\bullet$). [Reprinted with permission from: J. E. Bell and K. Dalziel, *Biochim. Biophys. Acta, 410*, 243–251 (1975).]

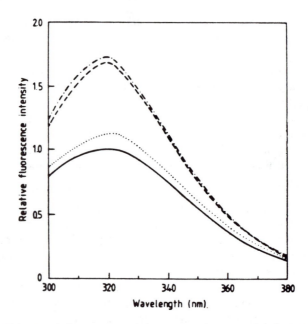

Figure 12-2   Protein fluorescence emission spectra of α_2-macroglobulin (α_2M) alone
(——), activated with methylamine ($\cdot\cdot\cdot$), interacting with trypsin (———), or activated
by methylamine and interacting with trypsin (—·—). (Reprinted with permission from:
I. Bjork, T. Lindblom, and P. Lindahl, *Biochemistry*, *24*, 2653–2660. Copyright 1985
American Chemical Society, Washington, D. C.)

not the case, indicating either that the premise that the subunits are initially identical
is not correct, or that a conformational change is occurring in the protein as the
saturation is increased, resulting in different quenching spectra for the different NAD
ligand molecules bound.

Protein fluorescence emission spectra have also been used to yield information
on conformational differences in proteins. Figure 12-2 illustrates the protein emission
spectra of the protease inhibitor bovine α_2 macroglobulin in several different states,
showing conformational differences between the protein as it is activated or interacts
with a target protease.

Noncovalent Ligand Fluorescence. Two types of noncovalent ligands can be
envisaged. In the first, a fluorescent molecule such as TNS [6,-(4-toluidino)-2-
naphthalenesulfonic acid] interacts with the protein at some site or sites unknown,
causing some change in the fluorescence properties of the ligand. When a second
ligand interacts with the protein, various situations can occur: (1) the bound TNS
fluorescence changes as the result of the environment of the bound TNS changing,
(2) the overall TNS fluorescence (bound plus free TNS) changes as a result of a change
in the distribution of bound and free TNS, or (3) there is no change in the overall
TNS fluorescence. In the first case it is clear that a conformational change is occurring
which affects the TNS fluorescence. In the second case, if there is an increase in the

amount of bound TNS it is clear that a conformational change in the protein must have occurred, resulting in either an increased number of sites or an increased affinity of the existing sites. If, however, the fluorescence change indicates that there is a decreased number of sites or lower affinity, it is possible that direct ligand competition is responsible, in which case conclusions about induced conformational changes are hard to draw. In both cases it is necessary to determine either the fluorescence properties of the bound ligand or the affinity of the protein for the ligand. These matters are discussed in detail in Chap. 17 and it is sufficient for the present discussion to say that the properties of the bound fluorophore can be established by titrating a fixed concentration of fluorophore with increasing concentrations of protein until no further changes in ligand fluorescence are detected; at this point *all* of the fluorophore is bound and the fluorescence properties reflect those of the bound fluorophore only.

In the third case, that where no change in overall TNS fluorescence occurs, the only definable conclusion that can be drawn is that if a conformational change in the protein is occurring, TNS fluorescence cannot be used to detect it. It is possible that there is no overall conformational change, but it is also possible that any conformational change is not detected by the particular fluorescent ligand used.

Proteins in general do not have specific binding sites for fluorophores such as TNS, and as a result such fluorophores are of the most use in detecting overall conformational changes. Figure 12-3 shows the fluorescence emission spectra of TNS bound to α_2 macroglobulin under circumstances similar to those in Fig. 12-2. Obviously, very similar conclusions can be drawn from this experiment as were drawn from the protein fluorescence measurements shown in Fig. 12-2.

The second type of ligand we consider is one that has a specific binding site on the protein. In many proteins a naturally occurring ligand is fluorescent and can be used as a probe of its own binding site if its fluorescence properties change upon binding. Figure 12-4 illustrates experiments using NADH fluorescence to follow conformational changes induced across subunit interfaces by NAD binding to the hexameric enzyme glutamate dehydrogenase.

In these experiments the fluorescence enhancement of the bound NADH is determined by titrating NADH with enzyme (see Chap. 17) in the presence of various amounts of NAD. The results clearly indicate that at half-saturation of the hexamer with NAD, the fluorescence enhancement of the bound NADH shows an abrupt decrease, suggesting that NAD induces a conformational change in the hexamer that is detectable by NADH fluorescence. In this particular instance, however, a more far-reaching conclusion can be drawn. Since NAD and NADH compete for the same site *on each subunit*, the NAD that is causing the conformational change in the hexamer (as detected by the trace amount of NADH used) *cannot* be bound to the same subunit as the fluorescent reporter NADH. The conformational change that is occurring must be induced across subunit interfaces. In addition, since the reporter fluorophore group (the NADH) is bound at the active site of the enzyme and the NAD also binds there (but to a different subunit under experimental conditions used), it is possible to reach the conclusion that ligand binding to one active site affects the conformation, as detected by the altered fluorescence properties of NADH, of a separate active site within the oligomer.

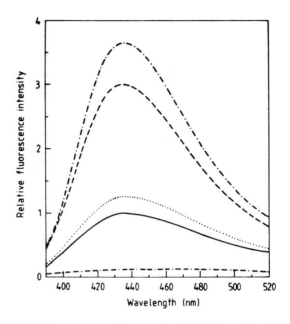

Figure 12-3   TNS fluorescence emission spectra for TNS bound to α_2M alone (——), activated with methylamine (· · ·), interacting with trypsin (– – –), or activated by methylamine and interacting with trypsin (–·–). Also shown is the fluorescence of free TNS in the presence of trypsin (– –·– –). (Reprinted with permission from: I. Bjork, T. Lindblom, and P. Lindahl, *Biochemistry*, *24*, 2653–2660. Copyright 1985 American Chemical Society, Washington, D. C.)

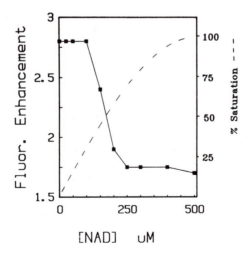

Figure 12-4   Fluorescence enhancement of NADH bound to glutamate dehydrogenase as a function of the saturation of the enzyme by NAD.

Even when a natural ligand of a protein is not fluorescent, it is often possible to use a ligand analog that is. In such instances it is important that control experiments be performed to show that the analog behaves the same way as the natural ligand; otherwise, it is possible that the interaction of the ligand analog with the protein will generate a different response in the protein conformation than in the natural ligand.

Covalent Fluorescent Probes. The most obvious potential problem with non-covalent fluorescent probes is the inability in some cases to experimentally determine the properties of the *bound* fluorophore. This is circumvented by using a covalently introduced fluorophore, making use for example of a reactive sulfhydryl group to introduce a fluorescent probe, as discussed in Chap. 7. Although this approach removes the problem of changes in the amount of fluorophore appearing to indicate conformational changes, it is important that the chemical modification used to intro-duce the fluorophore not result in changes in the normal properties of the protein that could alter its sensitivity to induced conformational changes.

Resonance Energy Transfer Measurements. In the discussion of fluorescence used to detect conformational changes we have concerned ourselves primarily with spectral measurements or intensities. We could equally as well have considered parameters such as polarization, lifetime, or ability to be quenched by collisional quenching. Somewhat different in approach, however, is the use of resonance energy transfer measurements to detect conformational changes. Resonance energy transfer between a donor and an acceptor depends on both the relative orientation of the donor and the acceptor, and the separation distance of the donor and acceptor. As such, it is exquisitely sensitive to conformational changes in a protein. If a suitable donor and acceptor are available, resonance energy transfer measurements between the two as a function of the concentration of a ligand thought to induce a conformational change in the protein can be used to study such a change. These measurements often involve resonance energy transfer from the protein aromatic residues to a covalent fluorophore acting as the transfer acceptor. Changes in energy transfer are easily interpreted. It is also possible to use either a noncovalent donor or acceptor, but changes in energy transfer must then be shown not to be due to changes in the binding of the noncovalent partner.

Circular Dichroism Measurements. Backbone CD measurements have been used to show that a protein changes its amounts of secondary structure in response to a ligand binding. As shown in Fig. 12-5, measurements of the far-UV CD of ATPase have been used to show that there is an increase in the α-helix content from 35% to 42% when MgATP binds to the enzyme in the presence of phosphate. In similar experiments (Fig. 12-6) it has been shown that calmodulin undergoes a conforma-tional change, as detected by CD measurements at 222 nm, when calcium binds to the protein.

In an approach exactly analogous to that described earlier using NADH as a conformational probe with glutamate dehydrogenase, CD measurements of blue

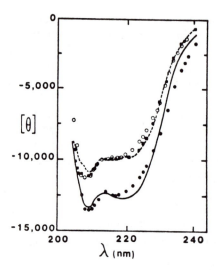

Figure 12-5 CD spectra of ATPase alone (○–○), in the presence of phosphate (■–■), or with phosphate plus MgATP (●–●). (Reprinted with permission from: B. Roux, G. Fellous, and C. Godinot, *Biochemistry*, *23*, 534–537. Copyright 1984 American Chemical Society, Washington, D. C.)

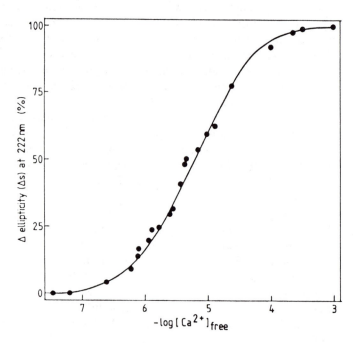

Figure 12-6 Titration of the Δ ellipticity at 222 nm of calmodulin as a function of the concentration of free calcium. (Reprinted with permission from: D. Burger, J. A. Cox, M. Comte, and E. A. Stein, *Biochemistry*, *23*, 1966–1971. Copyright 1984 American Chemical Society, Washington, D. C.)

dextran–Sepharose bound to the catalytic subunit of protein kinase have been employed to show that a variety of bound substrates induce similar conformational changes in the phosphoryl donor site. This enzyme, which catalyzes the cAMP-dependent phosphorylation of a variety of substrate proteins, consists of two regulatory and two catalytic subunits. The catalytic subunits are active in the absence of the regulatory subunits and phosphorylate a number of suitable peptide and protein substrates containing an exposed serine or threonine with one or more arginine residues to the N-terminal side. Kemptide is a synthetic substrate heptapeptide, with the sequence Leu-Arg-Arg-Ala-Ser-Leu-Gly.

Blue dextran has been shown to interact with a number of proteins containing nucleotide binding sites, and binds at the ATP site of the catalytic subunit. It has an absorbance maximum at 612 nm, and the low-energy transitions giving rise to this peak are extremely sensitive to alterations in the conformation of the chromophore. Blue dextran is used covalently linked to Sepharose to allow the removal of unbound enzyme or other ligands prior to the CD spectral measurements. Figure

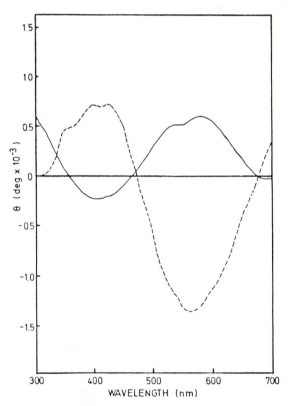

Figure 12-7 CD spectra of protein kinase catalytic subunit complexed with blue dextran Sepharose in the absence of substrate (——) or in the presence of a saturating amount of histone IIa (–––). (Reprinted with permission from: J. Reed and V. Kinzel, *Biochemistry*, *23*, 968–973. Copyright 1984 American Chemical Society, Washington, D. C.)

12-7 illustrates the CD spectrum of blue dextran bound to the catalytic subunit in the presence and absence of histone IIa.

The induced positive peak at approximately 400 nm and the large negative spectrum at about 560 nm are also seen when a variety of other substrates, such as glycogen synthase or kemptide, are bound to the kinase. Since the blue dextran is bound to the ATP site of the catalytic subunit, the induced CD spectral changes indicate that substrate binding induces a conformational change in the ATP site. When kemptide-induced changes are compared with those of the synthetic peptide **Arg-Gly-Tyr-Ala-Leu-Gly**, which has no phosphorylatable serine, different changes in the blue dextran are observed. These differences indicate that the substrate must contain the phosphorylatable serine as well as the arginine to induce conformational changes in the ATP site.

NMR Measurements. In some proteins it has been possible to uniquely assign proton NMR signals to individual side chains. In such instances NMR measurements can give detailed information on the conformation of those side chains and can detect conformational changes involving them. Four-hundred-megahertz proton NMR

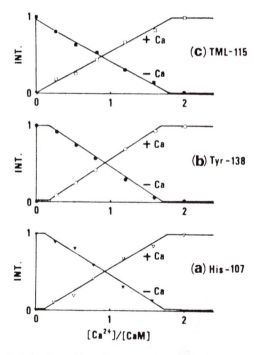

Figure 12-8 Relative intensities of proton resonances of His-107, Tyr-138, and Tml-115 as a function of the calcium concentration. −Ca indicates the Ca-free state; +Ca indicates the Ca-bound state. (Reprinted with permission from: M. Ikura, T. Hiraoki, K. Hikichi, M. Yazawa, and K. Yagi, *Biochemistry*, 22, 2573–2579. Copyright 1983 American Chemical Society, Washington, D. C.)

studies of calmodulin have shown that side-chain resonances fall into three categories: (1) those perturbed by calcium binding to its high-affinity site, which includes Tyr-138, trimethyllysine-115, His-107, and Tyr-99; (2) those associated with calcium binding to its low-affinity site; and (3) those associated with calcium binding at both sites. In the absence of calcium there is a hydrophobic region containing three phenylalanine residues (Phe-89, 92, and 141), a valine, and an isoleucine, all in the vicinity of the calcium binding sites. Titration of calmodulin with calcium and its effect on three of the group 1 resonances is shown in Fig. 12-8.

Nuclear Overhauser enhancement (NOE) techniques can be used to measure distances between atoms that are close enough to each other to provide a relaxation pathway. The NOE is obtained by irradiating one of the two coupled nuclei and measuring the resonance of the other. NOE difference spectra measurements suggest that upon calcium binding to the high-affinity site, a conformational change occurs in which one of the phenylalanine residues and the valine from the calcium-binding-site region approach Tyr-138, which is located away from the binding site.

Protein Chemistry Approaches

Any protein chemistry technique that can give information concerning the exposure of regions of the polypeptide chain to the general solvent environment can also be used to indicate whether the protein conformation has changed in such a way as to change the exposure. Where defined information concerning which region of the protein is so exposed is available, detailed information about regions of the protein involved in conformational changes can be obtained. Usually, however, such information is not sought and the experiment is limited to giving information only at the level of overall conformational changes.

In this section we consider side-chain reactivity to modification reagents, susceptability to proteolysis, the use of specific antibodies to detect conformation, and hydrogen-exchange studies.

Side-Chain Reactivity. In Chap. 7 a great many chemical modification reagents were discussed together with ways of quantitating their reactions with a protein. The exposure of a particular side chain together with any particular environmental effects on its reactivity are the major contributors when one considers how the side chain reacts toward a particular modification reagent. If a certain number of a particular type of residue reacts with a reagent (such as, for example, iodoacetic acid) when the protein is in a given conformation, changes in the protein conformation may affect the reactivity in several ways.

1. The protein conformation may change such that the number of groups modifiable does not change but the kinetics of modification does.
2. The conformational change may allow an altered number of side chains to react but not affect the rate of those side chains reactive in the original conformation.
3. Some combination of the first two possibilities may occur as a result of the conformational change.

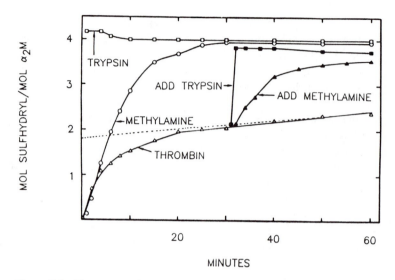

Figure 12-9   Time course of sulfhydryl generation when α_2M interacts with methyl-amine, thrombin, or trypsin. SH reactivity determined by DTNB titration. (Reprinted with permission from: J. P. Steiner, P. Bhattacharya, and K. Strickland, *Biochemistry*, *24*, 2993–3001. Copyright 1985 American Chemical Society, Washington, D. C.)

With each of these possibilities it is necessary to experimentally determine both the number of groups reacting and their rate (or rates).

Within each of these parameters there can be an increase or a decrease as a result of a ligand-induced conformational change of the protein. The conclusion is clear with increased rates of reactivity or numbers of modifiable groups; the confor-mation of the protein has changed. However, with decreased rates of reactivity or numbers of groups modifiable, it is possible that the ligand is sterically blocking access of the reagent rather than causing a conformational change.

If, in addition to these parameters the location of the groups involved (in terms of the primary sequence of the protein) can be ascertained, quite detailed information concerning conformational changes can be obtained.

The generation of reactive sulfhydryls in α_2 macroglobulin resulting from induced conformational changes has been followed using the DTNB reactivity of the exposed sulfhydryls. Figure 12-9 shows that the interaction of a variety of substrate ligands produces a conformational change resulting in the exposure of either two or four sulfhydryl groups.

Susceptibility to Proteolysis. This has been used in a number of systems to demonstrate ligand-induced conformational changes. The acid-induced unfolding of lactate dehydrogenase has been examined by limited proteolysis with pepsin and analysis of the resultant peptides by SDS-PAGE. The native enzyme is transferred to low pH and at various time intervals is pulsed with pepsin. The digestion products are separated on SDS-PAGE and undigested monomers and a variety of fragments

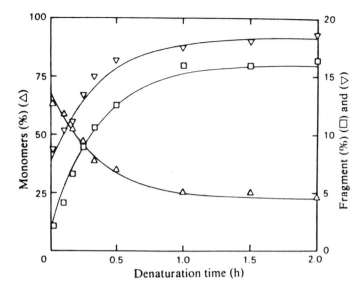

Figure 12-10   Correlation of loss of monomer and appearance of major fragments as a function of time of denaturation for lactate dehydrogenase. (Reprinted with permission from: R. Rudolf, *Biochem. Soc. Trans.*, *13*, 308–311. Copyright 1985 The Biochemical Society, London.)

ranging from 5 to 31 kDa are found. The decrease in the undigested monomer parallels the formation of two major fragments, as shown in Fig. 12-10.

This demonstrates that during the acid-induced unfolding of the lactate dehydrogenase monomer a conformational rearrangement results in the exposure of a specific interdomain cleavage site. From the molecular weights of the two major fragments (25 and 10.5 kDa) it would appear that cleavage occurs between Trp-225 and Lys-226.

Limited tryptic cleavage has been used to examine nucleotide- or phosphate-induced conformational changes in ATPase. Incubation with trypsin causes a rapid but slight activation, followed by a slower inactivation. Analysis of the proteolytic fragments by SDS-PAGE suggests that the initial rapid pulse is due to clipping of the α subunit, while the slower inactivation is due to subsequent proteolysis of α as well as β and γ subunits. The presence of ADP and ATP has no effect on the rapid phase but prevents the slower inactivation phase. Inorganic phosphate, on the other hand, slows both the initial phase and the subsequent inactivation phase, suggesting that it induces a quite different conformation in the protein than either ADP or ATP.

The Use of Specific Antibodies to Detect Conformation. Antibodies may recognize specific sequences of a polypeptide chain or a specific conformation of a region of polypeptide chain. In either case it is possible that the interaction of such an antibody with a protein can be used to detect conformational changes. In Chap. 9 we discussed the use of antibodies to detect folded structures of a protein, and emphasized the

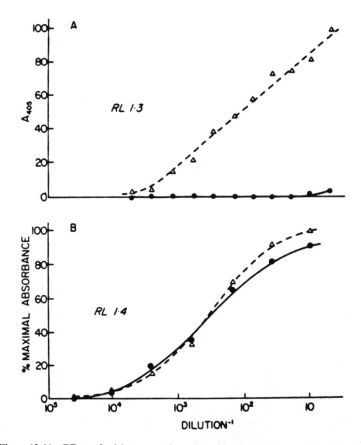

Figure 12-11   Effects of calcium on anti-prothrombin antibody interaction with prothrombin. Two different monoclonal antibodies, RL 1-3 and RL 1-4, were used in
an ELISA direct binding system. The amount of antibody bound, detected by absorbance measurements at 450 nm, is plotted as a function of the dilution of the anti-
body in the presence of calcium chloride ($\triangle$–$\triangle$) or EDTA ($\bullet$–$\bullet$). (Reprinted with
permission from: R. M. Lewis, B. C. Furie, and B. Furie, *Biochemistry, 22,* 948–954.
Copyright 1983 American Chemical Society, Washington, D. C.)

care that had to be taken in the interpretation of such experiments. Similar conditions hold with the use of antibodies to detect conformational changes. The experiments shown in Fig. 12-11 indicate that it is possible to prepare monoclonal antibodies
to prothrombin that are sensitive to conformational changes induced by calcium
binding.

   With the monoclonal antibody RL 1-3, the affinity of the interaction with prothrombin is clearly dependent on the presence of calcium. On the other hand, the
monoclonal antibody RL 1-4 binds to prothrombin in an essentially calcium-
independent manner. Monoclonal antibodies have also been produced which can
discriminate between the various forms of the chromoprotein phytochrome.

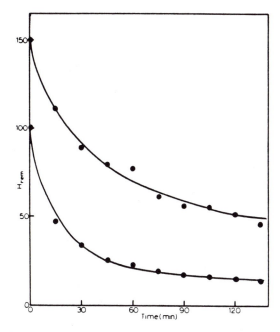

Figure 12-12 Exchange-out of tritium from labeled rhodanese in the *E* form (lower curve) and the *E_s* form (upper curve). [Reprinted with permission from: P. Horowitz and K. Falksen, *Biochim. Biophys. Acta, 747*, 37–41 (1983).]

Hydrogen-Exchange Kinetics. The accessibility of protons to solvent in many instances governs their behavior in hydrogen exchange with tritiated water. Clearly, conformational changes that alter this accessibility change the hydrogen-exchange kinetics of particular protons. Ligand-induced changes in hydrogen exchange have been used to follow conformational changes in many proteins. Allosteric regulators have been shown to increase the rates of hydrogen exchange in various proteins, including glutamate dehydrogenase and glutamine synthase. With the enzyme rhodanese, exchange out of tritium from pre-tritiated enzyme has been used to examine different conformational states of the protein. The experiments shown in Fig. 12-12 involve the free enzyme and a substrate-induced conformation of the enzyme.

In these studies the enzyme was prelabeled by incubation in tritiated water and the amount of tritium remaining with the protein determined after varying time intervals. The free enzyme, *E*, shows a faster overall back exchange of tritium than the substrate-induced conformation, E_s. Further analysis shows that the exchangeable protons fall into two classes and the major difference between the two conformations involves the number of protons in the faster class as well as the rate of exchange.

Where hydrogen-exchange kinetics can be followed by NMR measurements, it is possible to follow the exchange reactions of individual amino acids in a structure and obtain quite precise information on conformation. In bovine pancreatic trypsin inhibitor (BPTI), the β-sheet amide protons exchange extremely slowly. The effects of complex formation with trypsin on the hydrogen–deuterium exchange of these protons has been followed and it has been shown that whereas the NH proton of Tyr-35 is slowed by a factor of more than 1000, other NH protons are slowed only by a factor of 3 to 15. In a related system, streptomyces subtilisin inhibitor, the molecule exists as a dimer with extensive β-sheet regions at the interface. As in BPTI the exchange rates of NH groups in the sheet regions are extremely slow. This is in contrast to the two highly exposed regions of the molecule, one of which carries the recognition site for proteases, which are highly mobile. In this instance it has been suggested that the extremely slow rates of exchange of the β sheet protons result from an unusually rigid structure which compensates for the flexibility and mobility of other regions of the molecule.

Overall Shape Approaches

Various of the hydrodynamic techniques that were discussed in Chap. 4 in the context of molecular weight determination can also be used to obtain information about the overall shape of a macromolecule. In general, such approaches give no detail of any conformational change but in a variety of cases are used to indicate that a conformational change has occurred. In this section we briefly discuss the basis for using such methods.

Sedimentation Methods. If we consider a macromolecule as a prolate ellipsoid, we can consider two frictional coefficients: f_a, for rotation about the long a semiaxis, and f_b, for rotation about the short b semiaxis. The Perrin F factor is related to these frictional coefficients by

$$F_a = \frac{f_a}{f_{\text{rot}}} \tag{12-1}$$

$$F_b = \frac{f_b}{f_{\text{rot}}} \tag{12-2}$$

where f_{rot} is the rotational frictional coefficient and is related to the volume of the particle, V, and the viscosity of the solvent, η, by

$$f_{\text{rot}} = 6\eta V \tag{12-3}$$

As indicated in Chap. 4, the sedimentation properties of a macromolecule are related to the frictional coefficient, and as a result the sedimentation coefficient depends on the shape of the molecule. If we consider a monomer(m)–dimer(d) equilibrium, the sedimentation coefficients are related by

$$\frac{s_d}{s_m} = \frac{M_d/M_m}{f_d/f_m} = \frac{2f_m}{f_d} \tag{12-4}$$

When shape alone is the contributing factor to sedimentation coefficient changes, much smaller effects are observed. The theoretical ratio of sedimentation coefficients for a linear tetramer versus a tetrahedral tetramer is 0.944, indicating that even with such a drastic change in quaternary structure only small effects on the sedimentation coefficient are found. With such small changes it is vital that any molecular weight increase as a result of bound ligand be accounted for in assessing whether altered sedimentation properties are indeed the result of a conformational change.

Viscosity Methods. The intrinsic viscosity $[\eta]$ of a solute molecule is given in

$$[\eta] = \frac{v V_h N_0}{M} \qquad (12\text{-}5)$$

where v is the so-called Simha factor, N_0 the number of solute particles, M the molecular weight of the solute particles, and V_h the hydrated volume. The frictional coefficient, f, is related to the viscosity, η, by

$$f = 6\pi\eta \left(\frac{3}{4}\pi\right)^{1/3} V_h^{1/3} F \qquad (12\text{-}6)$$

Since as with sedimentation methods the frictional coefficient depends on the shape of the molecule, viscosity measurements can be used to assess shape changes, provided that the molecular weight is constant. Fluorescence polarization measurements are a sensitive way of determining viscosity and are useful in assessing shape changes of a fluorescently labeled macromolecule.

Gel Filtration Methods. In Chap. 4 we made use of the relationship $K_{av} = -A \log M + B$ to determine the molecular weight of a protein by gel filtration methods, and pointed out in Eq. (4-52) that in principle the molecular weight should be replaced by the effective hydrated radius, R_h: It is related to the frictional coefficient of a molecule by

$$f = 6\pi\eta R_h \qquad (12\text{-}7)$$

and as a result the elution behavior of a macromolecule in a gel filtration experiment is related to its shape. As before, changes in the elution behavior of a protein can be interpreted in terms of conformational changes only if molecular weight changes can be ruled out or taken into account.

Light-Scattering Methods. The radius of gyration of a macromolecule is obtained from light-scattering data (Chap. 4), and since it is defined as the root-mean-square distance of an array of atoms or groups from their common center of gravity, it is clear that light-scattering measurements can be used to detect conformational changes in a macromolecule if they are of sufficient magnitude to be reflected in the radius of gyration.

Summary

All the various methods discussed in this section *can* give indications of conformational changes in a macromolecule. The fact that in a particular system one or the other of these approaches fails to detect a conformational change does not mean that one is not occurring; it simply means that the experimental approach may not be sensitive enough to see the altered conformational state. With the approaches that give overall shape information, this is a particularly appropriate proviso which must be taken into account.

MECHANISMS OF CONFORMATIONAL CHANGES

Although we have discussed many experimental approaches that have been used to monitor conformational changes in proteins, not much attention has been paid to how such changes are transmitted through the protein. Except in a limited number of instances, little specific is known about how conformational changes occur. The most interesting exception is hemoglobin, where x-ray crystallographic studies of the protein in a number of conformations have allowed the mechanism of the cooperative interactions in oxygen binding to be understood in some detail. This particular example is discussed in Chap. 21. In the absence of such detailed information the general basis for the mechanism of conformational changes can be discussed.

Local changes such as might be involved in conformational adjustment of a binding site to optimize interaction with a substrate may involve only the reorientation of side chains of amino acid residues within the binding site. Depending on the stability of the local secondary structure, such reorientations may have no significant effect on the formal secondary structure. In Chap. 9 we examined the various forces involved in a particular secondary conformation: Small changes in one component (side-chain orientation) may not be sufficient to affect the overall conformational preference of a particular region of secondary structure.

Where the local secondary structure is poised delicately between two different conformations (e.g., a particular amino acid sequence may have almost equal preference for an α helix or a β sheet), apparently minor changes in side-chain orientation can cause a switch from one preference to the other. In this case the basis for the change in formal secondary structure can reside in the contribution of the side-chain orientation to the energy of a particular structure in terms solely of short-range interactions. However, long-range interactions that the side chain may make could well be altered, and changes in long-range interactions might trigger the conformational switch from one structure to another.

From this basis it is easy to understand how a conformational change in the protein can be triggered over long ranges. Because of the interplay of long- and short-range interactions in determining local structure of the protein, changes that affect either can be transmitted through the protein and result in conformational changes at points distant from the initial triggering point. Similarly, conformational changes can be transmitted across subunit interfaces, since interactions between poly-peptide chains at these interfaces are simply long-range interactions.

The considerations discussed here are also relevant to examining the effects of chemical modification on protein conformation. In Chap. 7 we looked at the use of side-chain chemical modification and stressed that the observation that a protein changes activity as the result of modification of a single defined residue does not indicate that the residue is directly present in the functional site of the protein. Conformational changes triggered by modification can occur across large distances and lead to alterations in the function of the protein.

13

Enzyme Kinetics: A Review

INTRODUCTION

In the preceding chapters we considered various aspects of the structure of a protein. In this and many of the succeeding chapters, we turn our attention to the function of the group of proteins known as enzymes. Enzymes are the largest class of proteins, and due to their ease of study, the most well characterized (as discussed in Chap. 1, there is often a simple way to follow their biological activity). All proteins share a common function: They interact with other molecules. In some cases the other molecule is unchanged by the interaction, while in others it is chemically altered by the interaction. Into the first category fall binding and structural proteins; into the second fall the enzymes. Proteins in both groups have specific binding sites recognized by the appropriate ligand. In some instances the ligand is a small molecule and in others it is another protein or some other macromolecule. In many cases proteins interact with either type of ligand at different binding sites.

Already we have considered some of the ways in which the chemical composition (i.e., the amino acid side chains) of a binding site can be determined. The major emphasis of the remainder of this book is to examine the area of protein–ligand interactions. We use the various characteristics of enzymes to discuss the information a protein chemist or enzymologist wishes to acquire and the experimental approaches used to acquire it. The methodology developed to study enzymes is often readily applicable to the questions that protein chemists explore concerning binding proteins or structural proteins.

In its simplest formulation, the action of an enzyme can be represented by the scheme in Fig. 13-1. In the first stage of the process the substrate ligand must recognize

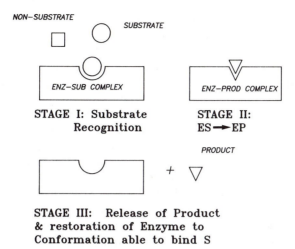

Figure 13-1 Schematic representation of the steps involved in a simple enzyme-catalyzed reaction.

and bind to its appropriate binding site. In addition to the chemical nature of the binding site, this interaction can be characterized by various kinetic and thermodynamic parameters.

The existence of a binding site on the enzyme for the substrate is the basis of one of the tenets of enzyme-catalyzed reactions; they are saturable. In its simplest terms, when all the catalytic sites of the enzyme are saturated with substrate the catalyzed reaction is proceeding at its maximum rate for that particular concentration of enzyme. A binding site, which makes specific interactions with various regions of the substrate, also confers another general property on enzymes; they are specific in their ability to recognize defined substrates. Recognition, by the binding site of the enzyme, of multiple points on the substrate confers stereo selectivity. When an enzyme binding site makes three distinct interactions with a substrate, then stereo-selective reaction is possible even in a symmetrical molecule such as citrate. This is the basis of the three-point attachment theory of enzyme specificity.

In the second stage an interconversion of substrate to product takes place, sometimes involving the formation of covalent intermediates between the substrate (or product) and some amino acid side chain or cofactor on the enzyme, and almost always involving the interaction of a variety of amino acid side chains with various stages of the substrate to product conversion.

Finally, in the third stage the product must be released from the binding site of the enzyme so that the cycle can begin again. As with the first stage, various kinetic and thermodynamic parameters describe this product release phase of the reaction.

In this simplified scheme we consider the interaction of a single ligand with a protein. In reality the majority of enzymes require two or more substrates to function and frequently a regulatory ligand as well. In some enzymes the protein molecule consists of a number of polypeptide chains: sometimes homopolymers, sometimes heteropolymers. In these instances the effects of protein–protein interactions must also be considered.

As soon as the situation becomes more complex than the simple case shown in Fig. 13-1, a number of other questions have to be answered. Consider briefly the possible effects of a second ligand on our example. We assume that the second ligand is *not* a substrate: it could affect some parameter related to any or all of the three stages mentioned. A *description* of the effects of this second ligand requires a knowledge not only of its binding site and related parameters, but also of the effects its binding has on the other stages of the reaction. An *understanding* of the effects of this second ligand requires a detailed knowledge of the protein structure connecting the various sites and the way in which the protein structure changes in response to ligand binding. In all but a few cases such an understanding has not been attained; the approaches described in the remainder of this book are those used to obtain a detailed description of ligand–protein interactions. Such descriptions are essential before the appropriate questions can be asked, questions whose answers lead to the understanding of enzyme structure and function relationships.

In this chapter some of the essential terminology and conceptual background relating to the topic "enzyme kinetics" are reviewed. The study of enzyme kinetics is often regarded simply as a description of the way a substrate or product interacts with an enzyme. As illustrated in later chapters, the various experimental approaches employed in the study of enzyme kinetics can give detailed insight into almost every phase of an enzyme-catalyzed reaction and its regulation.

CHEMICAL KINETICS

If we consider the reaction $S \rightleftharpoons P$, the rate of the reaction (which can be followed either as a decrease in $[S]$ or as an increase in $[P]$) is proportional to the concentration of S (i.e., $[S]$). This rate, v, at any given time, T, is given by

$$\frac{-d[S]}{dt} = \frac{d[P]}{dt} = k[S] = v \tag{13-1}$$

where k is the *rate constant* of the reaction. Integration of Eq. (13-1) gives

$$\ln \frac{[S_0]}{[S]} = kt \tag{13-2}$$

where $[S_0]$ is the initial concentration of S at $t = 0$. A half-time for the reaction, $t_{1/2}$, can be defined as the time taken for the concentration of S to decrease to half its value at any given point in time. From Eq. (13-2) we can obtain an expression for $t_{1/2}$ if we allow S_0 to decrease to $S_0/2$. Hence

$$\log 2 = \frac{kt}{2.303} \tag{13-3}$$

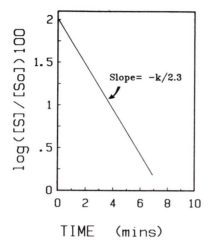

Figure 13-2 Determination of the rate constant k for a simple process where the concentration of S can be determined at various times during the course of the reaction.

and t (which is $t_{1/2}$) is equal to $0.693/k$. The half-time of a process is obtained directly from the rate constant of the process.

$$t_{1/2} = \frac{0.693}{k} \qquad (13\text{-}4)$$

Figure 13-2 graphically summarizes these points and indicates how k may be calculated for such a process from a plot of log $([S]/[S_0])$ versus t.

The reaction $S \rightleftharpoons P$ does not continue past a finite point: Not *all* of S is converted to P. The final concentrations of S and P are the *equilibrium concentrations* and are related by the equilibrium constant (K_{eq}) for the reaction

$$K_{eq} = \frac{P_{eq}}{S_{eq}} \qquad (13\text{-}5)$$

Thus we have two parameters that can be used to describe the reaction $S \rightleftharpoons P$: the rate constant, k, and the equilibrium constant, K_{eq}.

A chemical reaction proceeds through a *transition state*, and the reaction can be visualized in terms of an energy diagram. For the reaction we are considering, an energy diagram might resemble Fig. 13-3. Several important thermodynamic quantities are defined in this figure:

Standard Free-Energy Change

ΔG is related to the equilibrium constant, K_{eq}, for the reaction by

$$\Delta G^{\circ} = -RT \ln K_{eq} \qquad (13\text{-}6)$$

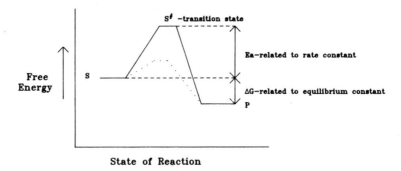

Figure 13-3   Energy diagram for the simple reaction S ↔ P.

The actual free-energy change of the reaction, ΔG, is related to $\Delta G°$ and the concentration of S and P by

$$\Delta G = \Delta G° + RT \ln \frac{[P]}{[S]} \qquad (13\text{-}7)$$

Thus a reaction can have a positive $\Delta G°$ but still proceed because the actual ΔG is negative. Since $\Delta G = \Delta H - T\,\Delta S$ the equilibrium position of a reaction can be regarded as a tendency toward a minimum of energy and a maximum of entropy.

Arrhenius Equation

This equation relates the rate constant, k, to the *activation energy*, E_a (defined in Fig. 13-4):

$$k = \frac{Ae^{-E_a}}{RT} \qquad (13\text{-}8)$$

which gives

$$\ln k = \frac{-E_a}{RT} + \ln A \qquad (13\text{-}9)$$

A plot of $\ln k$ versus $1/RT$ enables the activation energy, E_a, to be calculated, as illustrated in Fig. 13-4.

Transition-State Theory

Consider that an enzyme-catalyzed reaction proceeds through a transition state in much the same way as does the simple chemical reaction we just considered. The energy diagram for a simple one substrate–one product reaction, however, is more complex than that shown in Fig. 13-3. Apart from the chemical steps involved in converting substrate to product, one has also to consider the binding of substrate

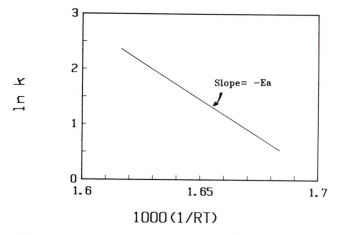

Figure 13-4   Determination of the activation energy, E_a, from the temperature dependence of the rate constant, k.

and product to the enzyme. Figure 13-5 shows a schematic energy diagram for such an enzyme system. As indicated, various thermodynamic parameters are associated with substrate binding and product release as well as the overall chemical equilibrium. Also shown is the hypothetical energy profile for the same reaction in the absence of enzyme. If the enzyme is acting as a true catalyst, it does not alter the chemical equilibrium position between substrate and product: It simply lowers the activation energy, allowing the equilibrium position to be attained more rapidly.

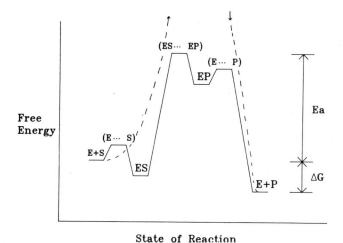

Figure 13-5   Energy diagram for a simple enzyme-catalyzed reaction. The dashed line shows the energy diagram for the non-enzyme-catalyzed reaction.

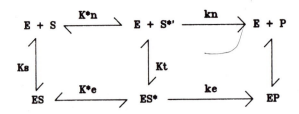

Figure 13-6 Simple scheme of a non-enzyme-catalyzed (top pathway) and an enzyme-catalyzed (bottom pathway) conversion of S to P.

If we assume that S*′ and S*, the respective transition-state structures of the nonenzymic and enzymic reactions, are similar in structure, we can derive several important points which are used in a variety of ways later.

Since the enzyme is acting as a catalyst (i.e., it is speeding a possible reaction), one must consider that the transition-state structure is in equilibrium with the reactant and can therefore write an equilibrium equation of the type $K^* = [S^*]/[S]$.

For the scheme shown in Fig. 13-6, we write a series of equilibrium equations,

$$K_s = \frac{[ES]}{[E][S]} \tag{13-10}$$

$$K_n^* = \frac{[S^{*\prime}]}{[S]} \tag{13-11}$$

$$K_e^* = \frac{[ES^*]}{[ES]} \tag{13-12}$$

$$K_t = \frac{[ES^*]}{[E][S^{*\prime}]} \tag{13-13}$$

where S*′ represents the transition state of the nonenzymic reaction and ES* represents the transition state of the enzyme-catalyzed reaction.

As before, the rate of either the nonenzymatic pathway or the enzymatic pathway is equal to the concentration of the reacting species (either S*′ or ES*) multiplied by a rate constant: k_n for the nonenzymatic path or k_e for the enzymatic path. From these relationships it can be shown that

$$\frac{K_t}{K_s} = \frac{K_e^*}{K_n^*} = \frac{k_e}{k_n} \tag{13-14}$$

and we can see that the rate enhancement of the enzymic reaction over the non-enzymic reaction (k_e/k_n) is a measure of the increased affinity the transition state has for the enzyme. We consider the ramifications of this conclusion further when we consider the chemistry of enzyme action.

INTRODUCTORY ENZYME KINETICS

For the simple reaction

$$E + S \underset{k_2}{\overset{k_1}{\rightleftharpoons}} ES \overset{k_{cat}}{\rightleftharpoons} E + P \tag{13-15}$$

one can derive an equation using either the steady-state assumption or the equilibrium assumption.

The *steady-state* assumption used by Briggs and Haldane is that the intermediates, in this case the ES complex, are always present in very small quantities compared to the reactants and products. Thus after an initial "*pre-steady-state*" phase, the concentrations of the intermediates remain constant, that is,

$$\frac{d[ES]}{dt} = 0 \tag{13-16}$$

The initial velocity, V_0, is defined by

$$V_0 = k_{cat}[ES] \tag{13-17}$$

Combining Eqs. (13-16) and (13-17), we find

$$\frac{d[ES]}{dt} = 0 = k_1[E][S] - [ES](k_2 + k_{cat}) \tag{13-18}$$

However, since the total enzyme concentration, e, is equal to $E + ES$ (the *enzyme conservation equation*), we get

$$E = e - ES \tag{13-19}$$

Substituting into Eq. (13-18) and rearranging, we find

$$k_1[ES][S] = [ES](k_2 + k_{cat}) \tag{13-20}$$

Therefore,

$$k_1[e][S] = (k_1[S] + k_2 + k_{cat})[ES] \tag{13-21}$$

and

$$[ES] = \frac{k_1[e][S]}{k_1[S] + k_2 + k_{cat}} \tag{13-22}$$

Combining Eqs. (13-22) and (13-17) results in the following equations:

$$V_0 = \frac{k_{cat}k_1[e][S]}{k_1[S] + k_2 + k_{cat}} \tag{13-23}$$

$$V_0 = \frac{k_{cat}[e][s]}{[S] + \left(\dfrac{k_2 + k_{cat}}{k_1}\right)} \tag{13-24}$$

and

$$V_0 = \frac{V_{max}[S]}{[S] + K_m} \qquad (13\text{-}25)$$

where $V_{max} = k_{cat}[e]$ and is the maximum rate of the enzyme-catalyzed reaction,

$$K_m = \frac{k_2 + k_{cat}}{k_1} \qquad (13\text{-}26)$$

where K_m is the Michaelis constant for the substrate.

In the *equilibrium assumption*, which can replace the steady-state assumption, k_{cat} is very much slower than k_2, and as a result the ES complex is in an effective equilibrium with E and S. Thus we can define an equilibrium constant

$$K = \frac{k_1}{k_2} = \frac{[ES]}{[E][S]} \qquad (13\text{-}27)$$

Substituting, the enzyme conservation equation [Eq. (13-19)], into Eq. (13-27), we get

$$K = \frac{[ES]}{[e][S] - [ES][S]} \qquad (13\text{-}28)$$

Rearranging Eq. (13-28), we find the equations

$$K[e][S] - K[ES][S] = [ES] \qquad (13\text{-}29)$$

and

$$K[e][S] = [ES] + K[ES][S] = [ES](1 + K[S]) \qquad (13\text{-}30)$$

Equation (13-30) can be put in terms of [ES]:

$$[ES] = \frac{K[e][S]}{1 + K[S]} \qquad (13\text{-}31)$$

As before, $V_0 = k_{cat}[ES]$, and using Eq. (13-31) we get

$$V_0 = \frac{k_{cat}K[e][S]}{1 + K[S]} \qquad (13\text{-}32)$$

Dividing by K we get an equation of similar form to Eq. (13-24):

$$V_0 = \frac{k_{cat}[e][S]}{[S] + 1/K} \qquad (13\text{-}33)$$

As in Eq. (13-25), $V_{max} = k_{cat}[e]$; thus

$$V_0 = \frac{V_{max}[S]}{[S] + K_m} \qquad (13\text{-}34)$$

where $K_m = 1/K = k_2/k_1$.

Equations (13-25) and (13-34) are the same: The difference lies in the definitions of K_m. In the earlier, steady-state case, K_m is equal to $(k_2 + k_{cat})/k_1$, whereas in the equilibrium case K_m is a true dissociation constant. This points to an important difference between steady-state and equilibrium assumptions. In the equilibrium assumption the Michaelis constant K_m is a true equilibrium constant, whereas in the steady-state assumption this is not the case. In steady-state mechanisms as k_{cat} gets slower (relative to k_2), we approach the *equilibrium* situation. At this point the difference may seem inconsequential; however, as is pointed out in Chaps. 14 and 15, this is important in establishing the formal kinetic mechanism of a multi-subunit enzyme.

The formula given in Eqs. (13-25) and (13-34) is the so-called Michaelis–Menten equation and describes the dependence of the initial velocity, V_0, on the substrate concentration. From this dependence two parameters, V_{max} and K_m, can be determined, as shown in Fig. 13-7.

In many cases it is not experimentally practical to plot kinetic data directly according to the Michaelis–Menten equation, and several linear transformations are popular. If we take the reciprocal of Eq. (13-34), we get

$$\frac{1}{V_0} = \frac{K_m + [S]}{V_m[S]} \tag{13-35}$$

$$\frac{1}{V_0} = \frac{K_m}{V_{max}[S]} + \frac{[S]}{V_{max}[S]} \tag{13-36}$$

$$\frac{1}{V_0} = \frac{K_m}{V_{max}} \frac{1}{[S]} + \frac{1}{V_{max}} \tag{13-37}$$

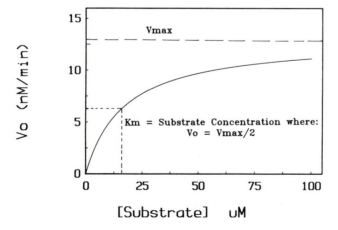

Figure 13-7 V_0 versus [S] plot of the Michaelis–Menten equation, indicating how V_{max} and K_m can be determined.

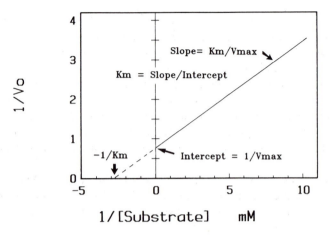

Figure 13-8   Double-reciprocal plot of kinetic data according to Eq. (13-37).

A plot of $1/V_0$ versus $1/[S]$ is linear, as illustrated in Fig. 13-8. This plot, which is the Lineweaver–Burk plot, allows extrapolation to the $1/V_0$ axis to give $1/V_{max}$ and extrapolation to the $1/[S]$ axis to give $-(1/K_m)$. Alternatively, K_m can be determined by dividing the slope of the plot by the intercept on the $1/V_0$ axis.

Alternatively, multiplying both sides by V_{max} and rearranging, we get

$$V_0 = -K_m \frac{V_0}{[S]} + V_{max} \tag{13-38}$$

and a plot of V_0 against $V_0/[S]$ is linear. This is the Eadie–Hofstee plot. As shown in Fig. 13-9, the two parameters V_{max} and K_m are obtained from the intercept on the V_0 axis or from the slope, respectively.

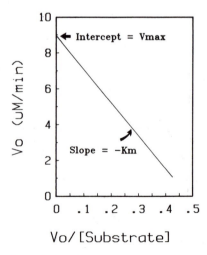

Figure 13-9   Eadie–Hofstee plot of enzyme kinetic data.

REVERSIBLE INHIBITION

The activity of an enzyme (or the bindiing activity of a protein) can be altered in a number of ways. In the realm of reversible inhibitors there are *two*. In the first, A, the inhibitor is considered to be a structural analog of the substrate and as such has an affinity for the active site. Once the active site has bound the inhibitor, subsequent binding of substrate is blocked. The interaction of the inhibitor, shown diagrammatically in Fig. 13-10A, can be described by a dissociation constant, K_i, where

$$K_i = \frac{[E][I]}{[EI]} \tag{13-39}$$

and where $[I]$ is the concentration of *free* inhibitor and $[EI]$ the concentration of the complex.

The second mode of interaction of an inhibitor with an enzyme, B, involves a site other than the active site (see Fig. 13-10B). As in the first case, interaction of inhibitor with protein can be described by a dissociation constant, K_i', which has a similar form to that of K_i. Since this does not directly involve the active site, we must consider how this type of inhibitor elicits its effect: (1) after inhibitor binding there may be a conformational change in the protein that blocks the active site from binding substrate; (2) as a result of a conformational change the ability of the active site to bind substrate may be altered but not prevented; (3) as a result of a conformational change the ability of the enzyme to *bind* the substrate may not be affected, but the ability of the enzyme to catalyze the reaction is affected; and (4) the site of inhibitor binding, although not at the active site, may be so close to it that there is

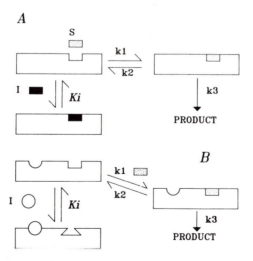

Figure 13-10 Modes of interaction of an inhibitor with an enzyme: (A) the inhibitor binds at the active site; (B) the inhibitor binds at a site other than the active site.

direct physical interaction between inhibitor and substrate binding, leading to distortion of the enzyme–substrate complex. Although all these mechanisms can lead to decreased activity (inhibition), the fourth type is conceptually quite different from the other three—no conformational change induced by inhibitor binding is required.

At this point it is appropriate to divide the modes of inhibitor interaction with enzyme into two categories; in category I is type A and type B-1. In both of these, binding of inhibitor *prevents* binding of substrate. Because inhibitor binding is reversible it follows that substrate binding in this instance can prevent inhibitor binding. Inhibition in these cases is therefore called *competitive inhibition.* In the presence of a fixed amount of inhibitor, the inhibition is completely overcome by saturating concentrations of substrate.

In the other types of inhibitor interaction that we have examined so far, the inhibitor *and* the substrate can be bound to the enzyme at the same time, giving an ESI complex. Where an ESI complex can be formed, the inhibitor is referred to as a category II inhibitor. In this complex the presence of the inhibitor can affect either the ability of the enzyme to bind substrate or the ability of the enzyme to catalyze the reaction once the enzyme has bound substrate. From the earlier discussion of basic kinetics in the absence of inhibitors, it is apparent that either K_m or $V_{\max}$ can be affected in these cases, whereas only K_m is affected in category I situations.

So far we have considered only cases where the inhibitor binds to free enzyme and has some effect on either substrate binding or catalysis. It is entirely possible, however, that the inhibitor may bind only *after* substrate has bound to the enzyme, giving an ESI complex (as in several of the situations already examined) that undergoes a conformational change leading to a decreased rate of product formation. As with the cases where an ESI complex is formed, in this instance the inhibition cannot be overcome by saturating substrate concentrations.

Rate equations describing the expected velocity in the presence of inhibitor in these cases can easily be derived using approaches similar to those described earlier for cases where no inhibitor is present. The only changes made are the inclusion of the appropriate terms describing the interaction of the enzyme and the inhibitor, which are manifested in the *enzyme conservation* equation.

Category I Inhibitors

This type of inhibitor, which causes the formation of EI at the expense of ES, can be described by the scheme shown in Fig. 13-11. From Eq. (13-39) it is apparent that $[EI] = [E][I]/K_i$. The enzyme conservation equation for this type of inhibition becomes

$$e = [ES] + [E] + [EI] \tag{13-40}$$

which, using the expression above for EI, becomes

$$e = [ES] + [E] + \frac{[E][I]}{K_i} = [ES] + [E]\left(1 + \frac{I}{K_i}\right) \tag{13-41}$$

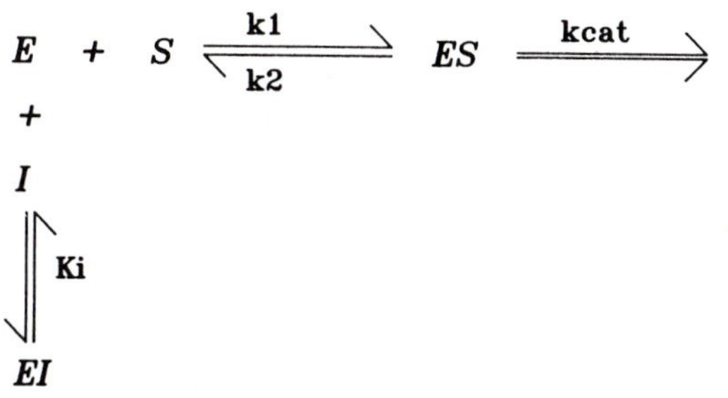

$$E \ + \ S \ \underset{k2}{\overset{k1}{\rightleftharpoons}} \ ES \ \overset{kcat}{\longrightarrow} \ E \ + \ P$$

$$+$$

$$I$$

$$\Big\updownarrow Ki$$

$$EI$$

Figure 13-11 Equilibria and catalytic steps involved with a category I type of inhibitor.

Applying the steady-state assumption, we get

$$\frac{d[ES]}{dt} = 0 = k_1[E][S] - [ES](k_2 + k_{cat}) \tag{13-42}$$

and it is apparent that

$$[E] = \frac{[ES](k_2 + k_{cat})}{k_1[S]} \tag{13-43}$$

Substituting this expression for $[E]$ into the enzyme conservation equation for this case [Eq. (13-40)], we get

$$e = [ES] + \frac{[ES] + (k_2 + k_{cat})}{k_1[S]}\left(1 + \frac{I}{K_i}\right) \tag{13-44}$$

which becomes

$$e = [ES]\left[\frac{1 + (k_2 + k_{cat})}{k_1[S]}\left(1 + \frac{[I]}{K_i}\right)\right] \tag{13-45}$$

As before, $V_0 = k_{cat}[ES]$, and from Eq. (13-45),

$$[ES] = \frac{e}{[1 + (k_2 + k_{cat})/k_1[S]][1 + [I]/K_i]} \tag{13-46}$$

and

$$V_0 = \frac{k_{cat}e}{[1 + (k_2 + k_{cat})/k_1[S]][1 + [I]/K_i]} \tag{13-47}$$

which, as before, can be rearranged to give

$$V_0 = \frac{V_{max}[S]}{[S] + K_m[1 + ([I]/K_i)]} \tag{13-48}$$

where K_m and V_{max} have the same meanings as previously.

Category II Inhibitors

Inhibitors Giving an ESI Complex. As described previously, the formation of an ESI complex is described by an equilibrium equation:

$$[ESI] = \frac{[ES][I]}{K_i'} \tag{13-49}$$

The enzyme conservation equation, as before, is

$$e = [ES] + [ESI] + [E] \tag{13-50}$$

Combining Eqs. (13-49) and (13-50), we get

$$e = [ES]\left(1 + \frac{[I]}{K_i'}\right) + [E] \tag{13-51}$$

which gives

$$e = [ES]\left\{\left(1 + \frac{[I]}{K_i'}\right) + \frac{k_2 + k_{cat}}{k_1[S]}\right\} \tag{13-52}$$

and

$$[ES] = \frac{e}{(1 + [I]/K_i') + (k_2 + k_{cat})/k_1[S]} \tag{13-53}$$

Since, as before, $V_0 = k_{cat}[ES]$, we get

$$V_0 = \frac{k_{cat}e}{(1 + [I]/K_i') + (k_2 + k_{cat})/k_1[S]} \tag{13-54}$$

Thus

$$V_0 = \frac{V_{max}[S]/(1 + [I]/K_i')}{[S] + K_m/(1 + [I]/K_i')} \tag{13-55}$$

Inhibitors Giving EI and ESI Complexes. In this case we have two equilibria involving the inhibitor: those involving EI and ESI. The equilibrium constants for these, K_i and K_i', respectively, are as described earlier. The enzyme conservation equation becomes

$$e = [E] + [ES] + [EI] + [ESI] \tag{13-56}$$

which, expressing each term in terms of ES gives

$$e = [ES]\left(1 + \frac{k_2 + k_{cat}}{k_1[S]} + \frac{k_2 + k_{cat}}{k_1[S]}\frac{[I]}{K_i} + \frac{[I]}{K_i'}\right) \tag{13-57}$$

which, in conjunction with $V_0/k_{cat}[ES]$, gives

$$V_0 = \frac{V_{max}[S]/(1 + [I]/K_i')}{[S] + K_m\left(\dfrac{1 + [I]/K_i'}{1 + [I]/K_i}\right)} \tag{13-58}$$

and when $K_i = K_i'$, this becomes

$$V_0 = \frac{(V_{max}[\text{S}])/(1 + [\text{I}]/K_i')}{[\text{S}] + K_m} \tag{13-59}$$

Linear Transformations

As with the equations derived in the absence of inhibitors, it is convenient to express these equations in double-reciprocal (Lineweaver–Burk) form and consider the effects in Lineweaver–Burk plots. Each of the three equations is shown in Lineweaver–Burk format in Table 13-1 and the appropriate plots in Fig. 13-12. From these double reciprocal equations it is apparent that in terms of Lineweaver–Burk plots, three types of inhibitors can be distinguished. Category I inhibitors affect *only* the slope, while category II inhibitors affect the intercept. The slope may also be affected in cases where an EI *and* an ESI complex are formed.

From Table 13-1 it is apparent that K_i or K_i' can be obtained from a plot of either slope or intercept from the Lineweaver–Burk plot versus [I], which gives a negative intercept on the abscissa of K_i or K_i', respectively. In instances where EI *and* ESI can exist, slope and intercept plots against I must be made; these are also illustrated in Fig. 13-12.

Since the enzyme kinetic data are also frequently represented by the Eadie–Hofstee plot, the effects of the three types of inhibitors on these plots are illustrated in Fig. 13-13. These three types of inhibition are known as competitive [Eq. (13-48)], uncompetitive [Eq. (13-55)], and noncompetitive or mixed [Eq. (13-59)]. In the latter case K_i and K_i' are not necessarily equal in Eq. (13-59). Equal values for K_i and K_i' indicate that the inhibitor binds to free enzyme and to the enzyme–substrate complex with equal affinity. It is quite possible that the inhibitor binds to either the free enzyme or the ES complex with higher affinity, and K_i is either smaller than or greater than K_i', respectively.

TABLE 13-1   Lineweaver–Burk equations[a]

inhibition	$1/V_0 = $ slope $\times 1/[\text{S}] + $	intercept
None	$\dfrac{K_m}{V_{max}}$	$\dfrac{1}{V_{max}}$
Category I	$\dfrac{K_m}{V_{max}}\left(1 + \dfrac{\text{I}}{K_i}\right)$	$\dfrac{1}{V_{max}}$
Category II (ESI)	$\dfrac{K_m}{V_{max}}$	$\dfrac{1}{V_{max}\left(1 + \dfrac{\text{I}}{K_i'}\right)}$
Category II (EI + ESI)	$\dfrac{K_m}{V_{max}}\left(1 + \dfrac{\text{I}}{K_i}\right)$	$\dfrac{1}{V_{max}}\left(1 + \dfrac{\text{I}}{K_i'}\right)$

[a] K_i values calculated from uninhibited parameter and the parameter obtained in the presence of a known concentration of I.

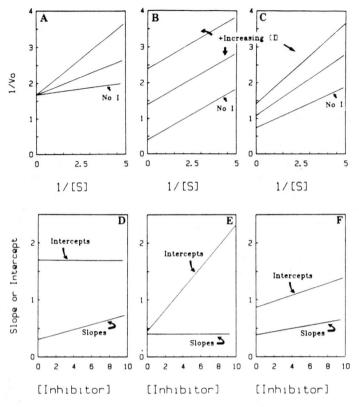

Figure 13-12 Lineweaver–Burk plots of various types of inhibition represented by (A) Eq. (13-48), (B) Eq. (13-55), and (C) Eq. (13-59). Also shown in parts (D) to (F) are the secondary plots of slope or intercept from the primary plot versus [I].

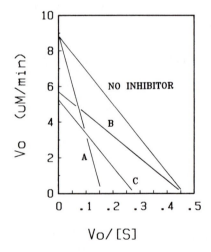

Figure 13-13 Effects of inhibitors on Eadie–Hofstee plots, according to (A) Eq. (13-48), (B) Eq. (13-55), and (C) Eq. (13-59).

330

The Dixon Plot

In an alternative approach to analyzing inhibition data to obtain a value for K_i, Dixon proposed the use of plots of $1/V_0$ versus [I] at several fixed concentrations of the substrate. Analysis using such Dixon plots requires a somewhat different experimental design to that used in Lineweaver–Burk or Eadie–Hofstee plot analysis, as the emphasis is on using a number of inhibitor concentrations at several discrete substrate concentrations. From Eq. (13-48) it is apparent that for a category I inhibitor, where only EI can exist, a plot of $1/V_0$ versus [I] at different values of [S] gives linear plots with an intersection point of $-K_i$ (see Fig. 13-14). Similarly, for category II inhibitors when K_i is equal to K'_i, such plots can be used. When K_i does not equal K'_i, or when only ESI complexes are formed, Dixon plots do not intersect in this manner.

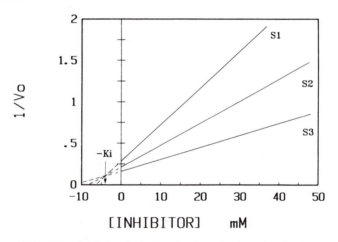

Figure 13-14 Use of a "Dixon" plot for the determination of K_i for a category I (competitive) inhibitor.

In this chapter we examined the basis for enzyme kinetics and demonstrated that for a simple one-substrate system, two quite distinct derivatives of the Michaelis–Menton equation [Eqs. (13-25) and (13-34)] can be employed. These derivatives lead to quite different physical significance for the Michaelis–Menton constant, K_m. Although this difference seems inconsequential when considering one-substrate systems, it is of much greater importance when dealing with multisubstrate systems.

14

Two-Substrate Kinetics

INTRODUCTION

Although some enzymes have only a single substrate, the vast majority have two or more. The equations developed in Chap. 14 describing the initial rate behavior of one-substrate enzymes and developing the ideas of reversible inhibitors must be modified to deal with the more common two-substrate enzyme. In this chapter the initial rate behavior of two-substrate enzymes is considered.

Before proceeding, the concept of "formal kinetic mechanism" for an enzyme-catalyzed reaction must be defined. This involves a description of the order in which the substrates bind to the enzyme as well as the sequence of product release after catalysis has taken place. Two basic classes of multisubstrate mechanisms exist. Those involving addition of *all* of the substrates prior to catalysis are (in the case of a two-substrate mechanism) referred to as *ternary complex mechanisms*. In the class where a formed product leaves the enzyme active site prior to the second substrate binding, the mechanism often involves a group transfer from one substrate to another, with that group forming a covalent intermediate with the enzyme prior to its transfer to the second substrate. This mechanism does not require formation of a ternary complex and is referred to as an *enzyme-substituted mechanism*.

In a two-substrate enzyme-substituted mechanism there is a required order of substrate addition to the enzyme: The substrate with the group to be transferred must bind first, react with the enzyme to form the substituted enzyme, and allow the first product to be released. However, in ternary complex mechanisms a number of options are possible. The enzyme may have to bind one substrate initially to induce a conformational change in the protein that allows the second substrate to bind. Such an enzyme would be described as having a *compulsory order* of substrate addition. Alternatively, it may make no difference to the enzyme that substrate binds first; the other substrate will then bind to give the ternary complex. This enzyme kinetic mechanism is referred to as *random order*.

With all of these a final component of the description of the formal kinetic mechanism involves knowledge of whether it is "*steady-state*" or "*rapid equilibrium.*" Both concepts were dealt with in Chap. 13 and are not reiterated here. It is the goal of an enzyme kinetic study to determine both the order and nature of substrate addition and, where appropriate, product release.

As a result of these considerations, for a two-substrate enzyme we have five formal kinetic mechanisms to consider. They are: compulsory order, steady state; compulsory order, equilibrium; random order, rapid equilibrium; random order, steady state; and enzyme substituted. In the next section equations for each of these are developed and, as will be seen, these equations (with one exception) fit the general form proposed by Dalziel:

$$\frac{e}{V_0} = \theta_0 + \frac{\theta_1}{A} + \frac{\theta_2}{B} + \frac{\theta_{12}}{AB} \tag{14-1}$$

This general rate equation is useful in two ways: It is easy to remember and it forms the basis for a way of analyzing initial rate data to give a series of parameters, whose properties give an indication of which formal kinetic mechanism is the most likely. The formal kinetic mechanisms examined in this chapter are not meant to form a comprehensive list, but a list of the most common. Each mechanism considered can be made more complex by the inclusion of other enzyme–substrate or enzyme–product complexes, and more complex equations describing such mechanisms derived. The more complex mechanisms usually still fit the generalized rate equation given in Eq. (14-1). Many enzyme kineticists adhere to the truism that one can never *prove* a formal kinetic mechanism for an enzyme; all one can do is to eliminate from consideration all reasonable alternatives, leaving the simplest formal mechanism that is consistent with the experimental data. It is the purpose of this chapter to show what types of initial rate data lead to this sort of description for the formal kinetic mechanism of a two-substrate enzyme-catalyzed reaction.

An understanding of the formal mechanism is important since it allows the use of the appropriate rate equation to give a variety of information on substrate (or product) affinities and/or various rate constants of binding or release. When in Chap. 21 we consider how enzyme reactions are regulated, information of this type will be essential to understanding the biological role of such regulation.

RATE EQUATIONS FOR TWO SUBSTRATE REACTIONS

Consider the reaction $A + B \rightleftharpoons P + Q$. There are a variety of kinetic mechanisms that can be considered for it, and each will be examined in turn to derive a rate equation. Once again, these are: (1) compulsory order, steady state; (2) compulsory order, equilibrium; (3) random order, rapid equilibrium; (4) random order, steady state; and (5) enzyme substituted.

Compulsory Order, Steady State

This mechanism, illustrated in Fig. 14-1, involves the obligatory addition of one substrate prior to the addition of the second. Together with a compulsory order of substrate addition, the products are released in a compulsory order.

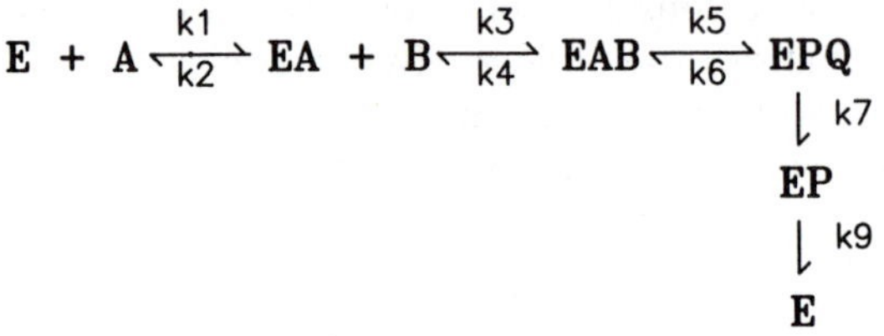

Figure 14-1   Reaction pathway for a compulsory-order, steady-state kinetic mechanism.

The enzyme conservation equation for this mechanism is given in

$$e = E + EA + EAB + EPQ + EP \tag{14-2}$$

There are five enzyme-containing species, and hence four steady-state equations are required:

$$\frac{dEA}{dt} = k_1 \cdot E \cdot A - k_2 \cdot EA - k_3 \cdot EA \cdot B + k_4 \cdot EAB = 0 \tag{14-3}$$

$$\frac{dEAB}{dt} = k_3 \cdot EA \cdot B - EAB(k_4 + k_5) + EPQ \cdot k_6 = 0 \tag{14-4}$$

$$\frac{dEPQ}{dt} = k_5 \cdot EAB - EPQ(k_6 + k_7) = 0 \tag{14-5}$$

$$\frac{dEP}{dt} = k_7 \cdot EPQ - k_9 \cdot EP = 0 \tag{14-6}$$

For convenience we can start the derivation with the assumption that the initial rate (V_0) is given by

$$V_0 = k_9 \cdot EP \tag{14-7}$$

However, other starting points can be used and give the same final equation. From Eq. (14-6)

$$EP = \frac{k_7 \cdot EPQ}{k_9}$$

Thus

$$EPQ = \frac{EP \cdot k_9}{k_7} \qquad (14\text{-}8)$$

From Eq. (14-5),

$$EAB = \frac{EPQ(k_6 + k_7)}{k_5} \qquad (14\text{-}9)$$

and hence

$$EAB = \frac{EP \cdot k_9(k_6 + k_7)}{k_5 \cdot k_7} \qquad (14\text{-}10)$$

From Eq. (14-4)

$$EAB(k_4 + k_5) = EPQ \cdot k_6 + EA \cdot k_3 \cdot B \qquad (14\text{-}11)$$

Thus

$$EA \cdot k_3 \cdot B = EAB(k_4 + k_5) - EPQ \cdot k_6 \qquad (14\text{-}12)$$

Using the expression for EAB in terms of EPQ, we get

$$EA \cdot k_3 \cdot B = EPQ \cdot \frac{(k_6 + k_7) \cdot (k_4 + k_5)}{k_5 - k_6} \qquad (14\text{-}13)$$

$$= \frac{EPQ}{k_5 \cdot (k_4 k_6 + k_4 k_7 + k_5 k_7)} \qquad (14\text{-}14)$$

$$= \frac{EP \cdot k_9}{k_5 k_7 \cdot (k_4 k_6 + k_4 k_7 + k_5 k_7)} \qquad (14\text{-}15)$$

Therefore,

$$EA = \frac{(EP \cdot k_9)(k_4 k_6 + k_4 k_7 + k_5 k_7)}{k_3 k_5 k_7 \cdot B} \qquad (14\text{-}16)$$

An expression for E in terms of EP is now obtained from either Eq. (14-3) or the steady-state equation for E,

$$\frac{dE}{dt} = EP \cdot k_9 + EA \cdot k_2 - E \cdot k_1 \cdot A = 0 \qquad (14\text{-}17)$$

Hence

$$E \cdot k_1 \cdot A = EP \cdot k_9 + EA \cdot k_2 \qquad (14\text{-}18)$$

Therefore,

$$E = EP \cdot \left(\frac{k_9}{kiA} + \frac{k_2 k_9(k_4 k_6 + k_4 k_7 + k_5 k_7)}{k_1 k_3 k_5 k_7 \cdot A \cdot B} \right) \qquad (14\text{-}19)$$

Using the enzyme conservation equation [Eq. (14-2)], the expression for the total enzyme concentration, e, becomes

$$e = EP\left(1 + \frac{k_9}{k_7} + \frac{k_9(k_6 + k_7)}{k_5 k_7} + \frac{k_9(k_4 k_6 + k_4 k_7 + k_5 k_7)}{k_3 k_5 k_7 \cdot B}\right.$$

$$\left. + \frac{k_9}{k_1 \cdot A} + \frac{k_2 k_9(k_4 k_6 + k_4 k_7 + k_5 k_7)}{k_1 k_3 k_5 k_7 \cdot A \cdot B}\right) \qquad (14\text{-}20)$$

Since from Eq. (14-7), $V_0 = k_9 \cdot EP$, we obtain an expression for e/V_0, given in

$$\frac{e}{V} = \frac{1/k_9 + 1/k_7 + (k_6 + k_7)}{k_5 k_7} + \frac{k_4 k_6 + k_4 k_7 + k_5 k_7}{k_3 k_5 k_7 \cdot B} + \frac{1}{k_1 \cdot A}$$

$$+ \frac{k_2(k_4 k_6 + k_4 k_7 + k_5 k_7)}{k_1 k_3 k_5 k_7 \cdot A \cdot B} \qquad (14\text{-}21)$$

In terms of Dalziel's θ parameters, this equation may be written:

$$\frac{e}{V_0} = \theta_0 + \frac{\theta_1}{A} + \frac{\theta_2}{B} + \frac{\theta_{12}}{AB} \qquad (14\text{-}22)$$

where the θ parameters are the appropriate groups of constants from the Eq. (14-21). If both substrates are considered to be saturating, Eq. (14-22) reduces to

$$\frac{e}{V_0} = \theta_0 \qquad (14\text{-}23)$$

Under these conditions the enzyme is operating at its maximum rate, V_{max}, which is related to θ_0 by

$$\frac{V_{max}}{e} = \frac{1}{\theta_0} \qquad (14\text{-}24)$$

If only one of the substrates is set at a saturating concentration, Eq. (14-22) becomes either Eq. (14-25) or (14-26), depending on the saturating substrate.

$$\frac{e}{V_0} = \theta_0 + \frac{\theta_1}{A} \qquad (14\text{-}25)$$

$$\frac{e}{V_0} = \theta_0 + \frac{\theta_2}{B} \qquad (14\text{-}26)$$

By rearrangement we obtain the Michaelis–Menten form of these equations:

$$V_0 = \frac{V_{max} A}{A + \theta_1/\theta_0} \qquad (14\text{-}27)$$

$$V_0 = \frac{V_{max} B}{B + \theta_2/\theta_0} \qquad (14\text{-}28)$$

Equilibrium Treatment of a Compulsory-Order Mechanism

As discussed for one substrate systems, when the more restrictive equilibrium assumption is applied the derivation of a rate equation is a simpler process. The rate-limiting step is determined by the equilibrium assumption as the catalytic step $EAB \rightarrow E + products$. Product release steps need not be considered, and the enzyme–substrate complexes can be described by the appropriate equilibrium equations.

$$E \ + \ A \xrightarrow{\ K1\ } EA \ + \ B \xrightarrow{\ K2\ } EAB \xrightarrow{\ k\ } EPQ$$

Figure 14-2 Reaction pathway for an equilibrium compulsory-order kinetic mechanism.

For the reaction shown in Fig. 14-2 we can write two equilibrium expressions:

$$K_1 = \frac{E \cdot A}{EA} \tag{14-29}$$

$$K_2 = \frac{EA \cdot B}{EAB} \tag{14-30}$$

The initial velocity V_0 is obtained by substitution from Eqs. (14-29) and (14-30) into the expression $V_0 = k \cdot EAB$, as in

$$V_0 = k \cdot EAB = \frac{k \cdot EA \cdot B}{K_2} = \frac{k \cdot E \cdot A \cdot B}{K_1 \cdot K_2} \tag{14-31}$$

The enzyme conservation equation for the reaction is given by

$$e = E + EA + EAB \tag{14-32}$$

which, using Eqs. (14-29) and (14-30) to substitute into Eq. (14-32) gives, after rearrangement,

$$e = E\left(1 + \frac{A}{K_1} + \frac{A \cdot B}{K_1 \cdot K_2}\right) \tag{14-33}$$

Thus

$$E = \frac{e}{1 + A/K_1 + AB/(K_1 \cdot K_2)} \tag{14-34}$$

and

$$V_0 = \frac{k \cdot e \cdot A \cdot B}{(K_1 \cdot K_2)[1 + A/K_1 + A \cdot B/(K_1 \cdot K_2)]} \tag{14-35}$$

$$= \frac{k \cdot e \cdot A \cdot B}{K_1 \cdot K_2 + A \cdot B + K_2 \cdot A} \tag{14-36}$$

and

$$\frac{e}{V_0} = \frac{K_1 \cdot K_2 + A \cdot B + K_2 \cdot A}{k \cdot A \cdot B} \tag{14-37}$$

$$= \frac{1}{k} + \frac{K_2}{K \cdot B} + \frac{K_1 \cdot K_2}{k \cdot A \cdot B} \tag{14-38}$$

which, in terms of Dalziel's θ parameters, becomes

$$\frac{e}{V_0} = \theta_0 + \frac{\theta_2}{B} + \frac{\theta_{12}}{AB} \tag{14-39}$$

It is apparent that for such a mechanism, θ, is equal to 0.

Random Order, Rapid Equilibrium

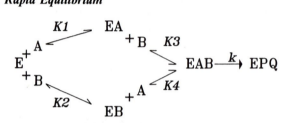

Figure 14-3   Reaction mechanism for a random-order, rapid-equilibrium kinetic mechanism.

As with the previous mechanism, derivation of the rate equation is quite simple. From the mechanism shown in Fig. 14-3, the four equilibrium constants K_1, K_2, K_3, and K_4 are defined by

$$K_1 = \frac{E \cdot A}{EA} \tag{14-40}$$

$$K_2 = \frac{E \cdot B}{EB} \tag{14-41}$$

$$K_3 = \frac{EA \cdot B}{EAB} \tag{14-42}$$

$$K_4 = \frac{EB \cdot A}{EAB} \tag{14-43}$$

The initial rate, V_0, is given by

$$V_0 = kEAB = \frac{k \cdot EA \cdot B}{K_3} = \frac{k \cdot E \cdot A \cdot B}{K_1 \cdot K_3} \tag{14-44}$$

The enzyme conservation equation for this mechanism is

$$e = E + EA + EB + EAB \tag{14-45}$$

which, using Eqs. 14-40 to 14-43, becomes

$$e = E\left(1 + \frac{A}{K_1} + \frac{B}{K_2} + \frac{A \cdot B}{K_1 \cdot K_3}\right) \tag{14-46}$$

Thus

$$V_0 = \frac{k \cdot e \cdot A \cdot B}{K_1 \cdot K_3[1 + A/K_1 + B/K_2 + A \cdot B/(K_1 \cdot K_3)]} \tag{14-47}$$

$$= \frac{k \cdot e \cdot A \cdot B}{K_1 \cdot K_3 + K_3 \cdot A + K_1 \cdot K_3 \cdot B/K_2 + A \cdot B} \tag{14-48}$$

and

$$\frac{e}{V_0} = \frac{K_1 \cdot K_3}{k \cdot A \cdot B} + \frac{K_3}{k \cdot B} + \frac{K_1 \cdot K_3}{K_2 \cdot k \cdot A} + \frac{1}{k} \tag{14-49}$$

However, $K_1 \cdot K_3 = K_2 \cdot K_4$ and therefore $K_1 \cdot K_3/K_2 = K_4$, and Eq. (15-49) becomes

$$\frac{e}{V_0} = \frac{K_1 \cdot K_3}{k \cdot A \cdot B} + \frac{K_3}{k \cdot B} + \frac{K_4}{k \cdot A} + \frac{1}{k} \tag{14-50}$$

which of course fits the generalized form of Dalziel's θ parameter equation,

$$\frac{e}{V_0} = \theta_0 + \frac{\theta_1}{A} + \frac{\theta_2}{B} + \frac{\theta_{12}}{AB} \tag{14-51}$$

Random Order, Steady State

If one does not apply the restrictive equilibrium assumption to the previous mechanism, in place of the equilibrium constants the individual rate constants for substrate or product binding and release must be used. A scheme for these rate constants is shown in Fig. 14-4.

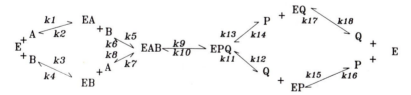

Figure 14-4 Reaction mechanism for a random-order, steady-state kinetic mechanism, showing rate constants for various ligand binding or release steps.

The enzyme conservation equation here includes two additional terms when compared to that for the compulsory order steady-state mechanism, and is given by

$$e = E + EA + EB + EAB + EPQ + EP + EQ \tag{14-52}$$

with EB and EQ being the additional terms.

In the absence of product (which eliminates the consideration of k_{12}, k_{14} and k_{16}), the steady-state expression, derived in an analogous manner to that for the compulsory order steady-state mechanism, is given by

$$\frac{e}{V} = \frac{\text{numerator}}{\text{denominator}} \tag{14-53}$$

where

$$
\begin{aligned}
\text{numerator} \quad &= k_2 k_4 [Y(k_6 + k_8) + k_9] \\
&+ A \cdot \{k_1 \cdot k_4 [Y \cdot (k_6 + k_8) + k_9] + k_2 k_7 (Y k_6 + k_9)\} \\
&+ B \cdot \{k_2 \cdot k_3 [Y \cdot (k_6 + k_8) + k_9] + k_4 k_7 (Y k_8 + k_9)\} \\
&+ A^2 \cdot k_1 k_7 (Y \cdot k_6 k_9) + B^2 \cdot k_3 \cdot k_5 (Y \cdot k_8 + k_9) \\
&+ A \cdot B \cdot \{k_1 \cdot k_5 [(Y + Z) \cdot (k_4 + k_7 \cdot A) + Y k_7]\} \\
&+ k_3 k_7 \cdot [(Y + Z) \cdot (k_2 + k_5 B) + Y k_6] + k_5 k_7 k_9
\end{aligned}
$$

$$\text{denominator} = A \cdot B \cdot k_9 [k_1 \cdot k_5 (k_4 + k_7 \cdot A) + k_3 k_7 (k_2 + k_5 \cdot B)]$$

and where

$$Y = 1 + \frac{k_{10}}{k_{13} + k_{11}}$$

$$Z = \frac{k_9 (k_{17} \cdot k_{15} + k_{17} \cdot k_{11} + k_{13} \cdot k_{15})}{k_{17} \cdot k_{15} (k_{13} + k_{11})}$$

This equation clearly does not fit the linear format of those describing the other mechanisms thus far considered, but rather is of the form

$$\frac{e}{V} = \frac{e + d \cdot A + f \cdot A^2}{a \cdot A + b \cdot A^2} \tag{14-54}$$

where one substrate, B, is held at fixed concentrations, and where a, b, c, d, and f, are functions of the fixed substrate, B, and various rate constants.

The first differential coefficient of Eq. (14-54) is

$$\frac{d(e/V)}{d(1/A)} = \frac{a \cdot c + 2b \cdot c \cdot A + (b \cdot d - a \cdot f)A^2}{a^2 + 2 \cdot a \cdot b \cdot A + b^2 \cdot A^2} \tag{14-55}$$

Under conditions where $b \cdot d < a \cdot f$, the slope of a Lineweaver-Burk plot of e/V versus $1/A$ is negative above some value of A. Apparent substrate inhibition is observed.

The second differential coefficient of Eq. (14-54) is

$$\frac{d^2(e/V)}{d(1/A)^2} = \frac{2F \cdot A^3}{(a + bA)^3} \tag{14-56}$$

where $F = a^2 \cdot f + b^2 \cdot c - a \cdot b \cdot d$ and may be positive or negative depending on the fixed substrate concentration and the relative values of the rate constants that make up a, b, c, d, and f.

If conditions are such that F is positive, the slope of the Lineweaver-Burk plot increases with increasing value of $1/A$, and the plot is concave upward. An equivalent plot of V versus $[A]$ is sigmoidal. If, on the other hand, F is negative, the Lineweaver–Burk plot is concave downward and resembles substrate activation.

Enzyme-Substituted Mechanism

In this mechanism (sometimes also referred to as double displacement or ping-pong) both substrates are not required to be bound to the enzyme together, in distinction to the previous four mechanisms where a ternary EAB complex must be formed. The formal scheme for this mechanism is shown in Fig. 14-5.

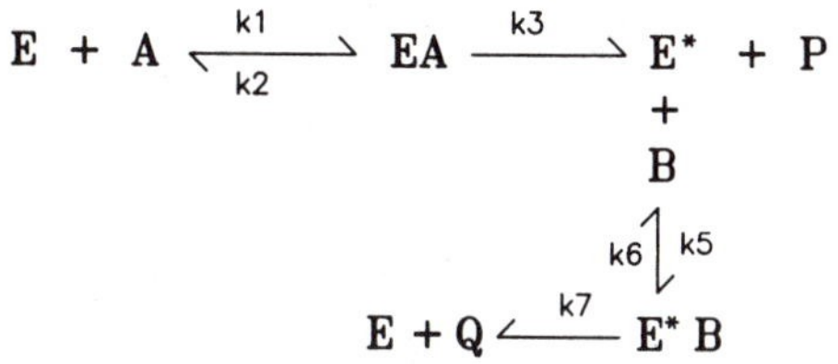

Figure 14-5 Reaction mechanism for an enzyme-substituted mechanism. E* is a covalent intermediate formed from E and A, with release of P.

As with other steady-state mechanisms, the product addition steps need not be considered, and the following steady-state equations are required:

$$\frac{dE}{dt} = EA \cdot k_2 + E^*B \cdot k_7 - E \cdot k_1 \cdot A = 0 \tag{14-57}$$

$$\frac{dE^*B}{dt} = E^* \cdot k_5 \cdot B - E^*B(k_7 + k_6) = 0 \tag{14-58}$$

$$\frac{dEA}{dt} = E \cdot k_1 \cdot A - EA(k_2 + k_3) = 0 \tag{14-59}$$

$$\frac{dE^*}{dt} = EA \cdot k_3 + E^*B \cdot k_6 - E^* \cdot k_5 \cdot B = 0 \tag{14-60}$$

From Eq. (14-58) we get an expression for E*:

$$E^* \cdot k_5 \cdot B = E^*B(k_7 + k_6)$$

Therefore,

$$E^* = \frac{E^*B(k_7 + k_6)}{k_5 \cdot B} \tag{14-61}$$

From Eq. (14-59) we get an expression for EA:

$$E \cdot k_1 \cdot A = EA(k_2 + k_3)$$

Therefore,

$$EA = \frac{E \cdot k_1 \cdot A}{k_2 + k_3} \tag{14-62}$$

For this particular mechanism, the net flow from E to EA, from EA to E*, and from E* to E*B must be the same. Therefore,

$$EA \cdot k_3 = E^*B \cdot k_7$$

Therefore,

$$EA = \frac{E^*B \cdot k_7}{k_3} \tag{14-63}$$

Substituting into Eq. (14-63) from Eq. (14-62) we get

$$E = \frac{E^*B \cdot k_7(k_2 + k_3)}{k_1 \cdot k_3 \cdot A} \tag{14-64}$$

The enzyme conservation equation for the reaction is given by

$$e = E^*B + E^* + E + EA$$

$$= E^*B\left(1 + \frac{k_7 + k_6}{k_5 \cdot B} + \frac{k_7}{k_3} + k_7 \frac{(k_2 + k_3)}{k_1 \cdot k_3 \cdot A}\right) \tag{14-65}$$

Since $V_0 = k_7 \cdot E^*B$, then

$$\frac{e}{V_0} = \frac{1}{k_7} + \frac{1}{k_3} + \frac{k_2 + k_3}{k_1 k_3 \cdot A} + \frac{k_6 + k_7}{k_5 \cdot k_7 \cdot B} \tag{14-66}$$

As with most of the other cases considered, this equation is in the general form of Dalziel's θ parameter equation, where the θ_{12} parameter is equal to 0.

$$\frac{e}{V_0} = \theta_0 + \frac{\theta_1}{A} + \frac{\theta_2}{B} \tag{14-67}$$

TABLE 14-1   Form of kinetic constants for two-substrate mechanisms

Mechanism[a]	V_{max} $1/\theta_0$	$K_m(A)$ θ_1/θ_0	$K_m(B)$ θ_2/θ_0
1	$\dfrac{1}{1/k_9 + 1/k_7 + (k_6 + k_7)/k_5 k_7}$	$\dfrac{1}{k_1[1/k_9 + 1/k_7 + (k_6 + k_7)/k_5 k_7]}$	$\dfrac{k_4 k_6 + k_4 k_7 + k_5 k_7}{k_3[(k_5 k_7/k_9) + k_5 + k_6 + k_7]}$
2	k	0	K_2
3	k	K_4	K_3
4	$\dfrac{1}{1/k_7 + 1/k_3}$	$\dfrac{k_2 + k_3}{k_1 k_3/k_7 + k_1}$	$\dfrac{k_6 + k_7}{k_5 + k_5 k_7/k_3}$

[a] 1, Compulsory order, steady state; 2, compulsory order, equilibrium; 3, random order, rapid equilibrium; 4, enzyme substituted.

TABLE 14-2 Summary of θ parameters for two-substrate mechanisms

Mechanism[a]	θ_0	θ_1	θ_2	θ_{12}
			Parameter	
1	$\dfrac{1}{k_9} + \dfrac{1}{k_7} + \dfrac{k_6 + k_7}{k_5 k_7}$	$\dfrac{1}{k_1}$	$\dfrac{k_4 k_6 + k_4 k_7 + k_5 k_7}{k_3 k_5 k_7}$	$\dfrac{k_2(k_4 k_6 + k_4 k_7 + k_5 k_7)}{k_1 k_3 k_5 k_7}$
2	$\dfrac{1}{k}$	0	$\dfrac{k_2}{k}$	$K_1 \dfrac{k_2}{k}$
3	$\dfrac{1}{k}$	$\dfrac{K_4}{k}$	$\dfrac{K_3}{k}$	$K_1 \dfrac{K_3}{k}$
4	$\dfrac{1}{k_7} + \dfrac{1}{k_3}$	$\dfrac{k_2 + k_3}{k_1 k_3}$	$\dfrac{k_6 + k_7}{k_5 k_7}$	0

[a] 1, Compulsory order, steady state; 2, compulsory order, equilibrium: 3, random order, rapid equilibrium; 4, enzyme substituted.

From these derivations it is clear that four of the five mechanisms considered give rise to the general form of a rate equation for two substrates as proposed by Dalziel. Only the random order, steady-state mechanism yields an equation that cannot, without further simplifying assumptions, give rise to linear Lineweaver–Burk plots. Over limited ranges of substrate concentrations even this mechanism may give an apparently linear Lineweaver–Burk plot, and "fit" to the Dalziel equation. Thus it is important to establish "linearity" over as wide a substrate concentration range as practical. Table 14–1 summarizes the values of the four θ parameters of the generalized rate equation for these four mechanisms. In Table 14–2 expressions for V_{max} and the Michaelis constants for the two substrates A and B are given.

EXPERIMENTAL DETERMINATION OF θ PARAMETERS

Experimentally, θ parameters are determined from a series of initial rate measurements with varied concentrations of the first substrate at several fixed concentrations of the second. The general rate equation may be written in two forms, depending on whether substrate A or substrate B is varied.

$$\frac{e}{V_0} = \left(\theta_0 + \frac{\theta_2}{B} \right) + \left(\theta_1 + \frac{\theta_{12}}{B} \right) \frac{1}{A} \qquad (14\text{-}68)$$

$$\frac{e}{V_0} = \left(\theta_0 + \frac{\theta_1}{A} \right) + \left(\theta_2 + \frac{\theta_{12}}{A} \right) \frac{1}{B} \qquad (14\text{-}69)$$

These two equations form the basis of the experimental determination of θ parameters from kinetic data. The slope and intercept of the appropriate Lineweaver–Burk plot is related to various constants and the concentration of the nonvaried substrate. Primary Lineweaver–Burk plots of data (Fig. 14-6) give slopes and intercepts

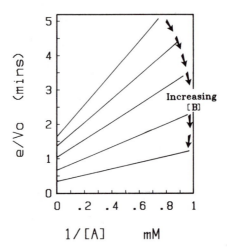

Figure 14-6   Primary Lineweaver–Burk plot of initial rate data according to Eq. (14-68). From this primary plot, slopes and intercepts of plots of e/V_0 versus $1/A$ are obtained at a series of fixed concentrations of B.

that are replotted versus the reciprocal of the concentration of the second substrate; for example, from Eq. (14-68), the slope of the primary plot with A as the varied substrate is given by the expression

$$\text{slope} = \theta_1 + \frac{\theta_{12}}{B} \tag{14-70}$$

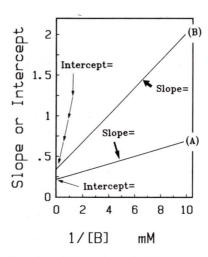

Figure 14-7   Secondary plots of (A) slopes, or (B) intercepts obtained from primary plots versus $1/B$. Individual θ parameters are obtained from the slopes and intercepts of the secondary plots as indicated.

and the intercept by the expression

$$\text{intercept} = \theta_0 + \frac{\theta_2}{B} \tag{14-71}$$

Clearly, as illustrated in Fig. 14-7, individual θ parameters are then obtained from the slopes and intercepts of such secondary plots, which are constructed by plotting either slopes or intercepts obtained from the primary plots versus $1/B$, according to Eq. (14-70) or Eq. (14-71), respectively.

Since the general rate equation is symmetrical [compare Eqs. (14-68) and (14-69)], it does not matter whether the primary plots are constructed with A or B as the varied substrate—equivalent values for the θ parameters are obtained.

The principal advantage of analyzing kinetic data with the Dalziel equation is that the initial rate parameters are obtained without any prior assumptions concerning the formal kinetic mechanism of the enzyme. It is also possible to assess the experimental accuracy of the parameters by suitable statistical analysis of the primary kinetic data rather than by using overall fit to one particular kinetic mechanism.

EXPERIMENTAL DISTINCTION BETWEEN VARIOUS TWO-SUBSTRATE KINETIC MECHANISMS

Based on the discussion to date and the rate equations derived for the various two-substrate kinetic mechanisms we are concerned with here, several methods for distinguishing among these mechanisms can be considered. Examination of the derived rate equations reveals two basic approaches, one involving the overall form of the rate equation and the other the physical significance of individual θ parameters in the generalized rate equations for various mechanisms.

Primary Plots of Kinetic Data

While all the mechanisms except the random-order, steady-state mechanism conform to the generalized rate equation proposed by Dalziel, two of the four are lacking one term in the rate equation. With the enzyme-substituted mechanism, the θ_{12} term is equal to 0. This has immediate implications for the pattern of Lineweaver–Burk plots obtained when one substrate concentration is varied at a series of fixed concentrations of the other. From the rate equation for the mechanism [Eq. (14-66)], it is evident that in Lineweaver–Burk plots with either A or B as the varied substrate, only the intercept changes as a function of the nonvaried substrate concentration. The slopes of the Lineweaver–Burk plots are either $(k_2 + k_3)/(k_1 k_3)$, when A is the varied substrate, or $(k_6 + k_7)/(k_5 k_7)$, when B is the varied substrate, and are of course independent of substrate concentration. The resultant series of parallel lines obtained are a quite distinctive feature of enzyme-substituted mechanisms.

The ternary complex mechanisms are all characterized by a positive value for θ_{12}, requiring that in such mechanisms slopes of Lineweaver–Burk plots change as a function of the concentration of the nonvaried substrate. With one of the three ternary complex mechanisms, whose rate equations do fit into the generalized Dalziel rate equations however, there is a distinctive feature. In the compulsory-order, equilibrium mechanism the θ_1 parameter is equal to 0. From the rate equation for the mechanism [Eq. (14-38)], it is apparent that in Lineweaver–Burk plots with B as the varied substrate, the intercept is equal to $1/k$ and is independent of the concentration of substrate A. With such a mechanism Lineweaver–Burk plots with B as the varied substrate intersect on the e/V axis for various fixed concentrations of the substrate A. With A as the varied substrate, of course, a normal pattern of plots is obtained: Both slope ($= K_1 \cdot K_2/B$) and intercept ($= 1/k + K_2/k \cdot B$) vary as the concentration of B is changed.

As we have seen in the previous discussion, several kinetic mechanisms have characteristic forms of the generalized rate equation which lead to distinctive patterns of Lineweaver–Burk plots. On the surface, compulsory-order, equilibrium, and enzyme-substituted mechanisms are easily distinguishable from the others considered. Care must be taken, however, in assigning a θ-parameter value equal to 0 and the experimental determination of θ parameters has to cover a sufficient range of substrate concentrations to allow the unequivocal demonstration of the presence or absence of a particular θ parameter. Consider the case of a random-order, rapid-equilibrium mechanism where one of the substrates binds extremely tightly to the enzyme in the presence of the other. The effect on the overall rate equation [Eq. (14-50)] of extremely tight binding of substrate A to EAB (i.e. $K_4 \to 0$) is that the term $K_4/(KA)$ tends to approach 0 and the mechanism may, if sufficiently low concentrations of A are not used in θ-parameter determinations, resemble a compulsory-order, equilibrium mechanism.

Such considerations aside, however, the major problem facing the enzymologist considering two substrate mechanisms is the distinction between a compulsory-order, steady-state and a random-order, rapid-equilibrium mechanism. A number of simple approaches are examined here (others are considered in Chap. 19) that are frequently used to assist in the distinction between these mechanisms. They can be grouped into three areas: (1) the use of alternate substrates, (2) comparison of kinetically derived parameters with independently derived parameters, and (3) the use of substrate analogs as reversible inhibitors.

Use of Alternative Substrates

Where several alternative substrates for a particular enzyme-catalyzed reaction are available it is often possible, from the results of simple initial rate measurements, to distinguish between a compulsory-order, steady-state mechanism and a random-order, rapid-equilibrium mechanism. The basis for such a distinction is the fact that from the rate equation for the compulsory-order, steady-state mechanism it is evident that $\theta_1 = 1/k_1$, where k_1 is the rate constant for the first substrate (A) binding to the enzyme. In the random-order, rapid equilibrium mechanism, this θ parameter does not have this simple physical significance. Consider then the effect of determining the four θ parameters in the generalized rate equation with a series of alternative substrates, B, B_1, B_2, and so on. If the mechanism is in fact compulsory order, steady state with B as the second substrate, the value of θ_1 determined for the different alternate substrates is constant. If the mechanism is really random order, rapid equilibrium or compulsory order, steady state, with B as the *first* substrate, θ_1 varies as the nature of B varies. To govern against this possibility it is necessary to use alternative substrates to A as well as to B. If θ_1 varies as a function of B, the determinations are repeated with a single B but a variety of alternative substrates for A. Constant values for the θ_2 parameter as the nature of A is varied indicate a compulsory order, steady-state mechanism with B as the first subtrate.

As is discussed in more detail in Chap. 19 alternative substrates can be used in a variety of other ways to assist in the establishment of kinetic mechanisms. The approach described here is a simple and readily interpretable one that has been used with a variety of enzymes to help distinguish between a compulsory-order, steady-state and a random-order, rapid-equilibrium mechanism.

Comparison of Kinetically Derived Constants with Independently Determined Values

As was outlined earlier and summarized in Table 14-1, the θ parameters of the generalized rate equation have specific physical significance depending on the kinetic mechanism. This allows a comparison to be made between various θ parameters (or ratios of θ parameters) and directly determined values for specific constants in the appropriate mechanism.

Consider the rate equation derived earlier for a random-order, rapid-equilibrium mechanism [Eq. (14-50)]. The Michaelis constants for the reaction (θ_1 and θ_2 divided by θ_0) give values for K_3 and K_4, the dissociation constants for A and B from the ternary EAB complex, respectively. Alternatively, if we take the θ_{12} parameter and divide by the θ_2 parameter, we get K_1, the dissociation constant for A from the EA complex. Since as discussed earlier, $K_1 K_3 = K_2 K_4$, the θ_{12} parameter divided by θ_1 gives K_2, the dissociation constant for B from the EB complex. Table 14-3 summarizes ratios between θ parameters for particular mechanisms.

While both the compulsory-order, steady-state and the random-order, rapid-equilibrium mechanism give values for K_1, the latter mechanism, as you would expect, is the only one that gives a value for K_2. Direct determination of K_2 by such

TABLE 14-3 Relationships between θ parameters for two-substrate systems

Mechanism[a]	Relationship			
	θ_1/θ_0	θ_2/θ_0	θ_{12}/θ_1	θ_{12}/θ_2
1	Complex	Complex	Complex	$= K_1$
2	$= 0$	Complex	$= 0$	$= K_1$
3	$= K_4$	$= K_3$	$= K_2$	$= K_1$
4	Complex	Complex	$- 0$	$- 0$

[a] 1, Compulsory order, steady state: 2, compulsory order, equilibrium: 3, random order, rapid equilibrium; 4, enzyme substituted.

approaches as equilibrium dialysis or spectroscopic titrations (both of which are considered in Chap. 17) and comparison with the ratio θ_{12}/θ_1 can give supportive evidence for a random-order, rapid-equilibrium mechanism.

The second facet of direct comparison with independently determined parameters involves the determination of individual rate constants for particular steps in the reaction. As we have discussed, in a compulsory-order, steady-state mechanism the θ_1 parameter $= 1/k_1$, where k_1 is the rate constant for substrate A binding to enzyme. Direct determination of k_1 by rapid reaction techniques such as stopped or continuous flow, or temperature jump (all of which are discussed in Chap. 18), and comparison with the θ_1 parameter determined via enzyme kinetics gives information about whether or not a compulsory-order, steady-state mechanism is applicable.

The comparison of directly and independently determined constants with those derived from initial rate kinetic studies by assuming particular formal kinetic mechanisms reiterates the truism stated in the introduction to this chapter. Agreement between such constants is merely consistent with the mechanism: Disagreement disproves the mechanism.

Use of Analog Inhibitors

The basic concepts and equations associated with the idea of a competitive inhibitor being an analog of one of the substrates of an enzyme were discussed in Chap. 13; with two-substrate systems, however, a more complex situation holds. A substrate analog of one of the substrates can bind in place of that substrate in either one or two sites, depending on whether a compulsory-order mechanism or a random-order mechanism exists. Consider an analog (BX) of substrate B in these mechanisms. In compulsory order, steady state, BX can bind only to an EA complex ($E \cdot A + BX \rightleftharpoons E \cdot A \cdot BX$) and the interaction can be described in terms of a dissociation constant K_i. Derivation of the rate equation for such a mechanism in the presence of BX leads to the generalized format

$$\frac{e}{V_0} = \theta_0 + \frac{\theta_1}{A} + \frac{\theta_2}{B}\left(1 + \frac{BX}{K_i}\right) + \frac{\theta_{12}}{AB} \tag{14-72}$$

With the random-order, rapid-equilibrium mechanism however, BX can combine to free enzyme as well as to the EA complex, and two additional steps must be considered in the derivation of the rate equation, $E + BX \rightleftharpoons E \cdot BX$ and $E \cdot A + BX \rightleftharpoons E \cdot A \cdot BX$, with dissociation constants of K_i' and K_i, respectively.

Derivation of the rate equation with the addition of these two steps yields

$$\frac{e}{V_0} = \theta_0 + \frac{\theta_1}{A} + \frac{\theta_2}{B}\left(1 + \frac{BX}{K_i}\right) + \frac{\theta_{12}}{AB}\left(1 + \frac{BX}{K_i'}\right) \tag{14-73}$$

Consider first Lineweaver–Burk plots with B as the varied substrate. With either mechanism only the slope is affected and the inhibition, as expected, is competitive, with the slope in the presence of the inhibitor being

$$(a) = \theta_2\left(1 + \frac{BX}{K_i}\right) + \frac{\theta_{12}}{A} \tag{14-74}$$

or

$$(b) = \theta_2 \left(1 + \frac{BX}{K_i}\right) + \theta_{12}\left(1 + \frac{BX}{K_i'}\right) \qquad (14\text{-}75)$$

depending on whether a compulsory-order mechanism (a) or a random-order mechanism (b) is operating.

When Lineweaver–Burk plots with A as the varied substrate are examined, however, we find that for either mechanism the intercept term is affected, but with the random-order mechanism the slope term is affected as well. With a compulsory-order, steady-state mechanism an analog of B is an uncompetitive inhibitor with respect to A. With a random-order, rapid-equilibrium mechanism, an analog of B is a noncompetitive inhibitor with respect to A.

For the compulsory-order mechanism, the intercept becomes

$$\text{intercept} = \theta_0 + \frac{\theta_2}{B}\left(1 + \frac{BX}{K_i}\right) \qquad (14\text{-}76)$$

while in the random-order mechanism the intercept and slope are given by

$$\text{intercept} = \theta_0 + \theta_2\left(1 + \frac{BX}{K_i}\right) \qquad (14\text{-}77)$$

$$\text{slope} = \theta_1 + \frac{\theta_{12}}{B}\left(1 + \frac{BX}{K_i'}\right) \qquad (14\text{-}78)$$

A further important conclusion concerning reversible inhibitors in two-substrate systems is also apparent from this discussion. That is, the determination of K_i values requires knowledge of the values of individual θ parameters in the presence and absence of the inhibitor.

EXAMPLES OF THE DETERMINATION OF KINETIC MECHANISMS

Histone Acetyltransferase

Reference: L.-J. Wong, and S. Wong, *Biochemistry*, 22, 4637–4642 (1983).

This enzyme catalyzes the acetylation of nuclear histones using acetyl-CoA as substrate. The second substrate is the side-chain amino groups of lysine residues close to the N termini of various histones. Initial rate studies with the enzyme purified from calf thymus can be conducted by following the transfer of tritiated acetyl groups from tritiated acetyl-CoA into histone, using ethanol precipitation of the protein substrate and product to separate the labeled product from remaining tritiated acetyl-CoA. Lineweaver–Burk plots with either histone or acetyl-CoA as the varied substrate are given in Fig. 14-8. Both primary plots obtained show a series of parallel lines indicating the absence of the θ_{12} parameter from the rate equation. These data clearly prove that histone acetyltransferase follows an enzyme-substituted mechanism.

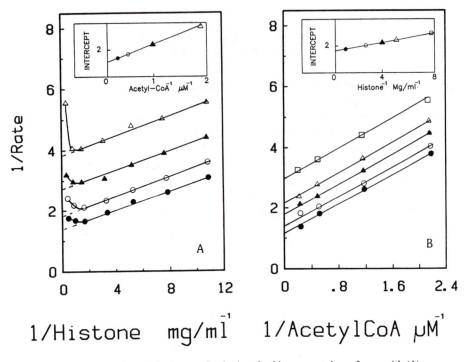

Figure 14-8 Primary Lineweaver–Burk plots for histone acetyltransferase with (A) histone concentrations varied at several acetyl-CoA concentrations, and (B) acetyl-CoA concentrations varied at several histone concentrations. The inserts are the secondary plots of intercepts from A or B versus the reciprocal of the concentration of the nonvaried substrate in the primary plots. (Reprinted with permission from: L.-J. C. Wong and S. S. Wong, *Biochemistry*, *22*, 4637–4641. Copyright 1983 American Chemical Society, Washington, D. C.)

Liver Alcohol Dehydrogenase

Reference: K. Dalziel, and F. Dickinson, *Biochem. J. 100*, 34–46 (1966).

This enzyme catalyzes the reversible oxidation–reduction of a variety of alcohols and aldehydes using NAD or NADH. The reaction is readily monitored either by following absorbance changes at 340 nm or by following fluorescence changes with emission at 450 nm and excitation at 340 nm (since the reduced coenzyme absorbs at 340 nm whereas the oxidized coenzyme does not). Using a wide range of substrate concentrations, the oxidation reaction has been studied with a variety of primary alcohols, and the reduction reaction has been studied with the corresponding aldehydes. From these it was found that all four parameters of the generalized equation were present, eliminating various mechanisms from further consideration. Table 14-4 shows the θ parameters obtained for both oxidation and reduction reactions with various alcohols and aldehydes.

TABLE 14-4   Parameters for reactions

Substrate	θ_0 (sec)	θ_1 (μM-sec)	θ_2 (μM-sec)	θ_{12} (mM^2-sec)
Ethanol	0.37	1.1	66	0.0072
Propan-1-ol	0.31	1.2	19	0.0012
Butan-1-ol	0.35	1.1	4	0.0004
2-Methylpropan-1-ol	0.34	1.7	40	0.0032
Acetaldehyde	0.0075	0.1	3.3	[a]
Propionaldehyde	0.0095	0.14	0.43	[a]
Butyraldehyde	0.0075	0.10	0.17	[a]
2-Methylpropionaldehyde	0.009	0.12	0.25	[a]

[a] No accurate value obtainable.

With both the oxidative and the reductive reactions, the θ_1 parameter is invariant with the nature of the alcohol or the aldehyde substrate, even though the θ_2 and the θ_{12} parameters change. These results are consistent with a compulsory order of substrate addition for both reactions with coenzyme as the required first substrate. Interestingly, when secondary alcohols or aldehydes are used as substrates, these relationships do not hold, suggesting that while the various primary alcohols all follow a steady-state compulsory order mechanism, the secondary alcohols may follow a different one.

Galactosyltransferase

Reference: J. E. Bell, T. Beyer, and R. Hill, *J. Biol. Chem.*, *251*, 3008–3013 (1976).

This enzyme catalyzes the transfer of galactose from UDP-galactose to an acceptor, which may be a glycoprotein or a small saccharide. The reaction is most conveniently followed using radioactive galactose in UDP-galactose and separating product from unused substrate chromatographically. Although the enzyme requires manganese, in the presence of saturating concentrations of $MnCl_2$, it can be regarded as a two-substrate enzyme. Initial rate studies with three different acceptor substrates showed that although the θ_1 parameter varied from 0.207 to 0.00068 depending on the nature of the acceptor substrate; all four parameters had positive values. These observations eliminated the enzyme-substituted mechanism and were inconsistent with a simple compulsory-order steady-state mechanism. Inhibition studies, using alternative substrates as analog inhibitors, are possible with this enzyme since the products obtained using, for example, N-acetylglucosamine and ovalbumin as acceptors, are readily separable. Figure 14-9 illustrates the results of experiments using N-acetylglucosamine as an inhibitor of the galactosylation of ovalbumin.

As expected, N-acetylglucosamine is a competitive inhibitor with respect to ovalbumin. When UDP-galactose concentrations are varied, N-acetylglucosamine is a noncompetitive inhibitor. Similar studies were performed with UDP-glucose,

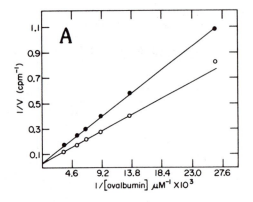
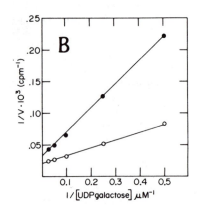

Figure 14-9   Inhibition of galactosyltransferase by *N*-acetylglucosamine with either ovalbumin (A) or UDP-galactose (B) as the varied substrate. ○, No *N*-acetylgluco-samine; ●, +10 m*M* *N*-acetylglucosamine. [From J. E. Bell, T. A. Beyer and R. L. Hill, *J. of Biol. Chem.*, *251*, 3003–3013 (1976). Reprinted with the permission of the copyright owner, The American Society of Biological Chemists, Inc., Bethesda, Md.]

which was a competitive inhibitor versus UDP-galactose and a noncompetitive in-hibitor versus ovalbumin. Table 14-5 summarizes the expected inhibition patterns for various possible two-substrate mechanisms for this enzyme.

The experimentally observed inhibition patterns are consistent only with the equilibrium random-order mechanism.

"Norvaline" Dehydrogenase

Reference: C. LiMuti and J. E. Bell, *Biochem. J.*, *211*, 99–107 (1983).

The enzyme glutamate dehydrogenase utilizes norvaline as a substrate for oxidative deamination, although the preferred substrate is glutamate. In detailed kinetic studies with norvaline as the varied substrate, Lineweaver–Burk plots could not be described

TABLE 14-5   Expected inhibition patterns[a]

	Compulsory order		
Inhibitor	UDP-galactose first	Acceptor first	Equilibrium random order
UDP-glucose			
versus UDP-Gal	C	C	C
versus ovalbumin	NC	UC	NC
N-Acetylglucosamine			
versus UDP-Gal	UC	NC	NC
versus ovalbumin	C	C	C

[a] C, Competitive inhibition; UC, uncompetitive inhibition; NC, noncompeti-tive inhibition.

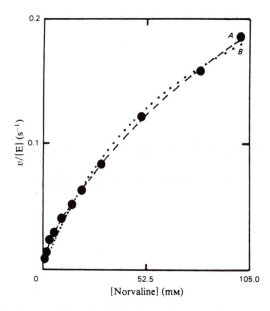

Figure 14-10 V_0 versus [norvaline] plot of data obtained with "norvaline" dehydrogenase showing theoretical curves for an equilibrium random-order mechanism ($\cdot \cdot$) and a steady-state random mechanism (——). (Reprinted with permission from: C. LiMuti and J. E. Bell, *Biochem. J.*, *211*, 99–107. Copyright 1983 The Biochemical Society, London.)

as linear when a wide range of substrate concentrations were used, suggesting the possibility of a steady-state random-order mechanism. Inhibition patterns with a variety of substrate analogs obtained using a limited range of substrate concentrations were consistent with a random order of substrate addition. As shown in Fig. 14-10, velocity versus [norvaline] plots are best described with an equation [in the form of Eq. (14-54)] containing second-order terms in the norvaline concentration.

From the theoretical fits to the data it is apparent that the steady-state mechanism is more appropriate than the equilibrium mechanism.

In this chapter a variety of enzyme initial rate kinetic approaches have been described that can yield information concerning the formal kinetic mechanism of an enzyme. In many cases initial rate kinetic studies must be used in conjunction with independently determined values for particular constants in the mechanism. This involves substrate equilibrium binding studies or studies of the rates of substrate binding. There are described in detail in Chap. 17 and 18, respectively.

15

Three-Substrate Kinetics
and Related Topics

INTRODUCTION

In Chap. 14 we extended the ideas of steady state and rapid equilibrium (first developed in Chap. 13) and applied them to various formal kinetic mechanisms for an enzyme with two substrates. The equations derived were put into the generalized Dalziel form of the initial rate equation. With the various two-substrate mechanisms we considered the use of experimentally determined initial rate parameters to distinguish among them, and came to the conclusion that to do this for certain mechanisms additional information is needed in the form of independently determined values of particular rate constants or equilibrium constants.

If instead of a two-substrate system we include a third substrate, it might intuitively be expected that this distinction is even more complex. As we will see, quite to the contrary, most three-substrate formal kinetic mechanisms can be identified on the basis of initial rate kinetic studies alone.

In this chapter we also consider two other situations where much information can be obtained from initial rate kinetic studies. Both are related to the general principles of three-substrate enzyme kinetics in that a third ligand is involved in an enzyme reaction with two substrates.

In the first situation, a regulatory ligand interacts with one or more of the enzyme complexes in a two-substrate ternary complex mechanism and alters the catalytic rate constant of the reaction. At nonsaturating concentrations of the regulatory ligand there are two forms of the enzyme ternary complex producing product, one with the regulatory ligand bound and the other with the normal ternary complex formed in the absence of the regulatory ligand. Initial rate studies can give a wealth of information concerning the interaction of the regulatory ligand with various enzyme complexes.

In the second situation we examine various kinetic aspects of metal-ion-dependent enzymes. In these systems one of the major questions that must be answered is: What are the true substrates of the enzyme? It is quite possible that one or another of the substrates can only interact with the enzyme in the form of a metal complex.

Finally, in this chapter we examine briefly several other enzyme kinetic approaches that have been used. However, as will be discussed, these techniques, which involve systems where products are added as inhibitors or build up significantly during the course of the reaction, do not find the general usage that the other techniques we have discussed in Chaps. 13, 14, and in this chapter have.

THREE-SUBSTRATE KINETICS

As with two-substrate enzyme kinetics we can separate possible formal kinetic mechanisms into two classes. In the first, all the substrates must be bound prior to the appearance of product and the mechanism is known as "quaternary complex". In the second class, one or more products appear before all the substrates have bound and, as with the two-substrate systems, the mechanism is said to be "enzyme substituted." With each of these classes there are considerations regarding the order of substrate addition and whether or not various steps are in rapid equilibrium or must be treated using the less restrictive steady-state assumption.

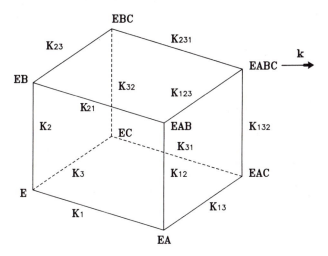

Figure 15-1 Random order of substrate addition in a three-substrate mechanism. The rapid-equilibrium assumption is made, allowing the various equilibrium constants to be included in the scheme. K_1, etc. are the equilibrium constants for A, etc. binding to free enzyme; K_{12}, etc. are the equilibrium constants for B, etc. binding to an EA-type complex; and K_{123}, etc. are the equilibrium constants for C, etc. binding to an EAB-type complex.

Formal Mechanisms

Five formal three-substrate mechanisms are considered. The first is the completely random-order addition of substrates in a quaternary complex mechanism, shown in Fig. 15-1. For the purposes of deriving a rate equation for this mechanism, the rapid-equilibrium assumption is made.

The second general mechanism considered is one involving a compulsory order of addition of each substrate, as shown in Fig. 15-2. Although this mechanism can be treated using either an equilibrium approach or the steady-state approach, for the purpose of this discussion we treat it using the steady-state assumption.

$$E+A \underset{k2}{\overset{k1}{\rightleftharpoons}} EA+B \underset{k4}{\overset{k3}{\rightleftharpoons}} EAB+C \underset{k6}{\overset{k5}{\rightleftharpoons}} EABC \underset{k'}{\overset{k}{\rightleftharpoons}} EPQ \underset{k8}{\overset{k7}{\rightleftharpoons}} Q+EP \underset{k10}{\overset{k9}{\rightleftharpoons}} P+E$$

Figure 15-2 Compulsory order of subbstrate addition in a three-substrate mechanism.

The final quaternary complex mechanism to be considered is illustrated in Fig. 15-3. In this mechanism there is an obligatory first substrate followed by a random-order addition of the second and third. The addition of the first substrate is usually treated using the steady-state approach, while the random order of the subsequent substrate additions is handled by the rapid-equilibrium assumption. As with two-substrate random-order mechanisms such treatment assumes that the steps indicated as k_8 and k_{10} are much faster than the subsequent rate of catalytic interconversion of the quaternary complexes.

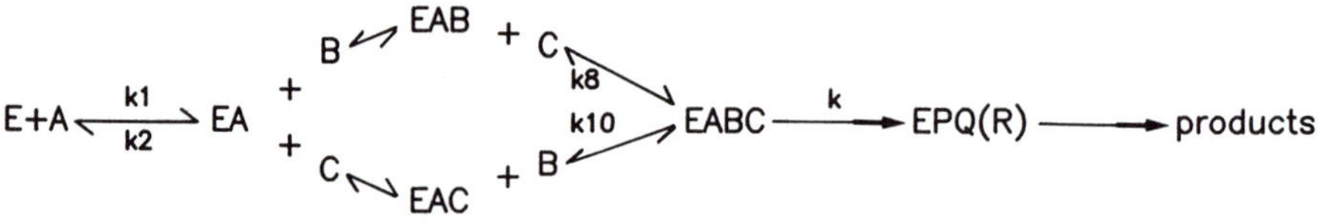

Figure 15-3 Quaternary complex mechanism with the compulsory addition of the first substrate followed by a random addition of the second and third substrates.

Two enzyme-substituted mechanisms are examined. In the first, shown in Fig. 15-4, there is a triple transfer involving two different enzyme-substituted species.

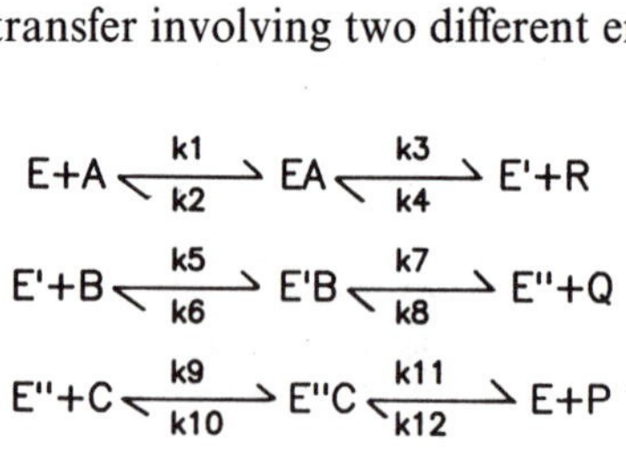

$$E+A \underset{k2}{\overset{k1}{\rightleftharpoons}} EA \underset{k4}{\overset{k3}{\rightleftharpoons}} E'+R$$

$$E'+B \underset{k6}{\overset{k5}{\rightleftharpoons}} E'B \underset{k8}{\overset{k7}{\rightleftharpoons}} E''+Q$$

$$E''+C \underset{k10}{\overset{k9}{\rightleftharpoons}} E''C \underset{k12}{\overset{k11}{\rightleftharpoons}} E+P$$

Figure 15-4 Enzyme-substituted mechanism for a three-substrate system involving a triple transfer.

In this mechanism there is an obligatory order of substrate addition and product release, and it is most simply treated by the steady-state approach.

In the other enzyme-substituted mechanism that we consider here, there is formation of a ternary EAB complex prior to formation of the enzyme-substituted intermediate (and release of the first product) and subsequent addition of the final substrate. The addition of the first two substrates can be ordered, as illustrated in Fig. 15-5, or random.

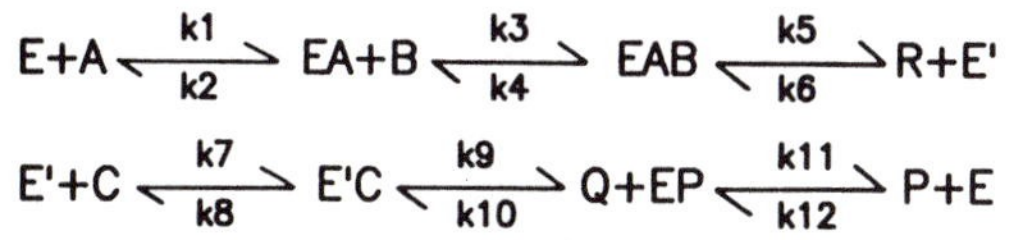

$$E+A \underset{k2}{\overset{k1}{\rightleftharpoons}} EA+B \underset{k4}{\overset{k3}{\rightleftharpoons}} EAB \underset{k6}{\overset{k5}{\rightleftharpoons}} R+E'$$

$$E'+C \underset{k8}{\overset{k7}{\rightleftharpoons}} E'C \underset{k10}{\overset{k9}{\rightleftharpoons}} Q+EP \underset{k12}{\overset{k11}{\rightleftharpoons}} P+E$$

Figure 15-5   Enzyme-substituted mechanism with an ordered formation of a ternary complex prior to generation of the enzyme-substituted intermediate.

If the ternary complex formation is random ordered, the equilibrium approach can be employed, provided that the "off"-velocity constants are sufficiently rapid to allow the equilibrium condition to be reached. The ordered and random equilibrium versions of this mechanism give the same generalized rate equation.

Generalized Rate Equations

Each of the three substrate mechanisms mentioned can be represented by a generalized rate equation of the form given in

$$\frac{e}{V_0} = \theta_0 + \frac{\theta_1}{[A]} + \frac{\theta_2}{[B]} + \frac{\theta_3}{[C]} + \frac{\theta_{12}}{[A][B]} + \frac{\theta_{13}}{[A][C]} + \frac{\theta_{23}}{[B][C]} + \frac{\theta_{123}}{[A][B][C]} \quad (15\text{-}1)$$

Equation (15-1) can be rearranged to give

$$\frac{e}{V_0} = \theta_0 + \frac{\theta_2}{[B]} + \frac{\theta_3}{[C]} + \frac{\theta_{33}}{[B][C]} + \left(\theta_1 + \frac{\theta_{12}}{[B]} + \frac{\theta_{13}}{[C]} + \frac{\theta_{123}}{[B][C]} \right) \frac{1}{[A]} \quad (15\text{-}1a)$$

which is in the form of a Lineweaver–Burk equation with A as the varied substrate at fixed concentration of the other two substrates, B and C. Equation (15-1) can, of course, be rearranged in a manner similar to that in Eq. (15-1a) to yield Lineweaver–Burk equations with either B or C as the varied substrate.

The only one of the five considered mechanisms to contain all eight terms of this equation is the rapid-equilibrium random-order mechanism of Fig. 15-1. The completely ordered mechanism (Fig. 15-2) lacks the θ_{13} term while the obligatory first substrate mechanism (Fig. 15-3) lacks the θ_{12} term and the θ_{13} term. Each of the quaternary complex mechanisms contains the θ_{123} term of Eq. (15-1).

In contrast, the various enzyme-substituted mechanisms (Figs. 15-4 and 15-5) each lack the θ_{123} term of Eq. (15-1). The triple transfer mechanism of Fig. 15-4 lacks four of the eight parameters; the θ_{12}, θ_{13}, and θ_{23} parameters are absent in addition to the θ_{123} parameter. The enzyme-substituted mechanism of Fig. 15-5 lacks the θ_{23} and θ_{13} parameters in addition to the θ_{123} parameter. In the discussion of Fig. 15-5 we considered the possibility that the ternary complex is formed by either

TABLE 15-1 Values of individual rate parameters for various three-substrate mechanisms

Mechanism[a]	Parameter							
	θ_0	θ_1	θ_2	θ_3	θ_{12}	θ_{13}	θ_{23}	θ_{123}
1	$\dfrac{1}{k}$	$\dfrac{K_{231}}{k}$	$\dfrac{K_{132}}{k}$	$\dfrac{K_{123}}{k}$	$\dfrac{K_{132}\cdot K_{31}}{k}$	$\dfrac{K_{123}\cdot K_{21}}{k}$	$\dfrac{K_{123}\cdot K_{12}}{k}$	$\dfrac{K_1\cdot K_{12}\cdot K_{123}}{k}$
2		$\dfrac{1}{k_1}$	$\dfrac{1}{k_3}$	$\dfrac{Y}{k_5}$	$\dfrac{k_2}{k_1\cdot k_3}$	0	$\dfrac{k_4\cdot Y}{k_3\cdot k_5}$	$\dfrac{k_2\cdot k_4\cdot Y}{k_1\cdot k_3\cdot k_5}$
3	$\dfrac{1}{k}$	$\dfrac{1}{k_1}$	$\dfrac{K_{132}}{k}$	$\dfrac{K_{123}}{k}$	0	0	$\dfrac{K_{123}\cdot K_{12}}{k}$	$\dfrac{K_2\cdot K_{123}\cdot K_{12}}{k_1\cdot k}$
4	$\dfrac{1}{k_3}+\dfrac{1}{k_7}+\dfrac{1}{k_{11}}$	$\dfrac{k_2+k_3}{k_1\cdot k_3}$	$\dfrac{k_6+k_7}{k_5\cdot k_7}$	$\dfrac{k_{10}+k_{11}}{k_9\cdot k_{11}}$	0	0	0	0
5(a)	$\dfrac{1}{k_5}+\dfrac{1}{k_9}+\dfrac{1}{k_{11}}$	$\dfrac{1}{k_1}$	$\dfrac{k_4+k_5}{k_3\cdot k_5}$	$\dfrac{k_8+k_9}{k_7\cdot k_9}$	$\dfrac{k_2(k_4+k_5)}{k_1\cdot k_3\cdot k_5}$	0	0	0
5(b)	$\dfrac{1}{k_5}+\dfrac{1}{k_9}+\dfrac{1}{k_{11}}$	$\dfrac{K_{21}}{k}$	$\dfrac{K_{12}}{k}$	$\dfrac{k_8+k_9}{k_7\cdot k_9}$	$\dfrac{K_1\cdot K_{12}}{k_5}$	0	0	0

$$Y = \frac{k_6\cdot k' + k_6\cdot k_7 + k\cdot k_7}{kk_7}$$

[a] 1, **Rapid equilibrium**, random order; 2, compulsory order, steady state; 3, compulsory first substrate, random second and third; 4, triple transfer, enzyme substituted; 5, ternary complex, enzyme substituted: (a) compulsory, (b) random.

an ordered or a random addition of the first two substrates. The generalized form of the rate equation is identical in either case. The actual values of the individual rate parameters for each of these mechanisms are given in Table 15-1.

Determination of θ Parameters

The eight initial rate θ parameters of Eq. (15-1) are determined in a manner analogous to that described in Chap. 14 for the parameters in a two-substrate system. Primary Lineweaver–Burk plots are constructed with one of the three substrate concentrations varied at fixed concentrations of the other two.

The slope and intercept of such a Lineweaver–Burk plot with A as the varied substrate are functions of four parameters and the concentrations of the other two substrates.

Primary plot:

$$\text{slope} = \theta_1 + \frac{\theta_{12}}{[B]} + \frac{\theta_{13}}{[C]} + \frac{\theta_{123}}{[B][C]} \tag{15-2a}$$

$$\text{intercept} = \theta_0 + \frac{\theta_2}{[B]} + \frac{\theta_3}{[C]} + \frac{\theta_{23}}{[B][C]} \tag{15-2b}$$

The concentration of one of the two remaining substrates is varied and at each concentration a primary plot obtained. The slopes and intercepts are plotted in secondary plots as a function of the reciprocal of the second varied substrate concentration. Two secondary plots are obtained, one of the primary plot slopes and one of the primary plot intercepts. The slopes and intercepts of these secondary plots are given by

Secondary plot of slopes (primary) versus $1/[B]$:

$$\text{slope} = \theta_{12} + \frac{\theta_{123}}{[C]} \tag{15-3a}$$

$$\text{intercept} = \theta_1 + \frac{\theta_{13}}{[C]} \tag{15-3b}$$

Secondary plot of intercepts (primary) versus $1/[B]$:

$$\text{slope} = \theta_2 + \frac{\theta_{23}}{[C]} \tag{15-4a}$$

$$\text{intercept} = \theta_0 + \frac{\theta_3}{[C]} \tag{15-4b}$$

Each of the secondary plot slopes and intercepts is a function of two parameters from Eq. (15-1) and the concentration of the third substrate. Two sets of secondary plots are obtained with varied concentrations of the third substrate and tertiary plots made of either slope or intercept from the appropriate secondary plot versus the reciprocal of the third substrate. Four tertiary plots are obtained and all eight initial rate parameters are determined from the appropriate slopes and intercepts.

Tertiary plots from secondary plots (intercepts):

(a) Secondary slopes:

$$\text{slope} = \theta_{23} \tag{15-5a}$$

$$\text{intercept} = \theta_2 \tag{15-5b}$$

(b) Secondary intercepts:

$$\text{slope} = \theta_3 \tag{15-5c}$$

$$\text{intercept} = \theta_0 \tag{15-5d}$$

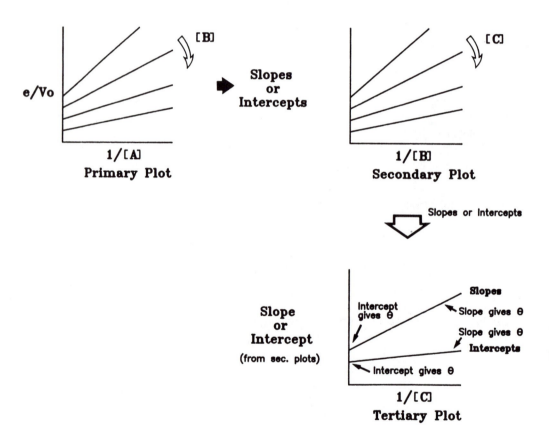

Figure 15-6 Outline of plots used to determine the initial rate parameters of Eq. (15-1).

Tertiary plots from secondary plots (slopes):

(a) Secondary slopes:

$$\text{slope} = \theta_{123} \tag{15-6a}$$

$$\text{intercept} = \theta_{12} \tag{15-6b}$$

(b) Secondary intercepts:

$$\text{slope} = \theta_{13} \tag{15-6c}$$

$$\text{intercept} = \theta_1 \tag{15-6d}$$

The plotting procedures used to obtain the initial rate parameters are illustrated in Fig. 15-6.

Distinguishing Mechanisms

From Table 15-1 it is clear that each of the mechanisms we are examining in this chapter have different parameters from Eq. (15-1) equal to 0. The exceptions are mechanism I, where all the parameters have positive values, and mechanisms V_i and V_{ii}, both of which have the same parameters equal to zero.

By determining the parameters of Eq. (15-1), each of the mechanisms I to V can be identified. To distinguish mechanisms V_i from V_{ii}, approaches similar to those described in Chap. 14 are used. In practice it is often possible to distinguish several of the mechanisms from the patterns obtained in the primary or secondary plots without the necessity of obtaining all the data needed to construct the tertiary plots. This can lead to a considerable saving of material and effort since the number of data points needed to construct the tertiary plots is n^3 (where n is the number of points on each line) compared to n^2 data points for the secondary plots. Inspection of Eqs. (15-2) to (15-6) and Table 15-1 shows that, for example, mechanism IV gives quite distinctive primary plots.

Lineweaver–Burk plots with any substrate varied yield a series of parallel lines at various fixed concentrations of either of the other two substrates. Both variants of mechanism V give primary Lineweaver–Burk plots with A as the varied substrate having slopes that are affected by the concentration of B but not C, while the intercepts are affected by the concentration of either B or C. Thus a pattern of either intersecting or parallel primary plots is obtained depending on whether the concentration of B or C is altered at a fixed concentration of the other.

With mechanisms I, II, and III intersecting patterns of primary plots are always obtained due to the presence of θ_{123}. If one substrate is fixed *at a saturating concentration*, however, various terms in the rate equation are eliminated. For example, saturation with substrate C leads to the elimination of all terms containing the concentration of C—that is, θ_3, θ_{13}, θ_{23}, and θ_{123}. Mechanism III is now readily distinguished from mechanisms I and II by the existence of parallel lines in primary plots with A as the varied substrate at fixed concentrations of B. Similarly, if the concentration of B is set at saturation, θ_2, θ_{12}, θ_{23}, and θ_{123} are eliminated and mechanisms I and II are easily distinguished.

The principal use of the determination of all the parameters in Eq. (15-1), apart from establishing the formal kinetic mechanism, is to find values for the various dissociation and rate constants involved in the mechanisms. As is apparent in Table 15-1, *all* the dissociation constants for mechanism I can be obtained from initial rate data and compared with independently determined values. Similarly, the rate constants of binding of the first and second substrates in the steady-state compulsory-order mechanism (mechanism II) can be obtained.

The Michaelis constants for the various substrates are given by the ratios θ_1/θ_0, θ_2/θ_0, and θ_3/θ_0 for substrates A, B, and C, respectively. In mechanism I these are the dissociation constants of the appropriate substrate from the quaternary complex EABC.

KINETICS OF TWO-SUBSTRATE SYSTEMS IN THE PRESENCE OF A REGULATOR

Many enzyme systems are subject to regulation by ligands other than the substrates. In such cases the regulatory ligand may activate or inhibit the reaction by binding at a site distant from the active site. An idea of the affinity the regulator has for the enzyme is often obtained by performing experiments at fixed substrate concentrations and varied regulator concentrations. Interpretation of this type of experiment assumes either that the regulator binds with equal affinity to all forms of the enzyme, or that it binds to only one form. Although these assumptions *may* be true in some systems, a far more complex situation is possible in a multisubstrate system.

The scheme illustrated in Fig. 15-7 represents a two-substrate random-order mechanism with a regulatory ligand, R, which can bind to both the binary complexes in the system. A rate equation for this scheme can be derived following the procedures outlined in Chaps. 13 and 14, with k_{cat}^1 as the rate constant for product production in the absence of regulator and k_{cat}^2 as the rate constant for product production in the presence of saturating amounts of the regulator. The regulator will be an activator if $k_{cat}^2 > k_{cat}^1$ and an inhibitor if $k_{cat}^2 < k_{cat}^1$. The rapid equilibrium rate equation derived making the rapid equilibrium assumption for this scheme is

$$\frac{e}{V_0} = \frac{K_{12} + K_{1R2} + K_{1R}[R]}{k_{cat}^1 \cdot K_{12} + k_{cat}^2 \cdot K_{1R2}[R]}$$

$$+ \frac{K_2(1 + K_{2R}[R])}{k_{cat}^1 \cdot K_{12} \cdot K_1 + k_{cat}^2 \cdot K_{1R2} \cdot K_{1R} \cdot K_1[R]} \frac{1}{A}$$

$$+ \frac{1 + K_{1R}[R]}{k_{cat}^1 \cdot K_{12} + k_{cat}^2 \cdot K_{1R2} \cdot K_{1R}[R]} \frac{1}{B}$$

$$+ \frac{1}{k_{cat}^1 \cdot K_{12} \cdot K_1 + k_{cat}^2 \cdot K_{1R2} \cdot K_{1R} \cdot K_1[R]} \frac{1}{[A][B]} \qquad (15\text{-}7)$$

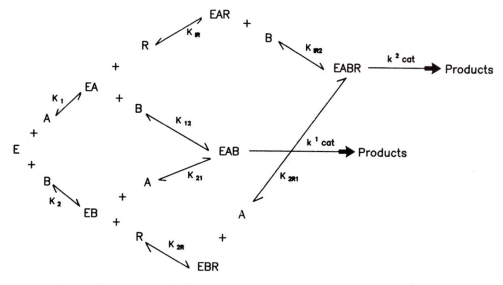

Figure 15-7 Rapid-equilibrium random-order scheme for an enzyme with two substrates and a regulatory ligand, R, which can bind to binary complexes.

From this equation, which fits the generalized rate equation for a two-substrate
system, it is apparent that various ratios of parameters are functions of the concentration of the regulator, R, and dissociation constants in the mechanism. Specifically,
θ_1/θ_{12} and θ_2/θ_{12} ratios are directly porportional to the concentration of R, as shown
by

$$\frac{\theta_1}{\theta_{12}} = K_2(1 + K_{22}[R]) \tag{15-8}$$

$$\frac{\theta_2}{\theta_{12}} = K_1(1 + K_{12}[R]) \tag{15-9}$$

From these it is obvious that a plot of θ_1/θ_{12} or θ_2/θ_{12} versus $[R]$ should be linear
and allow calculation of the dissociation constants K_{1R} and K_{2R} from Fig. 15-7.

The galactosyl transferase reaction represents an excellent example of this approach. In the absence of the regulatory molecule α-lactalbumin, galactosyl transferase
catalyzes the synthesis of N-acetyllactosamine from a sugar donor, UDD-galactose
(U), and an acceptor substrate, N-acetylglucosamine (A). The reaction is inhibited by
α-lactalbumin at high acceptor concentrations but activated at low acceptor concentrations. In the presence of α-lactalbumin the enzyme can also effectively utilize
glucose as a substrate to give lactose. Figure 15-8 shows plots of ratios of θ parameters
versus the concentration of α-lactalbumin which are clearly linear, as expected from
Eqs. (15-8) and (15-9). Values of the dissociation constants for α-lactalbumin binding
to the enzyme-acceptor or enzyme-donor complexes were also obtained.

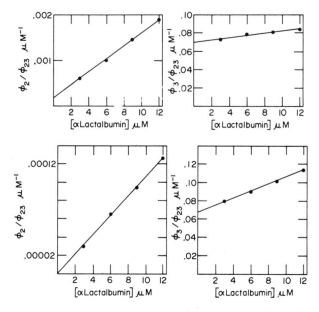

Figure 15-8   Plots of $\theta_2/\theta_{2,3}$ and $\theta_3/\theta_{2,3}$ versus [R] for galactosyl transferase using either N-acetylglucosamine (top) or glucose (bottom) as acceptor substrate. This enzyme has Mn^{2+} as a required independent first substrate. In the presence of saturating metal-ion concentrations, it can be regarded as a RERO two-substrate system.

    With galactosyl transferase, the possibility that the regulatory ligand could also bind to free enzyme and ternary complex was considered, as shown in Fig. 15-9. Kinetic analysis of this scheme using Eq. (15-7) and data obtained from a covalently cross-linked transferase-α-lactalbumin complex allowed direct estimation of 10 of the 12 dissociation constants in this mechanism. The remaining constants, K_{1R} and

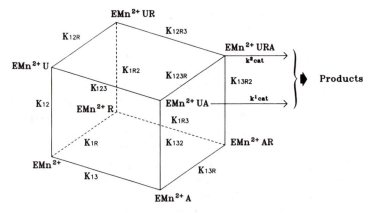

Figure 15-9   Scheme for galactosyl transferase-regulator interactions where the regulator can bind to free enzyme and the $E \cdot Mn^{2+} \cdot UA$ complexes.

Table 15-2 Dissociation constants for the mechanism shown in Figures 15-8 and 15-9

Constant	From initial rate data (μM)	From inhibition experiments (μM)	Independently determined (μM)
$K_{12}(\theta_{23}/\theta_3)$	13.7, 15.0, 10.1		14, 16
$K_{13}(\theta_{23}/\theta_2)$	6,600	4,800	5,000
$K_{123}(\theta_3/\theta_0)$	11,000	16,000	
$K_{132}(\theta_2/\theta_0)$	23		
K_{13R}	1.1	2.0	3.5
K_{12R}	50, 17.6	30, 35	17.7

K_{123R}, were calculated assuming overall equilibrium and were 555 μM and 0.3 μM, respectively. Table 15-2 shows values for some of the constants obtained from kinetic measurements together with comparable constants determined by independent methods. A comparison of the values of K_{1R}, K_{12R}, K_{13R}, and K_{123R} proves that the principal driving force for the regulator binding is the presence of the acceptor in the enzyme complex even though the regulator can bind to all the enzyme species given in Fig. 15-9.

Although this is a single example, it does illustrate the amount of information that can be obtained from initial rate kinetic studies of a system involving two substrates and a regulatory ligand. The equations here apply to the random-order equilibrium situation; this is the most complex situation that might be encountered. In compulsory-order systems, although the initial rate equations are more complex, the interpretation of the kinetic effects of a regulatory ligand are straightforward.

KINETICS OF METAL-DEPENDENT SYSTEMS

A number of enzymes have a required metal ion. In some instances it may be involved only in maintaining the structural integrity of the protein and is not involved directly in ligand binding or catalysis. In other cases the metal ion may be an absolutely required participant in the reaction; if so, and if the metal ion is reversibly bound to the protein, it must be considered as a substrate. The final possibility to consider is that the metal ion may not be required for reaction but acts as a regulatory ligand, and is considered as described in the preceding section.

When the metal ion is a reversibly bound, required participant in the reaction, several possibilities can be envisaged. The metal ion may bind only to the enzyme, in which case it can be regarded as a true third substrate and treated in exactly the manner described earlier for a three-substrate system. It is quite likely, however, that the metal can bind to one or other of the remaining substrates and the enzyme. In such a case the question that must be asked is: What is the true substrate of the enzyme, the free substrate or the metal–substrate complex? It is possible that the metal–substrate complex is the only form of the substrate to bind effectively to the enzyme, while the free substrate acts as an inhibitor (or vice versa). If the dissociation constants of the metal substrate complexes are known or can be determined, it is possible to resolve this question. Resolution depends on two assumptions: (1) the reaction velocity is represented by a first-degree equation in the concentrations of the true substrates, and (2) the metal–substrate complex is kinetically distinct from the free substrate. In a two-substrate plus metal-ion system where the metal ion interacts with only one of the substrates, it is a simple process to calculate the concentrations of free and metal-complexed substrate and plot the appropriate Lineweaver–Burk plots. If the metal ion complexes with both substrates, then changing any one of the participants affects the concentration of the remaining reactants and must be taken into account.

When initial rate data are plotted as a Lineweaver–Burk plot versus the total metal-ion concentration, an apparent substrate inhibition is seen if the true substrates are the free ligands, because increased metal-ion concentrations lead to an effective decrease of the true substrate concentration.

EFFECTS OF PRODUCTS

As indicated in the introduction to this chapter, there are various other initial rate kinetic techniques that we have not covered in detail in Chaps. 13 to 15. As emphasized, initial rate kinetics studies are best used in the context of eliminating possible kinetic mechanisms rather than in proving a particular mechanism. This is true because it is always possible to make a mechanism more complex to accommodate the experimentally obtained data. The two occasionally used initial rate studies described briefly in this section are inherently more complex than the other kinetic approaches described in this book, and as a result are not as easily interpreted.

As we have derived equations in Chaps. 13 and 14 to describe the dependence of the *initial* rate on substrate concentrations, it is assumed that product molecules are initially absent, and as a result, steps involving product *binding* (rather than release) can be ignored. Two types of experimental situations where this is *not* the case are sometimes employed. In the first, product molecules are added at fixed concentrations prior to reaction initiation by addition of enzyme. Since the product can bind to the enzyme to give E-P complexes, which are not on the pathway of substrates being converted to product, an inhibition of S-to-P conversion results. This is known as *product inhibition*, and its type depends on a variety of factors: Does the added product simply withdraw enzyme from otherwise productive catalysis, can the product act as a substrate analog (in which case it can be considered in the same way as described in Chap. 15 for substrate analogs), or can it form abortive E-S-P complexes? Rate equations for these various situations are easily obtained using the approaches described earlier, but of course are far more complex. This makes product inhibition studies of limited use in establishing a formal kinetic mechanism for an enzyme. Once a formal mechanism is known, however, these studies can be vital in establishing what types of "abortive" complexes exist in a mechanism. Such complexes are discussed further in Chap. 16.

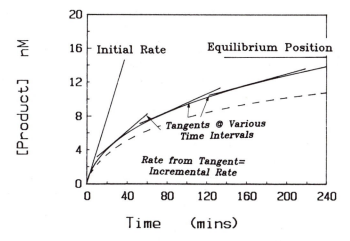

Figure 15-10   Full-time-course analysis of product accumulation. Product concentration is monitored as a function of time after addition of enzyme to a reaction mixture. The monitoring is continued until the equilibrium concentration of product is approached. Indicated on the graph is the initial rate and the incremental rate at some time point where product buildup effects are negligible. The incremental rate is determined solely by the decrease in substrate concentration. Also shown (---) is a projected full time course where product inhibition is seen.

The second experimental situation involves a *full time-course* analysis of an enzyme-catalyzed reaction rather than an initial rate study. In its simplest conception this analysis (where the amount of product is monitored over a much longer period than that in an *initial* rate study) offers the chance of obtaining a variety of parameters from a single experiment. Consider the situation in the reaction mixture once the reaction is initiated. At time $t = 0$, a true initial rate at a particular substrate concentration (the starting concentration) is obtained. At later time points the rate—as judged by the slope of the tangent to the time course (see Fig. 15-10)—is governed by a variety of considerations.

First, the substrate concentration at time point t is less than existed at $t = 0$, due to the reaction proceeding (as t increases, the substrate concentration continues to decrease until equilibrium is reached). Second, product molecules are now present and cause product inhibition and possibly abortive complex formation (as with decreasing substrate concentrations, the product concentration increases with time).

If product inhibition effects are insignificant a full-time-course analysis allows K_m and V_{max} to be obtained directly since the incremental rate shown in Fig. 15-10 is directly proportional to the incremental substrate concentration, as related by the Michaelis–Menton equation. In reality, of course, such a simple situation does not hold in the majority of cases and the analysis is far from simple since product inhibition with a continuously changing concentration of product must also be taken into account. At this point the complexity of the rate equation and the experimental accuracy of the data often preclude meaningful analysis.

OTHER NOMENCLATURES FOR KINETIC DATA

In this book we have used the double reciprocal form of the generalized initial rate equation first popularized by Dalziel. This form has a number of advantages in practical enzyme kinetics, not the least of which is the fact that it is easy to remember. With multisubstrate systems it yields a number of experimentally determined θ parameters whose accuracy can be determined independently of the fit of some proposed kinetic mechanism to the data. The various θ parameters are simply related to the V_{max} of the reaction and to the various Michaelis constants for substrates. As we have discussed in Chap. 14 and this chapter, each θ parameter has a distinct physical significance which is often employed as the basis of testing proposed mechanisms. The physical significance of individual parameters comes from the initial rate equation for the proposed mechanism. Chapters 14 and 15 discussed the various ways in which these equations are derived from the formal mechanism.

Over the years various nomenclatures for the constants that can be obtained from initial rate studies have been used. Apart from the Dalziel system two others are worthy of mention since they too are frequently employed in the literature—those of Alberty and Cleland. Table 15-3 lists these constants for a two-substrate system in terms of the Michaelis constant for each substrate, the dissociation constant for the first substrate in a ternary complex mechanism and the maximum velocity of the reaction. Although each of these constants has the same physical significance for a particular mechanism, the Dalziel parameters have the additional advantage that each K_m is the ratio of two initial rate parameters with individual physical significances as described in Chap. 14 and this chapter, whereas in the Alberty and Cleland nomenclatures the K_m values are individual parameters.

The final comments that must be made at this point are in regard to experimental design of enzyme initial rate kinetic studies. With one exception, all the formal kinetic mechanisms considered in Chaps. 13 to 15 provide initial rate equations predicting linear Lineweaver–Burk plots. In designing a kinetic study of an enzyme it is, however, important that as wide a range of substrate concentrations be used as is feasible.

TABLE 15-3   Kinetic constants for two-substrate reactions using the nomenclature of Dalziel, Alberty, and Cleland

Constant	Dalziel	Alberty	Cleland
K_m(first substrate)	$\dfrac{\theta_1}{\theta_0}$	K_A	K_a
K_m(second substrate)	$\dfrac{\theta_2}{\theta_0}$	K_B	K_b
K_d(first substrate)	$\dfrac{\theta_{12}}{\theta_2}$	$\dfrac{K_{AB}}{K_B}$	K_{ia}
V_{max}	$\dfrac{e}{\theta_0}$	V_f	V_1

This not only makes it possible to experimentally demonstrate linearity of Lineweaver–Burk plots, but it means that all the terms in the Dalziel generalized rate equation can be determined. To ensure this, it is necessary to make measurements at low substrate concentrations, where all the terms in the rate equation play significant roles in governing the rate.

At the other end of the concentration scale, many of the types of nonlinearity in Lineweaver–Burk plots that we discuss in Chap. 17 are clearly observable only at relatively high substrate concentrations—hence the need to use a wide range of substrate concentrations.

16

Deviations from Linear Kinetics

INTRODUCTION

In the consideration of enzyme kinetics in Chaps. 13 to 15 we examined a variety of systems and formal kinetic mechanisms. In each a common theme has been the development of equations to describe the initial rate behavior of the system. These (with one exception, which is considered again in this chapter) have all been, in the double-reciprocal form, equations describing linear Lineweaver–Burk plots. Although the theory discussed in Chaps. 13 to 15 was developed to describe linear kinetics, there are many instances of experimental observations that do not obey the simple theories. It is the purpose of this chapter to examine some of the types of nonlinear Lineweaver–Burk plots that have been reported and discuss the possible causes for such behavior. The experimental approaches that may be used in attempting to establish which of these various possibilities account for a particular deviation from linearity are also considered.

EXPERIMENTAL OBSERVATIONS

As has become apparent from Chaps. 13 to 15, a Lineweaver–Burk or Michaelis–Menten plot for an enzyme-catalyzed reaction is usually a straight line or a simple rectangular hyperbola, as illustrated in Fig. 16-1. Also shown are two often observed types of deviation from this simple behavior. In the first, the rate of the reaction appears to be more sensitive to changes in the substrate concentration at low substrate concentrations, while in the second the rate appears to be less sensitive to changes in substrate concentration at low substrate concentrations.

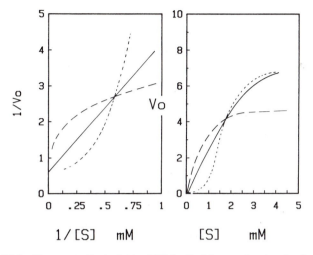

Figure 16-1 Lineweaver–Burk plot and Michaelis–Menten plot showing "normal" behavior (——) and deviations from normal behavior resulting in an increased response to changes in [S] at low concentrations (– – –) or a decreased response to changes in [S] at low concentrations (· · ·).

Another sort of deviation from normal behavior is shown in Fig. 16-2. In this instance as the substrate concentration increases past a certain point, the rate actually decreases, as indicated by the downward curvature of the V versus [S] plot or the sudden upward turn of the Lineweaver–Burk plot as 1/[S] decreases.

The final deviation is given in Fig. 16-3. This superficially resembles one of the cases illustrated in Fig. 16-1, but is distinguished based on the abruptness of the

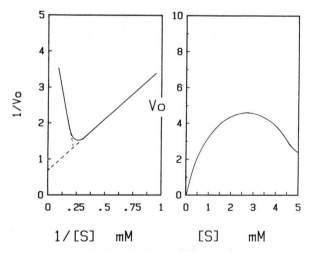

Figure 16-2 Deviations from normal behavior in a Lineweaver–Burk plot or a Michaelis–Menten plot, indicating substrate inhibition at high substrate concentrations.

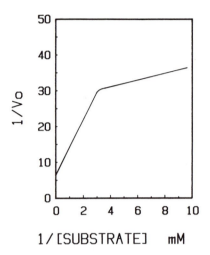

Figure 16-3   Lineweaver–Burk plot illustrating sharp transitions between apparently linear phases.

change in slope. In Fig. 16-1 the line is represented as essentially a continuous curve, while in Fig. 16-3 there is a sharper transition between two (or more) apparently linear regions.

POSSIBLE EXPLANATIONS

For each of the deviations in Figs. 16-1 to 16-3 there are at least two possible explanations, including substrate inhibition, substrate activation, the existence of isoenzymes, cooperative kinetics, and the existence of a steady-state random-order kinetic mechanism. The issue is further complicated by the fact that any given enzyme may display more than one of these phenomena, giving rise to quite complex Lineweaver–Burk or V versus [S] plots. Although in some systems it is possible that complex Lineweaver–Burk plots result from the experimental use of very wide, nonphysiological ranges of substrate concentrations, the complexity of an enzyme's kinetic behavior is more often related to its regulatory properties.

Substrate Inhibition

Substrate inhibition is the term used to describe behavior of the sort illustrated in Fig. 16-2, where at some particular substrate concentration the rate of the reaction actually decreases as the concentration of the substrate is further increased. If this type of kinetic behavior is observed, special precautions must be taken when initial rate parameters are being determined. The various substrate concentration ranges employed must fall outside the range where the onset of substrate inhibition occurs. Careful inspection of Fig. 16-2 shows that the deviation from linearity actually occurs

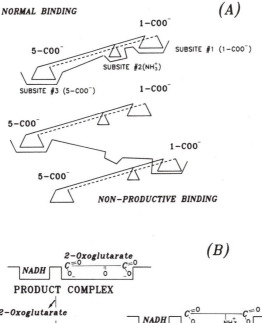

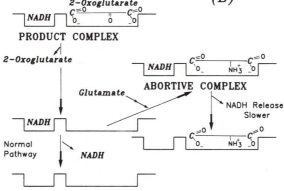

Figure 16-4 Hypothetical case of alternative substrate binding modes that can lead to substrate inhibition: (A) nonproductive binding of glutamate to a hypothetical binding site; (B) glutamate is illustrated forming an abortive complex in the glutamate dehydrogenase reaction once 2-oxoglutarate has vacated the binding site.

before the velocity starts to decrease. There are several possible causes for substrate inhibition.

Nonproductive Substrate Binding. The binding site for the substrate is usually made up of several subsites, each of which recognizes a specific structural feature of the substrate. If as the substrate concentration is increased, it is possible for these various subsites to interact with different substrate molecules (as outlined for a hypothetical case in Fig. 16-4), the resultant enzyme is catalytically inert, and the substrate binding is said to be nonproductive. The nonproductive nature of the binding results from the fact that unless all the required interactions of the various substrate moieties are made with their appropriate enzyme subsites, the catalytic groups on the enzyme and the reactive regions of the substrate cannot be correctly aligned.

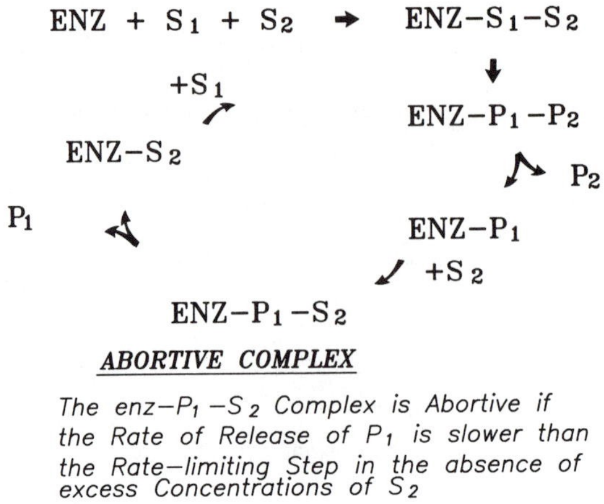

$$ENZ + S_1 + S_2 \;\; \longrightarrow \;\; ENZ\text{–}S_1\text{–}S_2$$

$$\begin{array}{c} +S_1 \\ \nearrow \end{array}$$

$$\downarrow$$

$$ENZ\text{–}S_2 \qquad\qquad ENZ\text{–}P_1\text{–}P_2$$

$$\begin{array}{c} \swarrow \end{array} P_2$$

$$P_1 \qquad\qquad\qquad\qquad ENZ\text{–}P_1$$

$$\begin{array}{c} \nwarrow \end{array} \qquad\qquad\qquad \begin{array}{c} \swarrow \end{array} +S_2$$

$$ENZ\text{–}P_1\text{–}S_2$$

ABORTIVE COMPLEX

The enz–P_1–S_2 Complex is Abortive if the Rate of Release of P_1 is slower than the Rate–limiting Step in the absence of excess Concentrations of S_2

Figure 16-5   Abortive complex formation in a multisubstrate enzyme.

Abortive Complex Formation.   In a multisubstrate enzyme system one of the substrates frequently resembles one of the products. Depending on the order of product release during the reaction cycle, it may be possible to form an enzyme complex that contains both some substrates and some products. A possible outline of this instance is illustrated in Fig. 16-5.

If the E-P-S complex releases product more slowly than the normal E-P complex, the presence of S in the E-P-S complex *may* result in a slowing of the overall reaction. In such an instance the E-P-S complex is said to be an abortive complex.

Even if product release in an E-P-S complex is slower than that from the E-P complex, substrate inhibition is not necessarily the result. The overall rate of the reaction (as measured by V_0) is decreased only if the rate of release of product from E-P is normally rate limiting or becomes rate limiting when E-P-S is formed. The appearance of substrate inhibition in, for example, a Lineweaver–Burk plot, is only apparent if the substrate has a lower affinity for E-P than it has for E. If substrate has the same affinity for E-P and E, an equal proportion of E-P-S is formed at all levels of saturation and the Lineweaver–Burk plot is linear.

Substrate Activation

A downward-curving Lineweaver–Burk plot, as shown in Fig. 16-1A or 16-3, indicates that as the substrate concentration increases, the rate of the reaction is disproportionally faster than is expected by simple extrapolation of the plot obtained at low substrate concentrations; this phenomenom is referred to as *substrate activation.* Such data could easily be obtained in an enzyme that contains a regulatory binding site for the substrate. If this site binds the substrate with lower affinity than the active site but after binding causes an activation of the reaction catalyzed by the active site, the resultant Lineweaver–Burk plot resembles the case we are considering. As the substrate concentration is raised, more of the regulatory site is bound and the enzyme is activated to a greater degree than at low substrate concentrations. If the regulatory site and the active site have equal affinities for substrate, linear Lineweaver–Burk plots result since both sites are saturated at the same rate.

Isoenzymes

At a number of points in this book we have discussed the existence of isoenzymes and briefly considered the fact that they can often be electrophoretically separated. It is possible, however, that in the course of protein purification and characterization, the existence of isoenzymes escapes detection. In such a case, if the isoenzymes have different substrate affinities, nonlinear Lineweaver–Burk plots are expected. At low substrate concentrations the isoenzyme having the higher affinity (i.e., lower K_m) preferentially binds substrate and catalyzes the reaction. As the substrate concentration is raised, the isoenzyme with the higher affinity becomes saturated at lower substrate concentrations than the other isoenzyme. The V versus $[S]$ plot at low substrate concentrations predominantly reflects the K_m of the isoenzyme having the higher affinity. At higher substrate concentrations the plot reflects predominantly the K_m of the isoenzyme with the lower affinity. In such a simple case with two isoenzymes, the Lineweaver–Burk plot shows a curvature, as illustrated in Fig. 16-1A. If the K_m values of the isoenzymes are sufficiently separated, the plot has a much sharper discontinuity and resembles the Lineweaver–Burk plot of Fig. 16-3. In the more complex situation of many isoenzymes all having different K_m values, the Lineweaver–Burk plot tends toward a more continuous curve.

A very important point that must be brought up in this context is the definition of an isoenzyme. As already indicated, isoenzymes are often thought of as being readily separable from one another. This implies that they differ in some biochemical or biophysical way, resulting in altered physical properties, but it is possible that they differ only in conformation (and thus do not have distinct physical differences). In many instances isoenzymes are multi-subunit, made up of varying proportions of different isoenzyme subunits. It is also possible that some multi-subunit enzymes, while they may contain chemically identical subunits, do not have them in the same conformations (i.e., they are conformationally asymmetric). Such a molecule clearly does not fall into the general concept of an isoenzyme. The type of kinetic behavior such an oligomeric protein might show is, however, little different from a mixture of isoenzymes. If the subunits behave independently within the system but are inherently asymmetric in terms of their ability to bind substrate, nonlinear Lineweaver–Burk plots of exactly the same types as would be expected for mixtures of isoenzymes are observed. As emphasized in Chap. 21, this is particularly important in light of the possibility that the observed nonlinear Lineweaver–Burk plots may be the result of cooperative interactions within an oligomer.

Cooperative Kinetics

Although in the section on isoenzymes the possible effects of oligomeric proteins were discussed, none of the explanations for nonlinear kinetics considered so far *require* the enzyme to contain subunits. The two types of deviations illustrated in Figs. 16-1 and 16-3 can also be explained by some sort of interaction between subunits in an oligomeric protein. The downward curvature of Figs. 16-1 and 16-3 can be the result of allosteric interactions between subunits, leading to decreased affinity for substrate as the saturation of the enzyme increases. The deviation shown in Fig. 16-1 can also result from subunit interactions, which lead to increased affinity for substrate as the saturation of the oligomer is increased. The various allosteric models that can generate these sorts of nonlinear kinetics are examined in detail in Chap. 21.

Steady-State, Random-Order Mechanisms

All of the mechanisms considered to explain nonlinear kinetic data invoke either heterogeneity (of enzyme molecules or of binding sites) or subunit interactions in an oligomeric protein. It is possible, however, for a single-subunit enzyme, with no isoenzymes or regulatory binding sites, to exhibit the deviations of either type shown in Fig. 16-1. As discussed in some detail in Chap. 14, a multi-substrate enzyme that follows a steady-state, random-order addition of substrates can, depending on the relative magnitudes of various rate constants in the mechanism, give rise to an initial rate equation that predicts nonlinear Lineweaver–Burk plots. Such nonlinearities can be of either sort illustrated in Fig. 16-1.

EXPERIMENTAL APPROACHES TO DISTINGUISH
AMONG POSSIBLE EXPLANATIONS

With such a diversity of possible explanations for nonlinear Lineweaver–Burk plots, experimental distinction among the alternatives can prove difficult. Table 16-1 summarizes the three basic types of deviation found and the possible explanations of each.

Although as shown in Table 16-1 there are at least two possibilities for each type of deviation, certain experimental approaches permit identification of the appropriate one. The simplest first step in distinguishing between them is to examine the substrate binding to the protein. A variety of techniques that can be used to study substrate binding are described in detail in Chap. 18. If nonlinear binding is also observed, the simple kinetic explanation of the existence of a steady-state, random-order mechanism can be eliminated, as in this case, simple, hyperbolic saturation to one binding site per enzyme molecule results. In the case of cooperative kinetics or isoenzymes, multiple classes of substrate binding to the active site are observed. These involve only the active-site binding, and since multiple classes of sites can exist by the same ligand being

TABLE 16-1 Deviations from linear kinetics

Type of deviation	Figures	Possible explanations
Downward curvature	16-1, 16-3	Substrate activation, isoenzymes, cooperative kinetics, steady state, random order
Upward curvature	16-1	Cooperative kinetics, steady state, random order
Substrate inhibition	16-2	Nonproductive binding, abortive complex formation

able to bind at a regulatory site as well as the active site, it is necessary to show that the observed "heterogeneity" results from active-site binding only. Accurate determination of the stoichiometry of binding is essential and it is helpful if a co-substrate can be shown to affect the affinity of all the putative active binding sites. In substrate activation it is possible to demonstrate binding of the substrate to its regulatory site in addition to the active site. As with the heterogenous classes of binding sites just discussed, the stoichiometry of binding must also be determined. Where substrate can bind to a second, nonactive site per molecule (or subunit), the total stoichiometry is twice that obtained in the previous case, where only active-site binding occurs. In both instances nonlinear binding is observed, emphasizing the requirement for stoichiometry determinations to distinguish between these possibilities. Binding studies are also useful in choosing between the possible explanations of substrate inhibition. For nonproductive binding a second binding site per subunit becomes apparent at high substrate concentrations. Abortive complex formation requires that the release of product from the E-P-S complex is slower (and becomes rate limiting) compared to release from the E-P complex. Such decreased "off"-velocity constants are often associated with decreased dissociation constants for the product ($K_d = k_-/k_+$: where k_- is the "off"-velocity constant and k_+ is the "on"-velocity constant), and it is possible to demonstrate in binding studies that the E-P-S complex is more stable than the E-P complex. It is possible that k_- and k_+ for the product are affected equally and thus a change in k_- is not reflected in K_d. The change in k_- still results in an overall slowing of the reaction in this case and must be demonstrated directly using rapid reaction techniques, which are discussed in Chap. 18.

Binding studies are an important contributor in understanding the basis of nonlinear kinetics. Binding studies, however, cannot discriminate between cooperative kinetics and isoenzymes as a possible explanation for observed downward curvature in a Lineweaver–Burk plot. The distinction between these two alternatives is examined further in Chap. 21.

17

Ligand Binding Studies

INTRODUCTION

Ligand binding studies play an important role in many aspects of protein chemistry and enzymology. As discussed in Chap. 1, many proteins must be assayed based on their specific ligand binding since they have no enzymatic activity. In Chap. 1 we briefly reviewed some of the experimental approaches used for determining specific ligand binding in the context of assaying nonenzymatically active proteins, and also referred to the use of the determination of the number of binding sites per molecular weight unit, and employing such information in establishing an extinction coefficient for a purified protein. The chapters on initial rate kinetic studies (Chaps. 13 to 15) have emphasized the importance of independently determined dissociation constants that can be used in conjunction with kinetically determined dissociation constants as a test of the validity of a proposed kinetic mechanism. In addition, as emphasized in Chap. 16, ligand binding studies are essential when examining systems that exhibit nonhyperbolic kinetics. In Chap. 18 we emphasize the importance of independently determined dissociation constants in studies of rapid kinetics and in the elucidation of mechanisms of enzyme regulation.

In any binding study, two parameters describe ligand binding: B_{max}, the maximum number of ligand molecules bound per mole of protein, and K_d, the dissociation constant of the reversible binding process. A variety of techniques have been developed to study ligand binding, although some cannot give a value for B_{max} because of the nature of the assumptions made in analyzing the data. Such methods have great use, however, often in terms of ease and accuracy, and can be employed where knowledge of K_d but not B_{max} is important. In other instances, B_{max} may be the parameter whose value is primarily required, and of course techniques must be chosen that will yield such information.

Approaches for studying ligand binding can be divided into two categories—direct and indirect—and we examine these separately. In addition to experimental methods for determining ligand binding, a number of ways of presenting ligand binding have been developed, and depending on the information being sought, different types of data presentation are more appropriate. In the context of the graphical representation of ligand binding data we examine the effects of systems that do not follow hyperbolic saturation.

METHODS TO STUDY LIGAND BINDING

Direct Methods

The various direct methods for estimating the amount of ligand bound to a protein have all evolved from equilibrium dialysis and depend on physically separating bound ligand from free. Assuming that there are direct methods for quantitating the amount of ligand bound and free in solution, these techniques all give unequivocal (except for experimental error) estimates of B_{max} and K_d. In equilibrium dialysis, a typical setup of which is shown in Fig. 17-1, the free ligand is allowed to reach an equilibrium across a semipermeable membrane that separates the protein from the bulk phase of the solution.

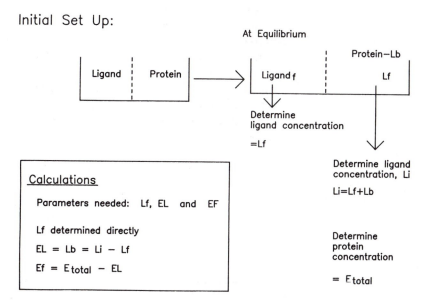

Figure 17-1 Scheme for an equilibrium dialysis experiment.

At equilibrium the concentration of ligand in the non-protein-containing compartment equals the concentration of *free* ligand (L_f) in the protein-containing compartment. If the *total* ligand concentration (L_t) in the protein-containing compartment is known (from experimental determination or, if this is not possible, by subtraction of the amount of ligand in the non-protein-containing compartment from the total amount of ligand initially added), the amount bound to the protein (L_b) can be calculated by subtracting the contribution of free ligand. From a single experiment one can find K_d by substitution into

$$K_d = \frac{[E_f] \cdot [L_f]}{[EL]} \tag{17-1}$$

where L_f is experimentally determined, $[EL]$ ($= L_b$) is calculated as described previously and E_f obtained by subtraction of $[EL]$ from E_{total}, which is presumably known. Application of this equation provides a value for K_d but does not give a value for B_{max}. In the simple situation where hyperbolic saturation of a single class of sites occurs, K_d is an accurate estimate. In a more complex situation, the dialysis experiment is repeated with a series of different total ligand or protein concentrations. Results are then plotted using one of the graphical methods described later.

Before moving on to some of the other direct methods for studying ligand binding, we examine in more detail the experimental procedures of equilibrium dialysis. A number of problems may be encountered:

1. True equilibrium of the free ligand may not be achieved. This is usually controlled for in two ways: (a) a duplicate experiment is set up with *no* protein in either compartment—if equilibrium is reached in the time period used in the experiment, the ligand concentrations in each compartment will be equal, and (b) duplicate determinations in the presence of protein are made, but in one the ligand is initially placed in the compartment containing the protein, while in the other the ligand is initially placed in the other compartment. If an effective equilibrium is established during the time course of the experiment, identical results will be obtained.

2. During the time required to reach equilibrium (often 10 to 15 hours), either the ligand or the protein may decay to inactive forms that interfere with the concentration determination. A more serious problem exists if the ligand decays to a form that competes with the true ligand at the protein binding site. The former artifacts can be controlled for by assaying the protein activity and the effective total ligand concentration before and after dialysis.

3. In many cases the free ligand concentration is conveniently assayed by spectrophotometric measurements. If the bound ligand has different spectral properties from the free ligand, the total ligand concentration in the protein-containing compartment will be incorrectly estimated. This can be overcome by denaturing the protein after equilibrium has been achieved and determining the concentration of the total ligand in that compartment when it is all free in solution. As will be discussed later, it is often possible to determine an extinction coefficient for the bound ligand

via spectral titrations, and if it is known, the concentration can be determined directly from the total absorbance of the protein-containing compartment after subtraction of the absorbance of the free ligand (which is known from the compartment lacking protein). These problems are overcome if a radioactive ligand is used.

4. In situations where the total ligand concentration is varied (usually the case), one reaches a situation as B_{max} is approached where the amount bound is calculated by subtraction of one large number (for free ligand concentration) from a slightly larger number (for [free] + [bound])—a situation that can lead to large experimental error.

Having considered these potential pitfalls, we now examine some of the other "direct" methods for following ligand binding together with some of their problems.

Forced Dialysis Methods. In a number of variants of equilibrium dialysis, there is physical separation of free from bound ligand via a semipermeable membrane. The ligand and protein are initially in the same compartment, separated from a collection vessel by the semipermeable membrane (see Fig. 17-2).

At the start of the experiment, pressure is applied to the protein-containing compartment to cause a "forced" dialysis: solvent, containing free ligand solute but *not* the protein solute, is forced through the membrane, collected, and the concentration of free ligand determined. In the early versions of such schemes the forced dialysis was continued until the protein (and its bound ligand) remained associated with the inside of the semipermeable membrane. The assumption was made that the solvent, containing free solute ligand, was in equilibrium with the protein-containing compartment, and hence the concentration of ligand in the forced dialysate was equal to the concentration of free ligand in the protein-containing compartment. To avoid problems that might arise from concentrating the protein, the experiment is terminated at a fixed point so that a constant protein concentration can be used. To

Initial Set Up:

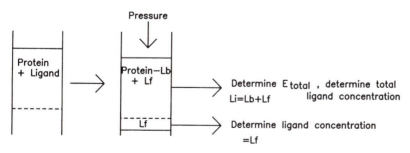

Calculations as in Equilibrium Dialysis

Figure 17-2   Scheme for forced-dialysis method of determining equilibrium binding.

achieve multiple determinations, buffer containing ligand is added to the protein-containing compartment at the termination of one run and the process repeated, allowing either increasing or decreasing total ligand concentrations to be used.

Protein Transport Methods. Several methods have been developed where protein is transported through a solution containing ligand, and the amount of ligand associated with the region containing protein is compared to regions lacking protein to give the amount of ligand bound to the protein. In particular, gel filtration and sedimentation have been successful in this approach.

The simplest way of using gel filtration chromatography to study ligand binding is exemplified by the *Hummel–Dreyer method*, the general principles of which are given in Fig. 17-3A; a gel filtration column is equilibrated with buffer containing a particular ligand at a defined concentration. Once the column is equilibrated (as determined by the eluent containing the same ligand concentration as the starting buffer), the protein sample (which is preequilibrated with the same ligand concentration as the equilibrating buffer) is introduced to the column and the column developed with the equilibrating buffer. Under circumstances where the protein binds the ligand, the sample solvent is depleted with respect to free ligand. As elution proceeds,

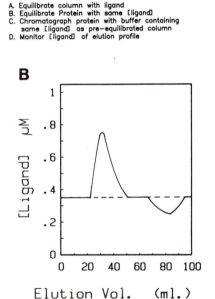

A   Use of Gel Filtration to Study
Ligand Binding

A. Equilibrate column with ligand
B. Equilibrate Protein with same [ligand]
C. Chromatograph protein with buffer containing
same [ligand] as pre-equilibrated column
D. Monitor [ligand] of elution profile

Figure 17-3   (A) Outline of steps in the use of gel filtration to study ligand binding; (B) typical experimental data obtained in Hummel–Dreyer method.

the protein is separated from ligand-depleted solvent and elutes as a peak with its bound ligand such that the *total* ligand concentration in the protein peak is higher than the concentration of the equilibration buffer: as a result, a corresponding trough follows this peak. When the *ligand* concentration is monitored as the column develops, the profile obtained looks like that in Fig. 17-3B. If the ligand does not bind to the protein, the measured concentration of *ligand* during elution does not deviate from the concentration of the equilibration buffer (the dashed line in Fig. 17-3B).

To achieve the best results, it is important to choose a gel matrix that gives maximum separation between the ligand and the protein. This is usually one that totally excludes the protein. It is also essential that the *ligand* elution profile has a plateau of constant ligand concentration between the peak and the trough. This indicates that for the flow rate used to elute the column, the bound and free ligand are in effective equilibrium. If this plateau is not observed, the binding equilibria are too slow relative to the flow rate of the column and the experiment must be repeated at a lower flow rate.

The simplest way to calculate the amount of bound ligand is to collect fractions and determine the ligand concentration per fraction. Although in theory the calculation of the amount of bound ligand can be made using either the peak or the trough, in practice it is easier to use the trough, as interference of the ligand concentration determination by the protein is minimized. Assuming that the ligand concentration is determined by absorbance measurements, the concentration of bound ligand is given by

$$\mu\text{mol bound} = \frac{\sum_i (\Delta A_i)(\text{ml}_i)}{\varepsilon_{mM}} \qquad (17\text{-}2)$$

where ΔA_i is the difference in the absorbance of the fraction, i, and the baseline absorbance (determined by the equilibrating ligand concentration), ml_i is the volume of fraction i and ε_{mM} is the millimolar extinction coefficient of the ligand. The summation is carried out for each fraction in the trough region of the chromatogram, giving the amount of ligand bound *by the amount of protein* in the initial sample.

Although the dissociation constant for the ligand binding process can be calculated from a single experiment, it is necessary to obtain additional data for more detailed analysis of the binding isotherm. Therefore, the experiment must be repeated at either different protein concentrations of the loaded sample or, preferably, at different concentrations of the equilibrating ligand. Essentially the same approach is used when bound and free ligand are separated by centrifugation. This method is outlined in Fig. 17-4.

Protein Precipitation Approaches. In a number of instances, especially where large molecule ligands are involved, it is convenient to study binding by precipitating the protein–ligand complex away from the free ligand and then determining the amount of protein and ligand in the complex; the free ligand concentration is then

Outline of Sedimentation Method of Studying Ligand Binding

1. Sedimentation cell contains ligand
2. Protein is sedimented through ligand. solut.
3. Ligand bound to sedimenting protein is estimated by one of two approaches:
   i. By change in mol.wt. of protein at a series of different [ligand] — this is quite successful even with relatively small igands (mol.wts *500—600)
   ii. Using absorbance measurements to give an estimate of bound ligand per sedimenting protein molecule

Figure 17-4 Outline of approach used to determine ligand binding by sedimentation methods.

determined in the supernatant. This method is outlined in Fig. 17-5. Although frequently used (out of necessity), it is subject to the criticism that the binding of ligand in the presence of the appropriate precipitant may not accurately reflect that found under normal circumstances. In addition, it is necessary to control for nonbound ligand that is included in the precipitated material. When a radioactive ligand is employed, this is often achieved by repeating the precipitation in the presence of a large excess of unlabeled ligand. In this case precipitated radioactive material is used as a control blank to be subtracted from the specifically bound ligand.

Indirect Methods

Spectroscopic Methods. These depend on either a change in the ligand's or the protein's spectral properties on complex formation, and are probably the most widely employed techniques for studying ligand binding. They have the advantages of being rapid, reproducible, and quite accurate. *However*, there are some potential sources of deception in what is an otherwise simple approach, and also (in general), these methods do not give a value for B_{max}, only for K_d.

Consider two cases, both involving fluorescence measurements (although the same arguments that we will use could be made for any other spectral parameter).

Ligand Binding By Complex Precipitation Methods

A. Protein and Ligand premixed at desired total [ligand] and pH etc.
B. Protein—Ligand complex is precipitated by addition of precipitating agent (eg Amm. SO4,Antibody, PEG etc.)
C. Bound ligand in precipitate is estimated

Figure 17-5 Scheme outlining the determination of ligand binding by complex precipitation.

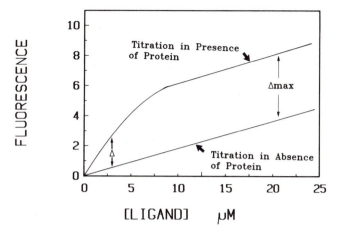

Figure 17-6   Fluorescence titrations in the presence and absence of protein.

In the first, we assume that the fluorescence of a ligand is enhanced upon binding to the protein. A typical experiment is shown in Fig. 17-6.

In the absence of protein, fluorescence intensity is linearly related to the ligand concentration. In the presence of protein the fluorescence intensity rapidly increases, but eventually becomes parallel to that obtained in the absence of protein. The maximum change in fluorescence is defined as Δ_{max}, as shown in Fig. 17-6. If experimentally the two titrations do not become parallel, a value for Δ_{max} can be obtained from a double reciprocal plot of $1/\Delta$ versus $1/[\text{ligand}]$, as in Fig. 17-7.

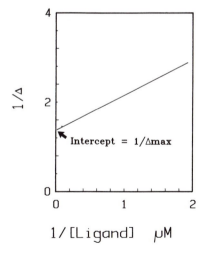

Figure 17-7   Experimental determination of Δ_{max} from double-reciprocal plots.

If it is assumed that Δ_{max} occurs when all the protein sites for the ligand are occupied, one can calculate the amount of ligand bound (L_b) at any point in the titration curve, if the initial concentration of binding sites (which is B_{max}) is known, from

$$L_b = \frac{\Delta}{\Delta_{max}} B_{max} \tag{17-3}$$

Knowing the total ligand concentration at any point in the titration allows calculation of L_f and subsequently a value for K_d. *This approach depends on prior knowledge of* B_{max}.

An alternative calculation is represented by the following: Let F_m be the experimentally determined fluorescence *in the presence* of protein at a given total ligand concentration T. L_b is the concentration of *bound* ligand at any given total ligand concentration. If we define F_b as the *specific molar fluorescence* of bound ligand and F_f as the specific molar fluorescence of free ligand, then at any given point in the titration,

$$F_m = F_b L_b + F_f(T - B) \tag{17-4}$$

$$= F_b L_b + F_f T - F_f B \tag{17-5}$$

However, since the experimentally measured fluorescence in the *absence* of protein $F_t = F_f T$, we get

$$F_m = F_b L_b + F_t - F_f L_b \tag{17-6}$$

Dividing both sides of Eq. (17-6) by F_t gives

$$\frac{F_m}{F_t} = \frac{F_b L_b}{F_t} + 1 - \frac{F_f L_b}{F_t} \tag{17-7}$$

Therefore,

$$\frac{F_m}{F_t} - 1 = \left(\frac{F_b}{F_t} - \frac{F_f}{F_t}\right) L_b \tag{17-8}$$

and we obtain an expression for L_b:

$$L_b = \frac{(F_m/F_t) - 1}{(F_b/F_t) - (F_f/F_t)} \tag{17-9}$$

Since $F_f T = F_t$,

$$L_b = \frac{(F_m/F_t) - 1}{(F_b/F_f)T - \dfrac{F_f}{F_f T}} \tag{17-10}$$

Multiplying top and bottom by T gives

$$L_b = T\frac{(F_m/F_t) - 1}{(F_b/F_f) - 1} \tag{17-11}$$

Defining a new constant, the *fluorescence enhancement*, FE, as FE $= F_b/F_f$, yields

$$L_b = T\frac{(F_m/F_t) - 1}{FE - 1} \tag{17-12}$$

which allows calculation of the amount of bound ligand, L_b, at any point in the titration, provided that the total ligand concentration (T) and the fluorescence enhancement (FE) are known. F_m and F_t are the experimentally determined fluorescences at any given ligand concentration in the presence and absence of protein, respectively. Provided that FE is experimentally determined, these titrations permit calculation of the amount of ligand bound without prior knowledge of B_{max}.

These two ways of approaching essentially the same type of experimental data illustrate an important point: Δ_{max} is a *protein-dependent parameter*, while FE is a *ligand-dependent parameter*. The experimental determination of ligand-dependent parameters allows B_{max} to be determined from indirect methods; however, if only a protein-dependent parameter can be followed, then only information about K_d can be obtained since B_{max} must be assumed.

Ligand-dependent parameters such as FE must be independently determined from an experiment of the type illustrated in Fig. 17-8. FE is determined by fluorescence titrations *at a fixed ligand concentration* with varied protein concentrations. Control titrations omit the ligand to allow the contribution (if any) of the protein to the fluorescence. σ_0 is the fluorescence of the chosen ligand concentration in the absence of protein. σ_i is the measured fluorescence at various protein concentrations (minus the protein fluorescence) and σ_{max} is obtained when the experimental and control titrations become parallel. At σ_{max} *all* the ligand is bound by protein and

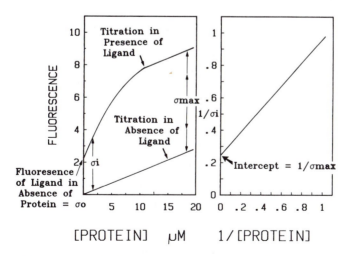

Figure 17-8 Outline of the determination of the ligand-dependent parameter of FE.

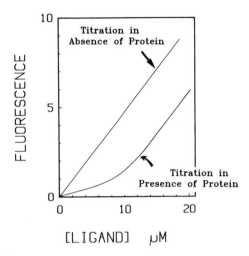

Figure 17-9   Fluorescence titration data obtained when the ligand fluorescence is quenched upon binding to protein.

hence FE is simply given by

$$FE = \frac{\sigma_{max}}{\sigma_0} \qquad (17\text{-}13)$$

If σ_{max} is not experimentally obtained from the titration, a double reciprocal plot of $1/\sigma_i$ versus $1/[\text{protein}]$ gives an intercept of $1/\sigma_{max}$ as in Fig. 17-8.

This example involves a fluorescent ligand that undergoes a fluorescence enhancement on binding to a protein. In some instances the ligand fluorescence may be quenched, and a titration of the type shown in Fig. 17-9 is obtained.

As with the case of enhanced fluorescence, equations for use in the calculation of the bound ligand concentrations at any point in the titration can be derived using either the protein-dependent parameter, Δ_{max}, or a ligand-dependent parameter, Q_c, defined by

$$Q_c = \frac{F_b}{F_f} \qquad (17\text{-}14)$$

where F_b and F_f are defined as previously. The equation for the protein-dependent parameter is as before, while the equation using Q_c is

$$L_b = T\,\frac{1 - (F_m/F_t)}{1 - Q_c} \qquad (17\text{-}15)$$

Although this discussion has centered on fluorescent ligands, any spectral property of the system (i.e., protein or ligand) that changes upon complex formation can be used to study ligand binding. Some of these are summarized in Table 17-1.

TABLE 17-1   Spectral properties used to study ligand binding

Parameter[a]	Dependence[b]	Comment
Protein fluorescence	Usually P	Usually quenching of protein fluorescence observed
Polarization	L	Bound ligand has higher polarization than free ligand
Absorbance	P or L	Can present problems if the ligand absorbs in the region 260–280 nm since protein absorbance changes may interfere with detection of ligand absorbance changes
ESR	Usually P	Useful with paramagnetic metal binding (e.g., Mn); otherwise, must introduce spin label to ligand
NMR	Usually P	In theory almost universally applicable; in practice is limited by experimental considerations such as protein concentration needed
NOE	L	Especially useful with weakly bound ligands

[a] ESR, Electron spin resonance; NMR, nuclear magnetic resonance; NOE, nuclear Overhausser effect.
[b] P, Protein dependent; L, ligand dependent.

In addition to the question of whether a protein-dependent or a ligand-dependent parameter can be determined, spectral methods of following ligand binding suffer from a fundamental limitation that can be resolved only through independently determining B_{max} using a direct method. Implicit throughout all of this is the assumption that in a multisite-per-molecule system (either two or more sites for the same ligand per polypeptide chain *or* a multi-subunit situation) *all* molecules of bound ligand contribute equally to the followed parameter. If one or more sites in a multisite system are spectrally unobservable, only binding of ligand molecules that do contribute to the signal are observed, and the true B_{max} cannot be determined, even using a ligand-dependent parameter.

Other Indirect Methods. Any property of a protein that changes upon the binding of a ligand can be used to study that ligand binding process. All such approaches depend on the experimental determination of a parameter that is equivalent to the Δ_{max} discussed previously. As a result, these methods must assume a value for B_{max}, and can only allow determination of K_d, but they are, however, quite useful, as they are often experimentally easy.

Some of the experimental properties that have been used to study ligand binding are given in Table 17-2.

TABLE 17-2   Protein properties that have been used to follow ligand binding

Property	Comment
Susceptibility to proteolysis	In all instances the presence of ligand in a protein–ligand complex
Denaturation by solvents	may either increase or decrease susceptibility to approach; in either
Heat stability	case the dependence of the change on ligand concentration can
Chemical modification	give binding information.

Affinity Chromatography

The general principles were described in some detail in Chap. 3. Because the retardation of passage of a specific protein through an affinity column is directly due to a specific ligand–protein interaction, it is natural that affinity chromatography can be used to study ligand binding.

If we consider a situation where a solution containing a protein, P, ligand, L, and complex, PL, are in equilibrium, we can write the equilibrium constant, K_{pl}:

$$K_{pl} = \frac{[PL]}{[P_f][L_f]} \tag{17-16}$$

If this mixture is now chromatographed on an affinity column with immobilized X, which is capable of interacting reversibly with *any* component of the equilibrium, we can write a series of equations for the appropriate equilibrium constants and calculate the product concentration,

$$P + X \longleftrightarrow PX \qquad [PX] = K_{px}[P_f][X_f] \tag{17-17}$$

$$L + X \longleftrightarrow LX \qquad [LX] = K_{lx}[L_f][X_f] \tag{17-18}$$

$$L + PX \longleftrightarrow LPX \qquad [LPX] = K_1 K_{px}[P_f][L_f][X_f] \tag{17-19}$$

$$P + LX \longleftrightarrow LPX \qquad [LPX] = K_2 K_{lx}[P_f][L_f][X_f] \tag{17-20}$$

$$X + PL \longleftrightarrow LPX \qquad [LPX] = K_3 K_{pl}[P_f][L_f][X_f] \tag{17-21}$$

where the designation f indicates the free species.

The volume (V_a) of the affinity column that is accessible to the protein is given by t:

$$V_a = V_0 + K_{av}^* V_s \tag{17-22}$$

where V_0 is the void volume, V_s the volume of the stationary phase, and K_{av}^* that fraction of the stationary phase accessible to the protein.

Affinity chromatography usually involves a gel matrix as the stationary phase, and retardation of the protein involves a gel filtering effect as well as specific interaction. The concentration of "immobilized" protein, $[P]_i$, at any point is given by

$$[P]_i = K_{px}[P_f][X_f] + Y[P_f][L_f][X_f] + [P_f](1 + K_{pl}[L_f])K_{av}^* \tag{17-23}$$

where $Y = K_1 K_{px} + K_2 K_{lx} + K_3 K_{pl}$.

The concentration of protein in the mobile phase, $[P]_m$, is given by

$$[P]_m = [P_f](1 + K_{pl}[L_f]) \tag{17-24}$$

and the ratio of the concentration of "immobilized" to "mobile" protein is given by

$$K_{av} = \frac{[P]_i}{[P]_m} \tag{17-25}$$

From Eqs. (17-23) to (17-25) it is apparent that

$$K_{av} = \frac{K_{av}^* + [X_f](K_{px} + Y[L_f])}{1 + K_{pl}[L_f]} \qquad (17\text{-}26)$$

K_{av} and K_{av}^* are converted into elution volumes by

$$V_p^* = V_0 + K_{av}^* V_s \qquad (17\text{-}27)$$

$$V_p = V_0 + K_{av} V_s \qquad (17\text{-}28)$$

where V_p and V_p^* are the elution volumes of the protein in the presence or absence, respectively, of interaction with the immobilized ligand, X.

From Eqs. (17-26) and (17-27), the equation

$$V_p - V_p^* = \frac{V_s[X_f](K_{px} + Y[L_f])}{1 + K_{pl}[L_f]} \qquad (17\text{-}29)$$

is obtained. $[X_f]$ may be written in terms of the *total* concentration of X, $[X]_t$, as shown by

$$[X_f] = \frac{[X]_t}{1 + K_{px}[P_f] + K_{lx}[L_f] + Y[P_f][L_f]} \qquad (17\text{-}30)$$

Using Eqs. (17-29) and (17-30), we get the general equation for $(V_p - V_p^*)$,

$$V_p - V_p^* = \frac{V_s[X]t(K_{px} + Y[L_f])}{(1 + K_{pl}[L_f])(1 + K_{lx}[L_f] + [P]t)(K_{px} + Y[L_f])} \qquad (17\text{-}31)$$

where $[P]_t$ is the total protein concentration.

Clearly, this is completely general in that it contains all six of the equilibrium constants defined by Eqs. (17-16) to (17-21). Specific equations for use in particular experimental systems are obtained from this general equation by setting certain equilibrium constants equal to 0. We consider here the two cases most likely to be encountered. They are as follows: (1) protein binds to either immobilized ligand or to free ligand B, but not to both, and (2) only the protein–ligand complex binds to the immobilized ligand. These two cases yield the following equations (obtained by setting other constants = 0):

Case 1:

$$\frac{1}{V_p - V_p^*} = \frac{K_p^l[L_f]}{V_s[X][K_{px}]} + \frac{1 + K_{px}[P]}{V_s[X]K_{px}} \qquad (17\text{-}32)$$

Case 2:

$$\frac{1}{V_p - V_p^*} = \frac{1}{(V_s[X]K_3 K_{pl}[L_f])} + \frac{1 + K_3[P]}{V_s[X]K_3} \qquad (17\text{-}33)$$

In case 2, $V_p^* = V_p$ and V_p can be substituted directly. In case 1, however, V_p^* is not equal to V_p, but it is possible to use V_p by subtracting the expression for $V_p - V_p^*$

[obtained by setting $[L] = 0$], in which case $V_p = V_p^*$, giving

$$(V_p - V_p^*) = \frac{V_s[X]K_{px}}{1 + K_{px}[P]}$$

and we get, for case 1,

$$\frac{1}{V_p - V_p^*} = -\left\{ \frac{(1 + K_{px}[P])^2}{V_s K_{px} K_{pl}[X][L_f]} + \frac{1 + K_{px}[P]}{V_s K_{px}[X]} \right\} \tag{17-34}$$

Now we have expressions in terms of V_p and V_p^* and where V_p is the elution volume of the protein under conditions where it interacts with the matrix and V_p^* is the elution volume where no interaction occurs.

These equations also involve the equilibrium concentration of the *free* ligand $[L_f]$, which is related to the total ligand concentration $[L]$ by

$$[L_f] = \frac{[L]}{1 + K_{pl}[P]} \tag{17-35}$$

which upon substitution into Eqs. (17-32) and (17-33), gives, for cases 1 and 2, respectively,

$$\frac{1}{V_p - V_p^*} = -\left\{ \frac{(1 + K_{px}[P])^2(1 + K_{pl}[P])}{V_s K_{px}[X]K_{pl}[L]} + \frac{1 + K_{px}[P]}{V_s K_{px}[X]} \right\} \tag{17-36}$$

and

$$\frac{1}{V_p - V_p^*} = \frac{1 + K_{pl}[P]}{V_s[X]K_3 K_{pl}[L]} + \frac{1 + K_3[P]}{V_s[X]K_3} \tag{17-37}$$

In *either* case a plot of $1/(V_p - V_p^*)$ versus $1/[L]$ is curvilinear, and the limiting slopes and intercepts (as $1/[L] \to 0$) are given in Table 17-3.

From these it is obvious that in either case, a single set of experiments at a fixed protein concentration yields two expressions in three unknowns. A third expression is required. Although several approaches can be used, the simplest is to perform a second set of experiments at a different protein concentration with a range of ligand concentrations, including 0, to allow determination of the new value of V_p. The dissociation constant K_{pl} for the protein-free ligand complex can now be determined.

TABLE 17-3   Values of slopes and intercepts from plots of $1/(V_p - V_p^*)$ versus $1/[L]$

Case	Slope	Intercept
1	$-\left\{ \dfrac{(1 + K_{px}[P])^2}{V_s K_{pl} K_{px}[X]} \right\}$	$-\left\{ \dfrac{1 + K_{px}[P]}{V_s K_{px}[X]} \right\}$
2	$\dfrac{1}{V_s[X]K_3 K_{pl}}$	$\dfrac{1 + K_3[P]}{V_s[X]K_3}$

REPRESENTATION OF LIGAND BINDING DATA

For the simple equilibrium $P + L \rightleftharpoons PL$, the dissociation constant, K_d, is given by

$$K_d = \frac{[P_f][L_f]}{[PL]} \tag{17-38}$$

where f denotes the concentration of free species at equilibrium. The total concentration of protein, $P_t = [PL] + [P_f]$, and the concentration of PL, may be expressed as a function of $[L_f]$ if $[P_f]$ is eliminated from Eq. (17-38), as shown in

$$[PL] = \frac{[P_t]}{1 + K_d/[L_f]} \tag{17-39}$$

If P_t is constant, as is usually the case, a plot of $[PL]$ versus $[L_f]$ is a rectangular hyperbola and, analogous to the Michaelis–Menten equation, the concentration of $[L_f]$ at which $[PL] = \frac{1}{2}[P_t]$ is equal to K_d. This is summarized in Fig. 17-10.

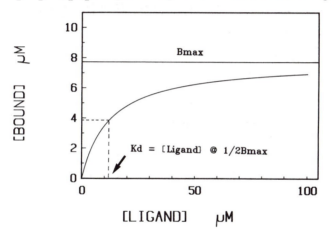

Figure 17-10   Saturation curves for ligand binding to protein.

As with enzyme kinetics, several linearized versions of this equation are used, most notably the Klotz equation,

$$\frac{1}{[PL]} = \frac{1}{[P_t]} + \frac{K_d}{[P_t]} \frac{1}{[L_f]} \tag{17-40}$$

and the Scatchard equation,

$$[PL] = [P_t] - \frac{K_d[PL]}{[L_f]} \tag{17-41}$$

The equivalent linear plots, the Klotz plot and the Scatchard plot, are illustrated in Figs. 17-11 and 17-12, respectively. All of these equations refer to a case where a single protein molecule has a single binding site for ligand.

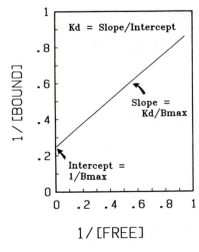

Figure 17-11 Klotz plot of equilibrium binding data.

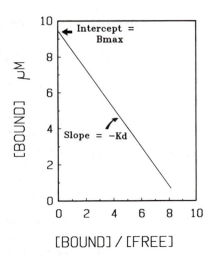

Figure 17-12 Scatchard plot of equilibrium binding data.

Identical, Independent Binding Sites

Many oligomeric proteins contain more than one binding site per molecule for a particular ligand, and in cases where the sites are independent *and* have the same microscopic dissociation constant, the interactions of the protein (P) with the ligand (L) can be characterized by the following equilibria:

$$P_0 + L \longleftrightarrow P_1$$
$$P_1 + L \longleftrightarrow P_2$$
$$P_{n-1} + L \longleftrightarrow P_n$$

where the number indicates the number of ligand molecules that are bound to the protein. Each site has the same microscopic dissociation constant, designated by mK. The *macroscopic dissociation constant* K_d, however, depends on the level of occupancy of the molecule, as indicated by

$$K_{d_1} = \frac{[P_0][L_f]}{[PL_1]} \tag{17-42}$$

$$K_{d_2} = \frac{[PL_1][L_f]}{[PL_2]} \tag{17-43}$$

$$K_{d_n} = \frac{[PL_{n-1}][L_f]}{[PL_n]} \tag{17-44}$$

The macroscopic dissociation constant K_{d_i} is related to the microscopic dissociation constant mK in such a system by the relationship given in

$$K_{d_i} = \left(\frac{\sum n, i - 1}{\sum n, i} \right) mK \qquad (17\text{-}45)$$

where $\sum n, i$ is the number of microscopic forms that make up PL_i. From Eqs. (17-42) to (17-45) we can write

$$[PL_i] = \frac{[PL_{i-1}][L_f]}{K_{d_i}} \qquad (17\text{-}46)$$

and since $\sum n, i = n!/(n-1)!i!$, we can write

$$[PL_i] = [PL_{i-1}] \frac{n - i + 1}{i} \frac{[L_f]}{mK} \qquad (17\text{-}47)$$

Similarly, from expressions for $PL(i-1)$, $PL(i-2)$, and so on, we get

$$PL_i = PL_0 \left(\prod_{j=1}^{i} \frac{n - j + 1}{j} \right) \left(\frac{[L_f]}{mK} \right)^i \qquad (17\text{-}48)$$

The moles of ligand bound per mole of protein, designated as L_b, is given by

$$L_b = \frac{\sum\limits_{i=0}^{n} i[PL_i]}{\sum\limits_{i=0}^{n} [PL_i]} \qquad (17\text{-}49)$$

and using Eq. (17-48) for $[PL_i]$, we get

$$L_b = \frac{\sum\limits_{i=1}^{n} i \left\{ \prod\limits_{j=1}^{1} [(n - j + 1)/j] \right\} ([L_f]/mK)^i}{1 + \sum\limits_{i=1}^{n} \left\{ \prod\limits_{j=1}^{i} [(n - j + 1)/j] \right\} ([L_f]/mK)^i} \qquad (17\text{-}50)$$

which, using the expression for $\sum n, i$, simplifies as follows:

$$\prod_{j=1}^{i} [(n - j + 1)j] = \frac{n!}{(n - i)! \, i!} \qquad (17\text{-}51)$$

Therefore,

$$L_b = \frac{\sum\limits_{i=1}^{n} i[n!/(n - i)! \, i!]([L_f]/mK)^i}{1 + \sum\limits_{i=1}^{n} [n!/(n - i)! \, i!]([L_f]/mK)^i} \qquad (17\text{-}52)$$

The denominator of this equation is the binomial expansion of $(1 + [L_f]/mK)^n$:

$$\left(1 + \frac{[L_f]}{mK} \right)^n = 1 + \sum_{i=1}^{n} [n!(n - i)! \, i!] \left(\frac{[L_f]}{mK} \right)^i \qquad (17\text{-}53)$$

which, upon differentiation with respect to $[L_f]/mK$ and multiplication by $[L_f]/mK$, yields

$$n\left(\frac{[L_f]}{mK}\right)\left(1 + \frac{[L_f]}{mK}\right)^{n-1} = \sum_{i=1}^{n} i\left[\frac{n!}{(n-i)!\,i!}\right]\left(\frac{[L_f]}{mK}\right)^i \qquad (17\text{-}54)$$

Substitution into the expression for $[L_b]$ gives

$$[L_b] = \frac{n[L_f]/mK}{1 + [L_f]/mK} \qquad (17\text{-}55)$$

Since by definition $L_b = [PL]/[P_t]$, we get, after multiplication of both sides by $[P_t]$,

$$[PL] = \frac{n[P_t]}{mK/[L_f] + 1} \qquad (17\text{-}56)$$

giving

$$[PL] + [PL]\frac{mK}{[L_f]} = n[P_t] \qquad (17\text{-}57)$$

which rearranges to

$$[PL] = n[P_t] - mK\frac{[PL]}{[L_f]} \qquad (17\text{-}58)$$

which is the Scatchard equation derived earlier for a single site per molecule, where n is the number of sites per molecule. As before, a plot of $[PL]$ versus $[PL]/[L_f]$ is linear with a slope of $-mK$ and an intercept of $n[P_t]$.

Multiple Classes of Independent Sites

A frequently encountered situation is the case where a Scatchard plot is non-linear, as illustrated in Fig. 17-13. Where there are n_i independent sites with intrinsic microscopic dissociation constants mK_i, we can write

$$[L_b] = \sum_i \frac{n_i[L_f]/mK_i}{1 + [L_f]/mK_i} \qquad (17\text{-}59)$$

which gives (following the same process as previously)

$$[PL] = \sum_i n[P_t] - \sum_i mK\frac{[PL]}{[L_f]} \qquad (17\text{-}60)$$

In the Scatchard plot of Fig. 17-13, the intercept on the $[PL]$ axis is, for the case shown (which involves *two* classes of sites) $n_1 + n_2$, and the intercept on the $[PL]/[L_f]$ axis is $n_1/mK_1 + n_2/mK_2$. The most realistic values of n_1, n_2, mK_1, and mK_2 are obtained by an iterative process.

Assuming initially that the x-axis intercept is dominated by the smaller mK value (i.e., mK_1), a tangent to the curve at regions approaching $[PL] = 0$ gives an intercept

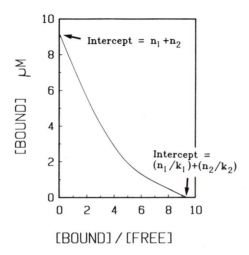

Figure 17-13   Nonlinear Scatchard plot as might arise from two independent but nonidentical binding sites.

of n_1/mK_1, and on the y-axis an initial value for n_1. With initial estimates of n_1 and mK_1, we can subtract the contribution of the high-affinity sites from the data obtained at higher degrees of saturation, which can then be plotted to give initial estimates of n_2 and mK_2. Once these estimates of n_2 and mK_2 have been obtained, the contribution of the *low*-affinity sites to the data at low degrees of saturation can be subtracted and new estimates of n_1 and mK_1 obtained. Throughout the procedure $n_1 + n_2$ must equal the observed $[PL]$ intercept, and the iterations are continued until $\sum_i (n_i/mK_i)$ equals the x-axis intercept.

Dependent Binding Sites

In a situation where two identical ligand molecules bind to a protein molecule (either one subunit with two sites or two subunits, each with a binding site), it may be that binding the ligand to the first binding site alters the affinity of ligand binding to the second site. Two cases are possible: In the first, ligand binding at the first site *increases* the affinity of the second site, while in the second case, ligand binding at the first site decreases the affinity of the second site. In either instance nonlinear Scatchard (Fig. 17-14) or Klotz (Fig. 17-15) plots result.

As discussed in the chapters on nonlinear kinetics and allosteric models (Chaps. 16 and 21, respectively) it is not possible to distinguish cases of independent nonidentical sites from cases where the first ligand decreases the affinity of the second ligand. The Scatchard plot in Fig. 17-13 resembles that of Fig. 17-14, curve B. Mechanistically, such binding can result from allosteric interactions giving negative cooperativity, or from direct steric interaction of the bound ligands. In contrast to this situation, Scatchard or Klotz plots indicating that the first ligand molecule to bind increases the affinity of the second ligand molecule can be explained only by allosteric models.

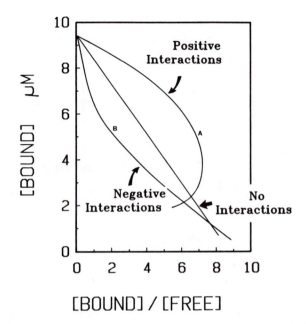

Figure 17-14   Scatchard plots for dependent binding sites: (A) increased affinity of second ligand; (B) decreased affinity of second ligand.

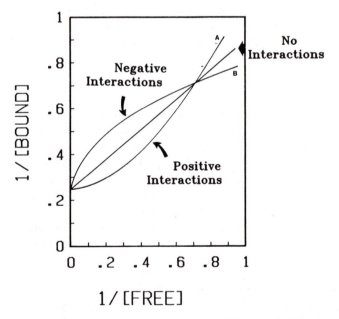

Figure 17-15   Klotz plots for dependent binding sites: (A) increased affinity of second site; (B) decreased affinity.

This chapter examined a variety of experimental approaches for determining the dissociation constant and maximum binding capacity of a ligand binding to a protein. It must be emphasized that in the treatments developed here the binding is assumed to be in a free equilibrium on the time scale at which the experiment is performed. As discussed in Chaps. 14 to 16, such equilibrium binding studies are of extreme importance in elucidating kinetic mechanisms of enzymes and establishing the basis of non-linear kinetics in systems that show non-Michaelis–Menton behavior. Similarly, as developed in Chap. 18, equilibrium binding studies complement the information obtained in rapid kinetic studies on the rates of ligand binding and release to and from protein complexes.

18

Rapid Kinetics

After the concept of "steady state" was introduced to enzymology it became of interest, both from a theoretical and an experimental standpoint, to examine events during the course of the "pre-steady state". As the rate equations of many potential two- and three-substrate enzyme mechanisms were developed, it also became apparent that in a number of instances (see Chaps. 14 to 16) a mechanism could be tested by correlating a rate constant derived from steady-state kinetics with an experimentally determined rate constant. In these cases a rate constant for a binding step was usually involved.

Since many enzymes have turnover numbers on the order of 10^3 per second and higher, it is apparent that simple mixing and observation spectrophotometrically, in a manner suitable for the study of steady-state rates, is not sufficient for the examination of pre-steady-state phenomena or for the direct observation of binding phenomena, which in many instances are faster than the steady-state rate. This experimental limitation led to the development of a variety of rapid reaction techniques.

Initially, Hartridge and Roughton introduced the continuous-flow method for studying the combination of oxygen with hemoglobin. Two solutions, one containing oxygen, the other hemoglobin, were kept in elevated containers (see Fig. 18-1). From these, two tubes ran into a crude mixing chamber and then into an observation tube. To start the experiment the valves from the two containers were opened and the reactants allowed to mix and flow into the observation chamber. Spectrophotometric observations were made at various positions in the long observation chamber, representing (at a constant flow rate) different time intervals after mixing. Alter-

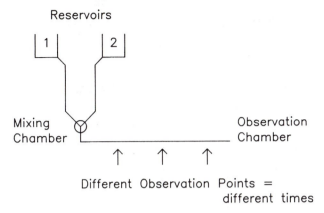

Reservoirs

1   2

Mixing   Observation
Chamber   Chamber

Different Observation Points =
different times

Figure 18-1   Outline of a continuous-flow method for following a rapid reaction.

natively, the pressure causing the flow could be changed, altering its velocity and allowing observations to be made at a fixed position. The time interval from mixing to the point of observation was determined (and if necessary altered) by the flow rate. Although this approach worked well, it used large amounts of material and had a relatively large "dead time" (which is the period of time between effective mixing and the start of observation).

This technique was later modified by Millikan, who used two syringes that could be coincidentally driven to deliver the reacting solutions to the mixing chamber and observation point. Chance further modified the approach, using an accelerated flow and a multijet mixing chamber. This, combined with the introduction of spectrophotometers, permitted dead times to be cut from several seconds to several milliseconds.

These developments led to the concept of "stopped" rather than continuous flow: In stopped flow the reactants are mixed and forced into an observation chamber and the flow stopped. The reacting mixture is then observed continuously as it ages. The main features of a stopped-flow rapid-mixing instrument are shown in Fig. 18-2, together with the way a stopped-flow experiment is usually performed. In many instances it is important to know the mixing, or premixing, order since premixing the enzyme with one or other of the reactants may alter the results observed or introduce experimental artifacts.

The major advantage of stopped flow is that very much smaller amounts of material are needed, and one continuously monitors the reaction rather than having to make a series of separate measurements at different time points. Since the mixing procedure is essentially the same as in the case of continuous flow, the dead time of a stopped flow experiment is of the order of 1 to 2 ms, depending on the exact design of the instrument. Modern stopped-flow instruments have pressure-driven syringes, permit observation of either fluorescence or absorbance, and have microprocessor control of data acquisition, allowing many replicates of the reaction to be rapidly collected and the data averaged and stored for later analysis.

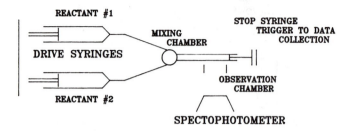

REACTION REPRESENTED AS:

ENZ + REACTANT #1

ENZ + REACTANT #2

Figure 18-2 Schematic representation of stopped-flow experiment. Reactions are usually indicated as shown.

The major limitation of such stopped-flow methodology resides in the dead time. As indicated, this is usually on the order of 1 to 2 ms. If, to collect sufficient data for accurate analysis we wish to observe 50% of the total reaction, the upper limit of the constant that can be estimated is obtained from the relationship for the half-time of the process, given by

$$t_{1/2} = \frac{0.693}{k} \qquad (18\text{-}1)$$

If we use a dead time of 1 ms, the upper limit of the rate constant is approximately $700 \ \mu s^{-1}$. In actual practice, for a single exponential process we need not follow 50% of the reaction and the practical upper limit is probably of the order of 1000 to $1500 \ s^{-1}$. However, it is most important when only a small portion of the reaction is monitored that the *amplitude* of the process followed is consistent with the initial and final levels of product. If two exponential processes are occurring with well-separated rate constants and only the later stages of the slower process are followed, an examination of the amplitude of the followed process compared to the overall measured amplitude reveals the presence of a faster process occurring within the dead time of the method. This limitation of the dead time of a rapid-mixing approach led to the development of equilibrium perturbation techniques, which have dead times on the order of 5 to 10 μs. Such a dead time gives an upper limit for an observable rate constant of about 10^5 per second. Although some enzyme processes do occur at faster rates than this, most binding rates can be satisfactorily measured with this amount of dead time.

Before considering some of the early examples of rapid reaction studies (which illustrate the range of potential applications), we must consider several fundamental points.

1. In these experiments the enzyme is used in quantities whose stoichiometric relation to the substrate is not of the order used in initial rate experiments, and must be considered as a participant in the reaction rather than as a catalyst.

2. The concentration of substrate cannot be *assumed* to be constant in such experiments (although in certain types of experiments it may effectively be a constant).

3. For rapid reaction techniques to be applied, there must be some parameter that can be experimentally followed—usually in a continuous manner, although as we will see when considering some examples, this is not always necessary.

Early Examples of Rapid-Mixing Studies

Lipoyl Dehydrogenase. On reaction with a reducing substrate such as NADH, this enzyme undergoes a two-electron transfer that yields a compound with a long-wavelength absorbance. The second electron is probably transferred to a sulfur-containing group in the protein. Using rapid reaction techniques it was shown that in reactions with NADH using oxidized lipoic derivatives as electron acceptors, the rate-limiting step in each case is the reaction of the intermediate (with long-wavelength absorbance) with the oxidized lipoic acid derivative. The catalytically active intermediate does *not* give an ESR signal, presumably as a result of spin pairing between the flavin and the sulfur radical. The fully reduced form of the enzyme is formed by reaction with NADH, but its rate of oxidation by substrate is much slower than that of the intermediate, with the result that the overall reaction is inhibited. In most situations this is prevented because the fully reduced form is rapidly converted to the intermediate form by interaction with NAD.

D-Amino oxidase also forms intermediates with long-wavelength absorbance that must contain substrate since the shape of the spectrum depends on the type of amino acid substrate used. The intermediate is formed and removed at rates consistent with its participation in catalysis and is converted to a fully reduced form that appears slowly and is not involved in ordinary functioning; this form has no ESR spectrum.

A semiquinone form of enzyme results from reduction with dithionite and has an ESR spectrum but does *not* react with either oxygen or amino acids.

Glucose Oxidase. This enzyme functions by alternating between fully oxidized and reduced forms. However, no trace of a change in long-wavelength absorbance is observed. One can obtain a semiquinone form by reduction with dithionite, which gives an ESR spectrum, but it does *not* react readily with glucose or oxygen.

Microsomal NADP–Cytochrome c Reductase. This enzyme alternates between an intermediate and a fully reduced form. At the end of the reaction the enzyme remains in the intermediate form. On mixing oxidized enzyme with NADH, both the fully reduced and the intermediate forms appear together, which suggests that during reduction the two molecules of flavin (which are known to be associated with the enzyme) may cooperate, giving an electronic exchange formally equivalent to a

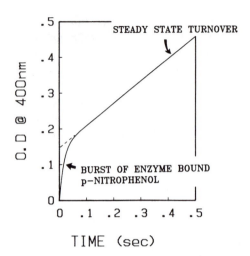

Figure 18-3 Outline of the experiment and experimental results obtained in stopped-flow experiments with trypsin and *p*-nitrophenyl acetate.

hydride-ion transfer, followed either by the full reduction of one flavin or by the appearance of an intermediate form of both flavins by intramolecular electron transfer.

Hydrolytic Enzymes. The reactions catalyzed by these enzymes are not associated with major spectrophotometric changes in enzyme and thus either the substrate or the product must be followed directly. Gutfreund followed the hydrolysis of *p*-nitrophenyl acetate, which gave a spectrophotometric change associated with hydrolysis that was markedly biphasic. The first phase was attributed to a relatively rapid formation of an enzyme–product complex, followed by slower turnover of enzyme as hydrolysis proceeded. This situation is illustrated in Fig. 18-3 and is important since it was one of the earliest demonstrations of a pre-steady-state "burst" of enzyme-bound product. We discuss the conditions that lead to the generation of a pre-steady-state "burst" of enzyme-bound product in more detail later in this chapter.

pH Changes. In reactions involving protons, Roughton and co-workers employed pH indicators to follow the rate of breakdown of carbonic acid in acid solution using a continuous-flow apparatus. Similar studies have been done to examine carbonic anhydrase and many other enzymes where a proton is involved.

Bioluminescent Reactions. Some bioluminescent systems require the simultaneous presence of several reactants before light emission takes place. Stopped-flow methods can give an indication of the order in which the various reactants enter the scheme. If three reactants are involved, one can use the various premixing steps shown in Fig. 18-4. If in the ordinary reaction the rate-limiting step is (b + c), and this reaction can take place in the absence of a, then on mixing (b + c) with a, light is observed much sooner than in the other premixing cases.

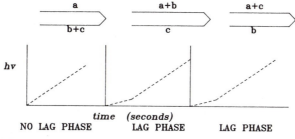

NOTES:

1. Combination of b+c involves a step that is rate limiting in the overall reaction.Pre–mixing b+c allows generation of luminescence immediately upon mixing with a. In the other premixing cases reaction of b+c must occur prior to interaction with a: hence a lag in appearence of luminescenceis observed.

Figure 18-4 Premixing scheme for a reaction involving three reactants leading to the generation of bioluminescence.

Quenching Methods

In the general principle of quenching methods, two reactants are driven through a conventional mixer and along an observation tube that terminates in a second mixer fed with a quenching reagent, as outlined in Fig. 18-5. The rapid reaction is initiated on mixing and terminated on quenching, and the effluent is collected and analyzed later.

A variation is the rapid-freezing method. Reactants are mixed and sprayed into a large volume of isopentane at $-140°C$. Since fine jets are used, rapid freezing occurs. This method is used primarily in conjunction with ESR measurements. The dead time of the apparatus is made up of three components: the time for cooling from room temperature to $0°C$, for freezing, and for cooling to $-140°C$. In addition,

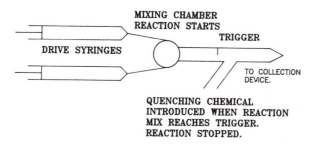

TIME INTERVAL CONTROLLED BY EITHER: a) Distance between mixing chamber and trigger for introduction of quencher, or b) Flow rate after mixing. Quenched sample is collected for latter analysis of amount of product.

Figure 18-5 Scheme of the setup of a rapid-mix-and-quench experiment.

any reaction continuing at $-140°C$ must be taken into account, especially if long periods are left between the freezing and later analysis. This approach has been used successfully in studies on xanthine oxidase and on the copper in cytochrome oxidase.

Equilibrium Perturbation Methods

As indicated earlier, the major limitation with rapid mixing experiments is the dead time. To overcome this problem, equilibration perturbation methods were developed. The principle is simple: An equilibrium mixture is set up under a particular set of thermodynamic conditions, which are rapidly changed and the "relaxation" of the equilibrium mixture to the equilibrium defined by the new thermodynamic conditions monitored. Since no rapid mixing of reactants is involved, the approach is limited instead by the time taken to change the thermodynamic conditions of the equilibrium. In the two most used methods, temperature jump and flash photolysis, the change requires on the order of 1 to 10 μs.

Temperature Jump. This procedure consists of taking a solution at one temperature containing an enzyme and substrate at equilibrium, heating the solution as quickly as possible, and following the exchange between enzyme and substrate as the equilibrium readjusts itself to the new value corresponding to the new temperature.

The temperature jump is produced by an electrical discharge between electrodes placed in the solution, resulting in a temperature rise of about 5°C. In most cases the reaction is followed by absorbance spectrophotometry, although fluorescence recording has been used (care must be taken because of temperature dependence).

The time resolution is limited by the rate of the temperature jump. This is normally on the order of 10^{-5} s, which represents an improvement of two orders of magnitude over flow methods. Although it seems that temperature-jump methodology offers many advantages over rapid-mixing methods, there are a few disadvantages. It is applicable only to systems in which an equilibrium is obtained with a reasonable amount of ligand present free in solution, and with an equilibrium that is perturbable. Also, there is the technical difficulty in producing a vigorous electric discharge in close proximity to a sensitive detection system.

Flash Photolysis. Energy, in the form of a light pulse, is delivered to the system under examination with a time constant of 5 to 10 μs for a flash energy of about 1000 J. The reaction is recorded photographically by using a second flash tube—the "spectroflash"—which records the spectrum of the solution that is being illuminated with the "photoflash." With electronic control of the interval between the photo and spectro flashes, the time resolution is about 1×10^{-5} s. Flash photolysis has many applications, including the study of the formation and spectra of triplet forms of chlorophyll in nonaqueous solvents, the formation and disappearance of long-wavelength absorbance derived from FMN coenzymes, and the interactions of heme-containing proteins with carbon monoxide. In hemoglobin and cytochrome oxidase, interactions with CO_2 were followed and it was found that the rate of reaction was about 60 times faster than had been determined by stopped-flow methods.

ASSOCIATION AND DISSOCIATION RATE CONSTANTS

A number of types of reactions illustrate the development of the theoretical basis for various rapid reaction techniques.

Simple Reactions

First examine a reactant, A, going to product, P, with a rate constant, k_1, and with the rate of the reaction given by

$$\frac{dA}{dt} = -k_1 A \tag{18-2}$$

Integration of Eq. (18-2) yields

$$A_t = A_0 e^{-k_1 t} \tag{18-3}$$

where A_0 is the initial concentration of A and A_t is the concentration at time $= t$. If the product P is followed, then, since $A_t + P_t = A_0$, we get

$$P_t = A_0(1 - e^{-k_1 t}) \tag{18-4}$$

The *half-time* of the reaction, $t_{1/2}$, where $A = P = A_0/2$, is given by substitution into Eq. (18-1), and we define a new parameter, τ, where $\tau = 1/k_1 = 1/k_{obs}$.

Reversible Reactions

If product P can react, with a rate constant k_2, to give A, we have a reversible reaction,

$$A \underset{k_2}{\overset{k_1}{\rightleftharpoons}} P$$

and can define an equilibrium constant K_e by

$$K_e = \frac{[P]}{[A]} = \frac{k_1}{k_2} \tag{18-5}$$

In this case $dA/dt = -k_1[A] + k_2[P]$, but since $A_t + P_t = A_0$,

$$\frac{dA}{dt} = -k_1[A] + k_2([A_0] - [A]) \tag{18-6}$$

This equation is integrated by separating the variables and multiplying each side by an exponential factor to give

$$\frac{dA}{dt} + [A](k_1 + k_2) = k_2 A_0 \tag{18-7}$$

$$\frac{dA}{dt} e^{k_1 + k_2} + [A](k_1 + k_2)e^{(k_1 + k_2)t} = k_2[A_0]e^{(k_1 + k_2)t} \tag{18-8}$$

Therefore,

$$\frac{d}{dt([A]\,e^{(k_1+k_2)t})} = k_2[A_0]e^{(k_1+k_2)t} \tag{18-9}$$

and

$$[A]e^{(k_1+k_2)t} = \frac{k_2}{k_1+k_2}\,[A]e^{(k_1+k_2)t} + C \tag{18-10}$$

where C is a constant. However, boundary conditions apply: namely, that at $t = 0$, $A = A_0$, and that at $t = $ infinity, the equilibrium concentration of A, $A_{eq} = A_0 k_2/(k_1 + k_2)$. Therefore,

$$A_t = \frac{A_0}{k_1+k_2}\,[k_1 e^{-(k_1+k_2)t} + k_2] \tag{18-11}$$

This expression can be divided into two terms: the exponential term with rate constant (k_{obs}), where in terms of relaxation kinetics, $k_{obs} = 1/\tau = k_1 + k_2$, and an amplitude factor, given by $k_1/(k_1 + k_2)$.

Two important points: (1) the measured rate constant for the approach to equilibrium is a simple exponential and is greater than either of the individual first-order rate constants—in fact, $1/\tau = k_1 + k_2$; and (2) k_1 and k_2 cannot be determined without a knowledge of the amplitude factor—however, if the concentrations of A and P at equilibrium are known, k_1 and k_2 can be estimated.

Enzyme–Substrate Association. If we consider the reaction $E + A \rightleftharpoons EA$, then if $A \gg E$, the reaction is effectively first order, and the system reduces to

$$E \underset{k_2}{\overset{k_1A}{\rightleftharpoons}} EA$$

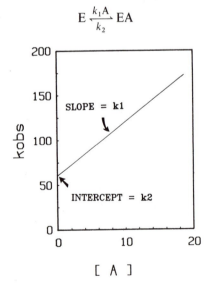

Figure 18-6 Dependence of k_{obs} on [A] for an enzyme–ligand association–dissociation reaction.

By analogy with the previous discussion, the relaxation time for this reaction is

$$\frac{1}{\tau} = k_2 + k_1 A = k_{obs} \qquad (18\text{-}12)$$

Since the rate constants are pseudo-unimolecular there is a concentration dependence of k_{obs}. Thus k_1 and k_2 can be obtained *without* the amplitude factor by following the dependence of k_{obs} on [A] as illustrated in Fig. 18-6.

Displacement Experiments. Consider the experimental situation shown in Fig. 18-7. A' is a displacing ligand present in excess such that its rate of binding is effec-

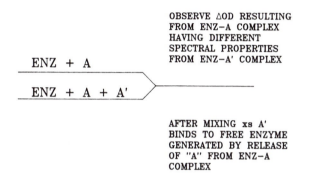

Figure 18-7   Scheme for a ligand displacement experiment.

tively infinitely fast and irreversible. The situation becomes, in effect,

$$A' + EA \xrightarrow{k_2} EA' + A$$

and resembles the simple irreversible reaction considered first. Thus $k_{obs} = k_2$, the off-velocity constant for A from the EA complex, and the analysis is quite simple.

Thermodynamic Principles of Relaxation

For a particular reaction the equilibrium concentrations of the reactants are dependent on some thermodynamic variable such as temperature.

Considering our earlier example, $E + A \rightarrow EA$, the equilibrium constant K can be written

$$K = \frac{(\bar{E})(\bar{A})}{\overline{EA}} = e^{-\Delta H^\circ/RT} + e^{\Delta S^\circ/R} \qquad (18\text{-}13)$$

where $\bar{E}$, $\bar{A}$, and $\overline{EA}$ are the equilibrium concentrations and ΔH° and ΔS° are the standard enthalpy and entropy changes for the reaction. To consider the effects of temperature, we must differentiate the expression for $\ln K$, to give

$$d(\ln K) = (\bar{E}^{-1} d\bar{E}) + (\bar{A}^{-1} d\bar{A}) - (EA^{-1} d\overline{EA}) \qquad (18\text{-}14)$$

But since $d\mathrm{EA} = -d\bar{\mathrm{E}} = -d\bar{\mathrm{A}}$, we get

$$d(\ln k) = (\bar{\mathrm{E}}^{-1} + \bar{S}^{-1} + \overline{\mathrm{EA}}^{-1})\, d\bar{\mathrm{E}} \tag{18-15}$$

From the van't Hoff relationship,

$$d(\ln K) = \frac{\Delta H^\circ}{RT^2}\, dT \tag{18-16}$$

Thus

$$d\mathrm{E} = \frac{1}{\bar{\mathrm{E}}^{-1} + \bar{\mathrm{A}}^{-1} + \overline{\mathrm{EA}}^{-1}}\frac{\Delta H^\circ}{RT^2}\, dT \tag{18-17}$$

and

$$d\mathrm{E} = \Gamma\,\frac{\Delta H^\circ}{RT^2}\, dT \tag{18-18}$$

where

$$\Gamma = \frac{1}{\bar{\mathrm{E}}^{-1} + \bar{\mathrm{A}}^{-1} + \overline{\mathrm{EA}}^{-1}}$$

Therefore, for a definite change in temperature, ΔT, we get

$$\Delta\bar{\mathrm{E}} = \Gamma\,\frac{\Delta H^\circ}{RT^2}\,\Delta T \tag{18-19}$$

Similarly, we can obtain expressions for the equilibrium shifts in $\bar{\mathrm{A}}$ and $\overline{\mathrm{EA}}$ from the expression

$$\Delta\bar{\mathrm{E}} = \Delta\bar{\mathrm{A}} = -\Delta\overline{\mathrm{EA}} \tag{18-20}$$

From Eq. (18-19) it is readily apparent that the perturbation of the equilibrium is dependent on two parameters: the equilibrium position itself, as expressed by Γ, and ΔH°.

If we consider the reaction

$$\mathrm{E} + \mathrm{A} \underset{k_2}{\overset{k_1}{\rightleftharpoons}} \mathrm{EA}$$

at equilibrium, when the system is suddenly heated, the concentrations [E], [A], and [EA], defined by the old equilibrium, are no longer the equilibrium concentrations of the new equilibrium, which may be symbolized by $[\bar{\mathrm{E}}]$, $[\bar{\mathrm{A}}]$, and $[\overline{\mathrm{EA}}]$, and

$$\frac{d[\mathrm{EA}]}{dt} = (k_1[\mathrm{E}][\mathrm{A}]) - (k_2[\mathrm{EA}]) \tag{18-21}$$

Equation (18-21) may be rewritten

$$\frac{d([\overline{\mathrm{EA}}] + \Delta[\mathrm{EA}])}{dt} = k_1([\bar{\mathrm{E}}] + \Delta[\mathrm{E}])([\bar{\mathrm{A}}] + \Delta[\mathrm{A}])$$
$$- k_2([\overline{\mathrm{EA}}] + \Delta[\mathrm{EA}]) \tag{18-22}$$

$$\frac{d[\overline{EA}]}{dt} + \frac{d\,\Delta[EA]}{dt} = k_1[\overline{E}][\overline{A}] + k_1([\overline{A}]\,\Delta[E] + [\overline{E}]\,\Delta[A])$$
$$+ k_1\,\Delta[E]\,\Delta[A] - k_2[\overline{EA}] + k_2\,\Delta[EA] \qquad (18\text{-}23)$$

From the equilibrium condition

$$\frac{d[\overline{EA}]}{dt} = k_1[\overline{E}][\overline{A}] - k_2[\overline{EA}] = 0 \qquad (18\text{-}24)$$

Hence

$$d\,\Delta[EA] = k_1([\overline{A}]\,\Delta[E] + [\overline{E}]\,\Delta[A]) + k_1\,\Delta[E]\,\Delta[A] + k_2\,\Delta[EA] \qquad (18\text{-}25)$$

If $\Delta[E]$, $\Delta[A] \ll [E]$ and $[A]$, we get

$$\frac{d\,\Delta[EA]}{dt} = k_1([\overline{A}]\,\Delta[E] + [\overline{E}]\,\Delta[A]) - k_2\,\Delta[EA] \qquad (18\text{-}26)$$

Adding the conditions $\Delta[E] = \Delta[A]$, and $\Delta[A] + \Delta[EA] = 0$, Eq. (18-26) becomes

$$\frac{d\,\Delta[EA]}{dt} = -(k_1([\overline{E}] + [\overline{A}]) + k_2)\,\Delta[EA] \qquad (18\text{-}27)$$

which yields

$$\Delta[EA] = \Delta[EA]^\circ e^{-(k_1([\overline{E}] + [\overline{A}]) + k_2)t} = \Delta[EA]^\circ e^{-1/\tau} \qquad (18\text{-}28)$$

From these final equations, it is quite apparent that k_{obs} ($= 1/\tau$) is equal to $k_1[\overline{E} + \overline{A}] + k_2$, and hence a plot of k_{obs} versus $[\overline{E} + \overline{A}]$ is linear, as shown in Fig. 18-8.

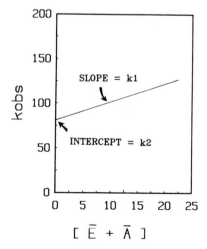

Figure 18-8 Dependence of k_{obs} on $[E + A]$.

So far we have examined a situation where an equilibrium is perturbed by a temperature change; however, the mathematical analysis is independent of *how* the perturbation is produced. In fact, one can combine the stopped-flow technique with the relaxation analysis and perform a "concentration-jump" experiment: the enzyme and ligand are preequilibrated in one syringe and then rapidly mixed with buffer plus any coligands contained in the other syringe. The decrease in spectral signal is monitored as the new equilibrium concentrations are established.

The concentration-jump technique is useful only if the equilibrium position is not too far to one side; otherwise, reequilibration upon dilution results in immeasurable concentration changes. Perturbation, in any case, results in relatively small concentration changes even with a favorable equilibrium. The analysis of the data obtained in such a concentration-jump experiment is identical to that used in a temperature-jump experiment.

In such an experiment, however, there are problems that may arise from the deliberate change in protein concentration. This is especially true in cases where the protein may undergo a concentration-dependent polymerization reaction. As in a regular stopped-flow experiment, this could contribute to the observed signal. Control experiments, where enzyme alone is subjected to the concentration jump, indicate such problems and provide "blank" reactions that can be subtracted from the experimental data in the presence of ligand. The case of glutamate dehydrogenase, which undergoes a concentration-dependent polymerization, illustrates this problem. Binding of the coenzyme NADPH can be followed by fluorescence changes, and data obtained in a typical concentration-jump experiment are given in Fig. 18-9.

Fluorescence is monitored as a function of time as enzyme–NADPH complex is subjected to rapid dilution. The continuous line shows the fit to an equation for two exponentials. When the experiment is repeated in the absence of coenzyme, one

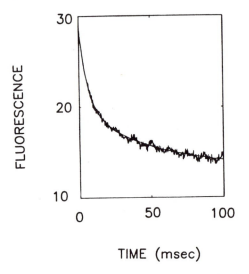

Figure 18-9 Fluorescence changes obtained in a concentration-jump experiment with glutamate dehydrogenase.

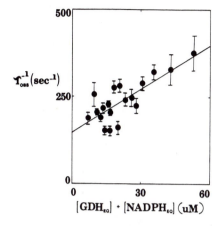

Figure 18-10   Relaxation times obtained from concentration-jump experiments. The bars represent estimated standard errors obtained from the fit. The straight line is a linear least-squares fit of the data.

of the exponentials is obtained, and is due to fluorescence intensity changes resulting from the depolymerization reaction of the enzyme as its concentration is decreased. The observed relaxation time for the other exponential is plotted (as its reciprocal) versus the calculated equilibrium concentrations of (enzyme + NADPH) in Fig. 18-10. Values for the association rate and the dissociation rate constant are obtained from the slope and intercept, respectively.

CATALYTIC TURNOVER EXPERIMENTS

So far we have considered ways to determine rate constants experimentally for the association and dissociation of ligand bound to enzyme. Rapid reaction techniques have also been applied to the analysis of enzymes undergoing catalytic reaction. Under conditions where the complete reaction is possible, rapid reaction techniques have the capability of examining various facets of the pre-steady-state phase. A variety of experiments are possible. If we consider the case where product formation is being monitored we observe a "pre-steady-state" phase only under conditions where product release is the overall rate-limiting step in the reaction cycle. This immediately gives information about the kinetic mechanism of the enzyme—it cannot obey a simple rapid-equilibrium random-order kinetic mechanism. If it did, no pre-steady-state phase could be observed in such an experiment: Product formation is the overall rate-limiting step. In situations where a pre-steady-state phase of the overall reaction is observed, however, we can examine a number of interesting questions.

Spectral Characterization of the Pre-Steady State

Some idea of the number and chemical nature of intermediates that contain product can be gotten from spectral studies on the pre-steady-state phase of the reaction. Such studies are performed by acquiring data over a period of, for example

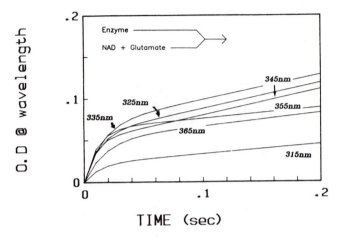

Figure 18-11   Stopped-flow traces obtained at the indicated observation wavelengths
for the reaction shown in the inset.

1 s, at a series of different wavelengths covering the potentially interesting range.
Although somewhat tedious, it is usually reasonable to obtain data at intervals of
2.5 to 5 nm over a 50 to 100 nm range. From the accumulated data, spectra can
be constructed at whatever time intervals are desired. The data are usually represented as "time-difference" spectra, using two different time points in the reaction.
Such spectra allow spectral characterization of species formed between the two time
points. For example, if glutamate dehydrogenase is rapidly mixed with NAD and
glutamate at pH 8.0, a pre-steady-state burst is observed when the reduced coenzyme
product is monitored by absorbance measurements. This indicates that a product
release step is rate limiting in the overall reaction. When a series of identical reactions are performed using different wavelengths of observation, a series of traces are
obtained, as shown in Fig. 18-11. From such data, time-difference spectra can be
calculated.

Typical results are given in Fig. 18-12, where the spectra are calculated from the
points indicated by the arrows in Fig. 18-11. In this particular case a clear "blue-
shifted" species is observed at 25 ms into the burst phase of the reaction. This species
shifts to a "red-shifted" spectrum (curve 2) at approximately 100 ms, before finally
resolving into the unshifted spectrum of free NADH (curve 3) at the end of the burst
phase. For clarity it is convenient to present the data in the manner shown here, as
this avoids multicomponent spectra that would be hard to interpret.

Derivation of Equations for Burst Amplitude and Analysis
of Rate Constants

Apart from the spectral characterization of transients occurring in the pre-steady-
state phase of a reaction, two other parameters can be measured in addition to the
steady-state rate. These are the "burst" size and rate. The burst size is usually ob-

PRE-STEADY STATE SPECTRA

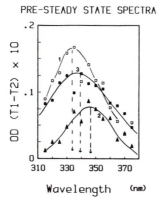

Figure 18-12 Time-difference spectra obtained from the data shown in Fig. 18-11; curve 1, $T_1 = 25$ ms, $T_2 = 0.5$ ms; curve 2, $T_1 = 100$ ms, $T_2 = 35$ ms; curve 3, $T_1 = 200$ ms, $T_2 = 100$ ms.

tained by extrapolation of the steady state to $t = 0$, as in Fig. 18-13, while the burst rate is the rate of the exponential process leading to the steady-state region. For the reaction

$$E + S_1 + S_2 \rightleftharpoons ES_1S_2 \rightleftharpoons EP_1P_2 \rightleftharpoons E + P_1 + P_2$$

under V_{max} conditions (i.e., when $[S_1] \to$ infinity and $[S_2] \to$ infinity), the simplest model for the formation of products can be written as

$$ES_1S_2 \underset{k_2}{\overset{k_1}{\rightleftharpoons}} EP_1P_2 \overset{k_3}{\longrightarrow} E + P_1 + P_2$$

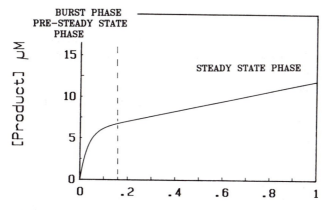

Figure 18-13 Phases of an enzyme reaction observed in a typical stopped-flow experiment.

and the total enzyme concentration

$$ET = [ES_1S_2] + [E] + [EP_1P_2]$$

Hence using E_S to represent ES_1S_2 and EP to represent EP_1P_2, we write

$$\frac{d(EP)}{dt} = k_1[ES] - (k_2 + k_3) \cdot [EP] \tag{18-29}$$

$$= k_1[ET - EP] - (k_2 + k_3) \cdot [EP] \tag{18-30}$$

$$= k_1 ET - (k_1 + k_2 + k_3) \cdot [EP] \tag{18-31}$$

Therefore,

$$\frac{d[EP]}{dt} + (k_1 + k_2 + k_3)[EP] = k_1 ET \tag{18-32}$$

and the concentration of EP is given by

$$[EP] = \frac{k_1 ET}{k_1 + k_2 + k_3} + Be^{-(k_1 + k_2 + k_3)t} \tag{18-33}$$

when $t = 0$, $[EP] = 0$, and

$$B = \frac{-k_1 ET}{k_1 + k_2 + k_3}$$

Therefore,

$$[EP] = \frac{k_1 ET}{k_1 + k_2 + k_3} (1 - e^{-(k_1 + k_2 + k_3)t}) \tag{18-34}$$

when t approaches infinity,

$$[EP] = \frac{k_1 ET}{k_1 + k_2 + k_3}$$

and is the steady-state concentration of EP under V_{max} conditions. If $k_1 \gg k_2 + k_3$, then EP = ET and the amount of EP is approximately equal to the total enzyme active-site concentration.

Similarly, from the formal mechanism we can write an expression for the formation of free product P, which is the product whose concentration can be monitored. $dP/dt = k_3[EP]$, hence

$$[EP] = \frac{1}{k_3} \frac{dP}{dt} \tag{18-35}$$

From Eqs. (18-33) and (18-35) we get

$$\frac{d[EP]}{dt} = \frac{1}{k_3} \frac{d^2P}{dt^2} = [k_1 ET - (k_1 + k_2 + k_3)] \frac{1}{k_3} \frac{dP}{dt} \tag{18-36}$$

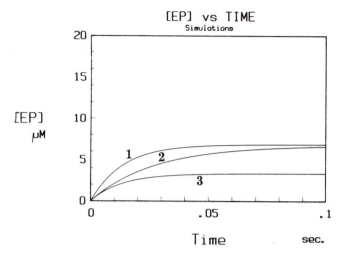

Figure 18-14 Computer simulations of [EP] versus time curves. k_1 is the rate constant for ES $\rightarrow$ EP, k_2 is the rate constant for EP $\rightarrow$ ES, and k_3 is the rate constant for product release, EP $\rightarrow$ E + P.

Constant	Curve 1	Curve 2	Curve 3
k_1	50	25	25
k_2	20	10	25
k_3	3	2	25

Hence

$$\frac{d^2P}{dt^2} + (k_1 + k_2 + k_3)\frac{dP}{dt} = k_1 k_3 ET \tag{18-37}$$

Therefore

$$[P] = \frac{k_1 k_3 ET}{k_1 + k_2 + k_3}\, t - \frac{k_1 k_3 ET}{(k_1 + k_2 + k_3)^2}\left(1 - e^{-(k_1 + k_2 + k_3)t}\right) \tag{18-38}$$

Figures 18-14 and 18-15 show the dependence of the formation of EP or P on t, as described by Eqs. (18-34) and (18-38), respectively. When t is large, the exponential term is negligible and a linear P versus t plot results, with

$$\text{slope} = \frac{k_1 k_3 ET}{k_1 + k_2 + k_3}$$

which is equal to V_{max} by steady-state treatment of the mechanism.

From these simulations it is apparent (Fig. 18-14) that k_2 has a major effect on the rate of attainment of the steady-state concentration of EP, and that the steady-state concentration of EP is markedly affected by the relative rate of k_3. Under the various conditions in Fig. 18-14, the steady-state concentration of EP is *not* stoichiometric with the active-site concentration.

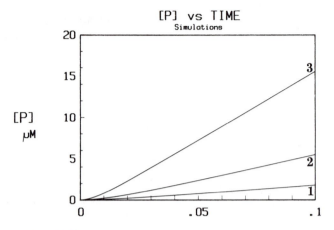

Figure 18-15

Constant	Curve 1	Curve 2	Curve 3
k_1	50	50	50
k_2	20	20	50
k_3	3	10	50

Other conditions are the same as in Fig. 18-14.

Figure 18-15 illustrates that under conditions where the concentration of P alone can be followed, there is a lag (as expected, since EP must be produced first) and that in the situation shown in Fig. 18-15, the production of P at its steady-state rate is attained after 20 to 30 ms. As expected, the intercept on the time axis $[= 1/(k_1 + k_2 + k_3)]$ can be affected by any of the three rate constants.

If the linear region is extrapolated to $P = 0$, the intercept is given by

$$0 = \frac{k_1 k_3 ET}{k_1 + k_2 + k_3} t - \frac{k_1 k_3 ET}{(k_1 + k_2 + k_3)^2} \tag{18-39}$$

and $t = 1/(k_1 + k_2 + k_3)$.

This formulation is of use if (and only if) one can measure P alone. However if, as is usually the case, the sum of $EP + P$ is measured experimentally, we get, from Eqs. (18-34) and (18-35),

$$[EP] + [P] = \frac{k_1 k_3 ET}{k_1 + k_2 + k_3} t + \frac{k_2 ET}{k_1 + k_2 + k_3} - \frac{k_2 k_3 ET}{(k_1 + k_2 + k_3)^2}(1 - e^{-(k_1 + k_2 + k_3)t}) \tag{18-40}$$

When t is large, the exponential term is negligible and the equation is linear in t.

$$
\begin{aligned}
[EP] + [P] &= \frac{k_1 k_3 ET}{k_1 + k_2 + k_3} t + \left[\frac{k_1 ET}{k_1 + k_2 + k_3} - \frac{k_1 k_3 ET}{(k_1 + k_2 + k_3)^2} \right] \\
&= \frac{k_1 k_3 ET}{k_1 + k_2 + k_3} t + \frac{(k_1 + k_2)k_1 ET}{(k_1 + k_2 + k_3)^2}
\end{aligned} \tag{18-41}
$$

Hence the slope of a plot of ($[EP] + [P]$) versus t at large t is the steady-state rate, $k_1 k_3 ET/(k_1 + k_2 + k_3)$. Such plots are given in Fig. 18-16.

Several points can be made based on the simulated curves of Fig. 18-16. The steady-state rate of the reaction measured on a macroscopic time scale is not reached for 50 to 60 ms, and as expected from Fig. 18-14, the rate constant k_2 has a major effect on the rate at which the final steady-state rate is attained. In all the cases shown in Fig. 18-16A, a burst is observed: The rate of product release, k_3, is much smaller than the rate of initial product formation, k_1; also, the size of the burst is *not* stoichiometric with the active-site concentration in these cases. Figure 18-16B shows that a burst can be obtained even though the rate of initial product formation in EP, k_1, is equal to the rate of product release, k_3, although the burst size is diminished under such conditions. In curve 3 of part (B) where k_1 is truly rate limiting ($k_1 = k_3/100$), a linear ($[EP] + [P]$) versus time plot is observed. The reaction rate is governed solely by the rate of production of P in the EP complex, and the reaction is a true rapid equilibrium.

The intercept of such plots, by extrapolation of the linear part of the plot to $t = 0$, is

$$\text{intercept} = \frac{(k_1 + k_2)k_1 ET}{(k_1 + k_2 + k_3)^2} \tag{18-42}$$

The relationship of the burst size to various rate constants is illustrated in Fig. 18-17. This figure shows the effects of the magnitude of k_1 compared to k_2 and k_3 on the size of the burst phase. As k_1 approaches the situation where $k_1 \gg k_2 + k_3$, the burst size nears the stoichiometric concentration of the active sites. Despite the fact that all three curves have the same k_3, the steady-state rate (attained within 50 ms) is different for all three. k_3 does not dominate the steady-state rate unless k_1 is at least $5 \times k_3$.

When $k_1 \gg k_2 + k_3$ this expression reduces to ET, and as before under this condition, the burst size is equal to the total enzyme active-site concentration. Without this restriction, however, the intercept *is not* the steady-state concentration of EP. The steady-state concentration is $k_1 ET/(k_1 + k_2 + k_3)$. The intercept from the plot of ($[EP] + [P]$) versus t is, however, the steady-state concentration of EP multiplied by the factor $(k_1 + k_2)/(k_1 + k_2 + k_3)$, as indicated by

$$\text{intercept} = \frac{(k_1 + k_2)k_1 ET}{(k_1 + k_2 + k_3)^2} = \frac{k_1 ET}{k_1 + k_2 + k_3} \frac{k_1 + k_2}{k_1 + k_2 + k_3} \tag{18-43}$$

Thus, depending on the magnitude of k_3 relative to k_1 and k_2, the intercept (the apparent burst size) is *smaller* than the steady-state concentration of EP. The complete equation may be written as

$$([EP] + [P]) = At + B(1 - e^{-k_{obs}t}) \tag{18-44}$$

where A and B are obtained from the slope and intercept of the *linear* region of (EP + P) versus t, and $k_{obs} = k_1 + k_2 + k_3$.

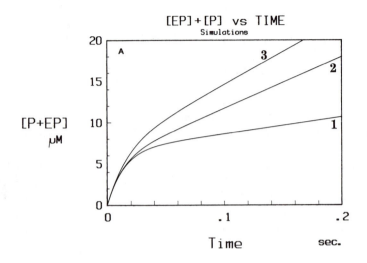

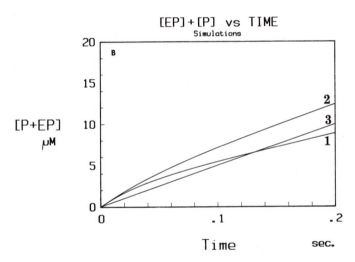

Figure 18-16 Computer simulations of $([EP] + [P])$ versus time curves. k_1 is the rate constant for ES → EP, k_2 is the rate constant for EP → ES, and k_3 is the rate constant for product release, EP → E + P.

	Constant	Curve 1	Curve 2	Curve 3
(A)	k_1	50	50	50
	k_2	20	20	3
	k_3	3	10	10
(B)	k_1	10	10	10
	k_2	10	0.1	1000
	k_3	10	10	1000

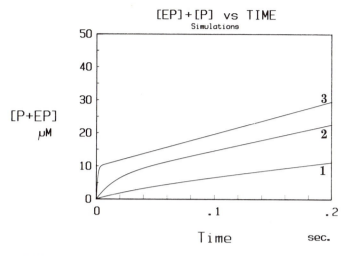

Figure 18-17 Rate constants are the same as in Fig. 18-16. Saturating concentrations of substrate are assumed. The active-site concentration is $10 \ \mu M$.

Constant	Curve 1	Curve 2	Curve 3
k_1	10	50	1000
k_2	3	3	3
k_3	10	10	10

From this known value of A, At can be calculated during the exponential phase and subtracted to give the true exponential.

$$([EP] + [P]) - At = B(1 - e^{-k_{obs}t}) \qquad (18\text{-}45)$$

From Eq. (18-45) we get

$$[([EP] + [P]) - At] - B = -Be^{-k_{obs}t} \qquad (18\text{-}46)$$

Therefore,

$$\log \{B - [([EP] + [P]) - At]\} = \log B - \frac{k_{obs}}{2.3} t \qquad (18\text{-}47)$$

or

$$\log \{B + At - ([EP] + [P])\} = \log B - \frac{k_{obs}}{2.3} t \qquad (18\text{-}48)$$

In other words, since $B + At$ is the calculated OD obtained by extrapolation of the steady-state phase of the reaction to $t = 0$ (i.e., to B) and $([EP] + [P])$ is obtained from the observed OD we can rewrite Eq. (18-48) as

$$\log (OD_t^{calc} - OD_t^{obs}) = \log OD_{t_0}^{calc} - \frac{k}{2.3} t \qquad (18\text{-}49)$$

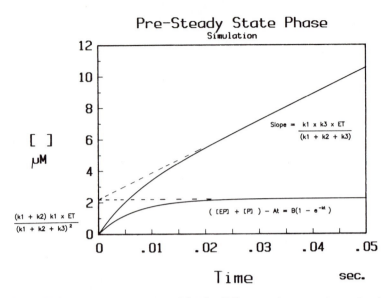

Figure 18-18 Computer simulation of ($[EP] + [P]$) versus time course (curve 1) and calculated exponential curve for the burst phase of the reaction. k_1 ($= 50$) is the rate constant for ES → EP, k_2 ($= 50$) is the rate constant for EP → ES, and k_3 is the rate constant for product release, EP → E + P. Saturating concentrations are assumed. The active-site concentration is 10 μM.

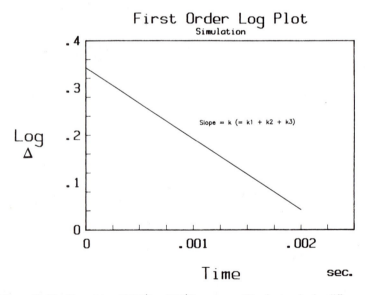

Figure 18-19 Plot of log ($OD_t^{calc} - OD_t^{obs}$) versus t. The former is the difference between the observed concentration of products and the concentration calculated by extrapolation of the steady-state rate to $t = 0$ in Fig. 18-18. The slope of the line is $-k_{obs}$, where k_{obs} is the rate constant of the exponential phase of the reaction (the burst rate). $k_{obs} = k_1 + k_2 + k_3$.

and from a plot of log $(OD_t^{calc} - OD_t^{obs})$ versus t we get a value for k_{obs}, as shown in Figs. 18-18 and 18-19.

It is apparent that the quantities obtained from experimental (EP + P) versus t curves are (1) the *apparent* burst size (B) from extrapolation of the linear region to $t = 0$; (2) the slope, A, of the linear part of the curve, $k_1 k_3 ET/(k_1 + k_2 + k_3)$; and (3) $k_{obs} = k_1 + k_2 + k_3$ from the exponential phase of the curve. As a result, summarized in Fig. 18-19, we can calculate the true steady-state concentration of EP from the apparent burst size (B), A, and k_{obs}.

$$[EP] = \frac{B - A}{k_{obs}} \tag{18-50}$$

From points 2 and 3 we can calculate $k_1 k_3 ET$ and (since ET is a known parameter) $k_1 k_3$.

$$k_1 k_3 ET = A k_{obs} \tag{18-51}$$

which also gives us k_3 as a function of k_1, and since $k_{obs} = k_1 + k_2 + k_3$ we obtain relative values of k_1 and k_2. From the $([EP] + [P])$ versus t curve it is possible to separate $[P]$ versus t or $[EP]$ versus t time courses for the reaction (Fig. 18-20).

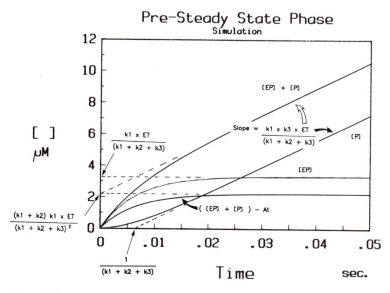

Figure 18-20   Summary of pre-steady-state phase for a reaction ES ↔ EP → E + P. Rate constants are the same as in Fig. 18-19. The intercept of the $[P]$ versus t plot is $1/(k_1 + k_2 + k_3)$, and is the reciprocal of the rate constant, k_{obs}, of the burst phase. $k_{obs} = k_1 + k_2 + k_3$. If free product can be monitored, this gives an alternative method to that outlined in Figs. 18-18 and 18-19 to obtain a value for k_{obs}.

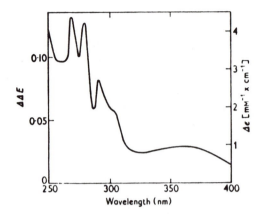

Figure 18-21   Difference spectrum of the complex of yeast GA-3P-DH with NAD at pH 8.0 and 20°C. Difference spectra measured at various concentrations of NAD between 0.02 and 1.00 mM were evaluated at various wavelengths to determine the apparent dissociation constant $K = 1.0 \times 10^{-5}$ and to extrapolate the measurements to 100% saturation. (Reprinted with permission from K. Kirschner, E. Gallego, I. Schuster, D. Goodall, *J. Mol. Biol.*, *58*, 29–50. Copyright 1971 Academic Press, Inc., New York.)

Examples of the Use of Rapid-Reaction Techniques

Cooperative Binding of NAD to Yeast Glyceraldehyde-3-Phosphate Dehydrogenase (GA-3P-DH)

When NAD binds to GA-3P-DH an absorbance at 360 nm specific for the co-enzyme bound to the enzyme is formed, as shown in Fig. 18-21. This difference spectrum has been used to study the equilibrium binding of NAD to the enzyme and serves as a convenient spectral signal to follow in rapid-reaction studies of coenzyme binding.

In a temperature-jump study, relaxation processes were detected spectrophotometrically only in mixtures of enzyme and NAD; three were observed with relaxation times τ_1, τ_2, and τ_3. The processes characterized by τ_1 and τ_2 are both clearly second order, as evidenced by an increase of rate with increasing ligand concentration. At low concentrations $1/\tau_1$ and $1/\tau_2$ are separated only by a factor of 2 to 3.

The rate of the slow relaxation process, $1/\tau_3$, was measured by temperature jump at low [NAD] and by stopped flow at high [NAD]. $1/\tau_3$ is constant at very high ligand concentrations, as shown in Fig. 18-22. The independence of $1/\tau_3$ on enzyme concentration was extended to low NAD concentrations. Clearly, $1/\tau_3$ must be identified with a slow first-order process involving the enzyme–NAD complex.

The first relaxation process can be identified unequivocally with a second-order reaction,

$$ \text{E} + \text{NAD} \underset{k_2}{\overset{k_1}{\rightleftharpoons}} \text{E} \cdot \text{NAD} $$

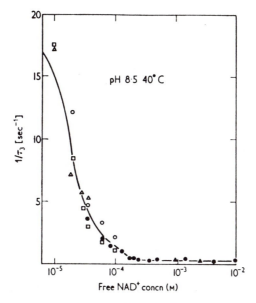

Figure 18-22 Dependence of $1/\tau_3$ on [NAD]. (Reprinted with permission from K. Kirschner, E. Gallego, I. Schuster, and D. Goodall, *J. Mol. Biol.*, *58*, 29–50. Copyright 1971 Academic Press, Inc., New York.)

Since it is well separated from the following step at high NAD concentrations, it can be regarded as an isolated process. The concentration dependence can thus be evaluated on the basis of

$$\frac{1}{\tau_1} = k_2 + k_1(\text{E} + \text{NAD}) \qquad (18\text{-}52)$$

and at high concentrations of ligand the concentration of free binding sites ($\bar{\text{E}}$) becomes negligible and Eq. (18-52) is reduced to

$$\frac{1}{\tau_1} = k_2 + k_1(\text{NAD}) \qquad (18\text{-}53)$$

From plots of $1/\tau_1$ versus NAD, as in Fig. 18-23, an estimate of the thermodynamic binding constant, $K_d = k_2/k_1 = 8 \times 10^{-5}$ M, is obtained. This is smaller than the concentration of free NAD at half-saturation of approximately 2.2×10^{-4} M (from direct binding studies), and thus the binding site involved in the fast relaxation process can be identified as a high-affinity site. The fact that the reciprocal relaxation time is an almost linear function of free NAD down to 3% saturation suggests that the concentration of free binding sites possessing high affinity are, under all conditions, small compared to the concentration of free NAD.

The second relaxation process is also clearly of second order. However, it can no longer be considered in isolation because it must be coupled to the faster process.

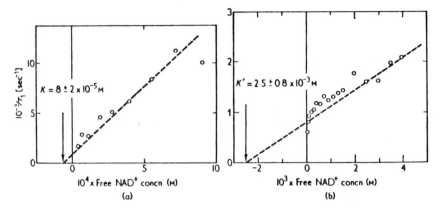

Figure 18-23 Dependence of $1/\tau_1$ and $1/\tau_2$ on the free [NAD]. (Reprinted with permission from: K. Kirschner, E. Gallego, I. Schuster, and D. Goodall, *J. Mol. Biol.*, *58*, 29–50. Copyright 1971 Academic Press, Inc., New York.)

If the binding sites exist independently, coupling is brought about by the common ligand (i.e., NAD), and the two reactions can be written as

$$E + NAD \underset{k_2}{\overset{k_1}{\rightleftharpoons}} E \cdot NAD \qquad \text{fast } (\tau_1)$$

and

$$E' + NAD \underset{k_2'}{\overset{k_1'}{\rightleftharpoons}} E' \cdot NAD \qquad \text{slow } (\tau_2)$$

Coupling can also occur via a common enzyme species, leading to the compulsory order of addition of ligands.

$$NAD + E + NAD \rightleftharpoons NAD \cdot E + NAD \rightleftharpoons NAD \cdot E' + NAD \rightleftharpoons NAD \cdot E' \cdot NAD$$

The two different types of binding sites are symbolized by differentiating between the left- and right-hand sides of E. Some unspecified isomerization of NAD $\cdot$ E to NAD $\cdot$ E' is required to generate the second binding site.

In attempting to explain only two binding reactions in the millisecond time range, the isomerization must necessarily have a comparable rate, and one of the intermediates (e.g., NAD $\cdot$ E') must be present at an undetectable concentration.

The compulsory-order addition mechanism is unnecessarily complex and neglects the fact that the *only* first-order process observed is slow compared to the binding steps. The data thus favor the independent-binding-sites scheme. In this case, the second relaxation process would obey the relation shown by

$$\frac{1}{\tau_2} = k_2' + k_1'\left(\bar{D} + \bar{E}\frac{K + \bar{D}}{K + \bar{D} + \bar{E}}\right) \tag{18-54}$$

where $\bar{D}$ is the concentration of free NAD.

At a high concentrations of NAD (i.e., $\bar{D} > \bar{E}, \bar{E}'$), decoupling of the two binding reactions occurs, reducing Eq. (18-54) to

$$\frac{1}{\tau_2} = k_2' + k_1'(\bar{D}) \tag{18-55}$$

The limiting slope of the values of $1/\tau_2$ at high NAD concentrations can now be used to obtain a rough estimate of the pertinent dissociation constant, $K_d' = 2.5 \times 10^{-3}$ M, which is far greater than the concentration of free NAD at half-saturation, and thus the binding site can be characterized as a low-affinity binding site.

Significance of Relaxation Times to Allosteric Models

The simplest sequential mechanism of Koshland, Nemethy, and Filmer, and the concerted mechanism of Monod et al., can be distinguished in favorable cases by such studies.

The sequential mechanism illustrated in Fig. 18-24 is characterized by five conformational hybrids of the enzyme, which are interrelated by four consecutive binding reactions. The implications for the relaxation kinetics of such a system are that only the superposition of second-order processes (a maximum of four) can be observed. Although the progressive changes of affinity could be due to induced conformational changes, the simplifying assumption converts the sequences of binding followed by a conformational change into an overall binding process. That is, isomerizations do not occur as independently observable elementary steps. It is clear that the simplest sequential mechanism cannot account for the existence of τ_3, no matter what the values of the individual rate and equilibrium constants are.

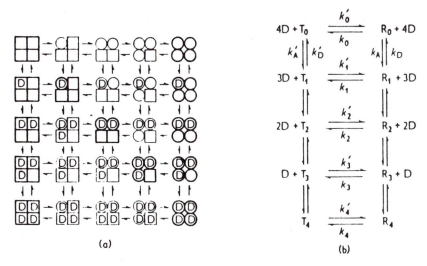

(a)   (b)

Figure 18-24 Sequential and concerted allosteric models for ligand binding cooperativity. (Reprinted with permission from K. Kirschner, E. Gallego, I. Schuster, D. Goodall, *J. Mol. Biol.*, 58, 29–50. Copyright 1971 Academic Press, Inc., New York.)

The concerted mechanism, however, explicitly allows independent isomerization reactions. For a tetrameric enzyme this mechanism is characterized by nine relaxation times. In the special case where isomerization is much slower than binding, and where the rates of binding are sufficiently separated, it is possible to derive closed analytical expressions for all nine relaxation times.

To the extent that the condition of identity and independence of the binding sites within each one of the two states is fulfilled, six out of nine relaxation processes have zero amplitudes. This degeneracy results from the fact that the only physical process associated with a finite enthalpy change is the binding of the ligand to the active site in either the R or the T state. The remaining relaxation times are related to normal modes involving the redistribution of ligand between equivalent binding sites at constant ligand concentration. These processes have zero enthalpy changes by definition. Under such conditions only three relaxation times are detectable, two of which show a dependence on ligand concentration while the third is independent of the ligand concentration. In the case of NAD binding to yeast GA-3P-DH, the observed results are consistent with a concerted model.

Binding of NADH to Yeast Glyceraldehyde-3-
Phosphate Dehydrogenase

NADH, on binding to the enzyme, quenches protein fluorescence ($_{290}F_{350}$) and also shows an increase in the polarization of the reduced nicotinamide fluorescence ($_{340}F_{450}$). Either of these parameters can be used to study NADH binding to the enzyme, in addition to monitoring changes in the NADH fluorescence upon binding. In rapid-reaction studies only one relaxation process was detected in mixtures of enzyme and NADH in the range $\tau = 10$ μs to 500 ms. Neither the apoenzyme nor the NADH solution showed measurable relaxation spectra when studied separately.

The characteristic relaxation time τ_1 did not vary with the specific method of observation: Protein fluorescence excited at 290 nm and measured between 300 and 400 nm, NADH fluorescence excited at 290 or 340 nm and observed at above 400 nm, or the polarization of NADH fluorescence excited at 340 nm. In each case the temperature jump resulted in rapid initial changes in the observed parameters that could not be resolved within the time resolution of the instrument, $\tau_{min} = 10$ μs, and are probably due to the intrinsic dependence of fluorescence quantum yield and polarization on temperature.

The observed signs of the amplitudes corresponding to the subsequent relaxation process are in the direction of increasing dissociation of NADH with increasing temperature, as predicted from equilibrium measurements at 20 and 40°C.

The relaxation experiments were done with two different total enzyme concentrations (20 and 0.2 mg/ml) and various methods of observation. An increase of $1/\tau_1$ with increasing concentration of free NADH is characteristic of a simple binding process at each of the four apparently identical and independent binding sites of the tetrameric enzyme.

The data obtained (see Fig. 18-25) for the lower enzyme concentration obey a linear relationship down to very low concentrations of NADH. At higher enzyme concentrations $1/\tau_1$ deviates at low concentrations of NADH. At the high NADH

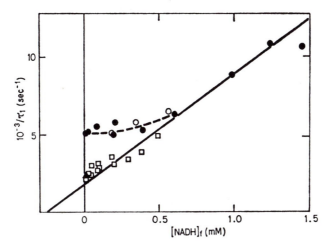

Figure 18-25 Dependence of $1/\tau_1$ on free [NADH] for NADH binding to yeast GA-3P-DH. [Reprinted with permission from: G. von Ellenrieder, K. Kirschner, and I. Schuster, *Eur. J. Biochem.*, *26*, 220–236 (1972).]

concentrations it merges into the straight line extrapolated from the measurements at low enzyme concentration.

This behavior is that expected for a simple binding reaction where the reciprocal relaxation time $1/\tau_1$ is related to the equilibrium concentrations of free binding sites $(\bar{E})$ and free NADH $(\overline{NADH})$ by the expression given in

$$\frac{1}{\tau_1} = k_2 + k_1(\bar{E} + \overline{NADH}) \qquad (18\text{-}56)$$

For $\overline{NADH} \gg \bar{E}$, $1/\tau_1$ is nearly related to $\overline{NADH}$. An upward deviation of $1/\tau_1$ with decreasing concentration of $\overline{NADH}$ is to be expected only when $\bar{E}$ and $\overline{NADH}$ are comparable (i.e., at a high level of total enzyme concentration).

When the data are plotted in the form $1/\tau_1$ versus $\bar{E} + \overline{NADH}$, a strictly linear plot is obtained at both enzyme concentrations, as seen in Fig. 18-26. The values of k_2 and k_1 obtained from the intercept and slope, respectively, of such a plot are $k_2 = 1.9 \times 10^{-3}\ s^{-1}$ and $k_1 = 7.9 \times 10^6\ M^{-1}\ s^{-1}$. Within the error limits, the value of the thermodynamic dissociation constant, $k_d = k_2/k_1 = 0.24\ mM$, agrees well with the value of $0.2\ mM$ obtained from direct binding studies by gel filtration.

Multiple Turnover Studies of the Reaction Catalyzed
by Glutamate Dehydrogenase

Glutamate dehydrogenase catalyzes the reversible oxidative deamination of glutamate. The reductive reaction, using NADH as coenzyme, is conveniently monitored by following absorbance at 340 nm. When enzyme is rapidly mixed with NADH, 2-oxoglutarate, and ammonia, no evidence of a pre-steady-state burst is seen (Fig. 18-27), consistent with the proposed rapid-equilibrium random-order addition of substrates in the reductive amination reaction.

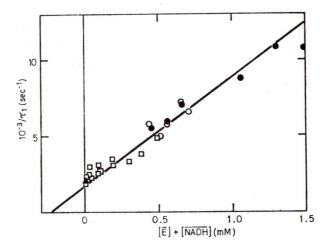

Figure 18-26   Dependence of $1/\tau_1$ on ($[E] + [\overline{NADH}]$) for NADH binding to yeast GA-3P-DH. [Reprinted with permission from: G. von Ellenrieder, K. Kirschner, and I. Schuster, *Eur. J. Biochem.*, **26**, 220–236 (1972).]

When the oxidative deamination reaction is monitored at pH 8.0 with varying amounts of glutamate, a distinct burst phase is seen at 10 mM glutamate and 250 μM NAD or NADP, as shown in Fig. 18-28. As indicated, three parameters—the steady-state rate, the burst size, and the burst rate—are obtained from such data. The existence of the burst clearly indicates that under these conditions a product release step is the overall rate-limiting step. The burst size, as discussed earlier, is less than the steady-state concentration of the enzyme–product complex, and in this instance

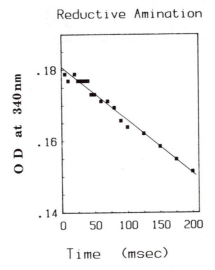

Figure 18-27   Results of stopped-flow study of the reductive amination reaction catalyzed by glutamate dehydrogenase.

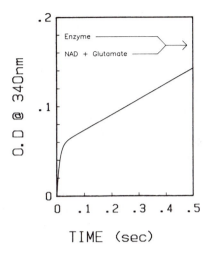

Figure 18-28 Oxidative deamination of glutamate dehydrogenase at pH 8.0 with 10 mM glutamate and 250 μM NAD.

is less than half the total active-site concentration. When the glutamate concentration is raised, a number of changes occur in the observed results: The steady-state rate decreases above 50 mM glutamate, the burst size approximately doubles, and as shown in Fig. 18-29, a distinct lag phase is observed between the pre-steady-state burst and the steady-state production of NADH.

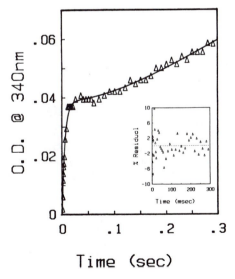

Figure 18-29 Oxidative deamination of glutamate dehydrogenase at high glutamate concentrations. The solid line is a best fit of the data to Eq. (18-57), and the insert is a residuals plot of the fit.

Time-difference spectral studies of the burst phase, the lag phase, and the steady-state phase of the reaction show that during the burst an enzyme–NADH-2-oxoglutarate complex is formed, and that during the lag phase 2-oxoglutarate is slowly released, presumably as the result of a rate-limiting conformational change, while, as would be expected, free NADH is rapidly produced during the steady-state phase. The doubling of the burst size, the appearance of the lag at high glutamate, and the apparent substrate inhibition seen in the steady-state rate led to the postulate of a reciprocating subunit mechanism for this hexameric enzyme. At low glutamate concentrations three or fewer of the subunits bind NAD and glutamate and react to produce NADH and 2-oxoglutarate bound to the enzyme. NAD and glutamate bind to the three vacant sites per hexamer and induce a conformational change that causes the release of NADH and 2-oxoglutarate from the original three sites. This process is repeated in a reciprocating subunit manner, with the two halves of the hexamer alternating in the production of NADH and 2-oxoglutarate. At high glutamate concentrations more than three sites are initially filled with NAD and glutamate and react to give NADH and 2-oxoglutarate and, as observed, the burst size increases. However, at this point there are essentially no empty sites for NAD and glutamate to bind to, so that product release is enhanced. As a result, there is a very slow release of 2-oxoglutarate, leading to the observed lag phase. Once 2-oxoglutarate has been released from a sufficient number of sites, glutamate can bind and reciprocation take place, leading to the steady-state rate of product accumulation. This model is shown in Fig. 18-30 and is described by

$$P_{\text{total}} = A(1 - e^{-kt} + mt(1 - e^{-lt})) \qquad (18\text{-}57)$$

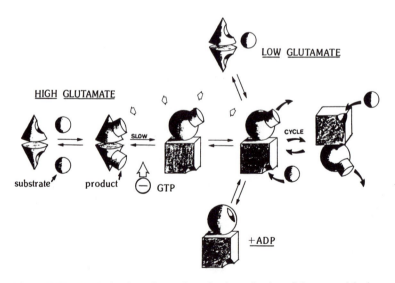

Figure 18-30   Model for the reciprocating subunit mechanism of glutamate dehydrogenase proposed on the basis of stopped-flow studies.

where P_{total} is the total amount of product NADH (free or enzyme-bound) produced at any time t, A is the burst amplitude, k the burst rate, m the steady-state rate of NADH production, and l the rate constant of the rate-limiting conformational change that allows 2-oxoglutarate to be released from the enzyme in the absence of binding energy input from glutamate. When the enzyme is able to operate via a reciprocating subunit mechanism at low glutamate concentrations, the rate constant l is fast and no lag is observed. As shown in the insert in Fig. 18-29, Eq. (18-57) provides an excellent description of the experimental data obtained at high glutamate concentrations, providing support for the model shown in Fig. 18-30.

From these examples it is apparent that information from rapid-reaction studies is invaluable in studying ligand binding steps as well as the complete catalytic reaction. In particular, data from such studies have proved useful in discerning among various allosteric models that have been proposed and, as shown by the last example, have allowed more complex allosteric models to be explored.

19

Use of Isotopes in Enzyme
Kinetic Mechanism Analysis

INTRODUCTION

The use of isotopically labeled compounds has led to a number of advances in enzyme kinetic analysis. These can be grouped into four categories, one of which was discussed in Chap. 1: the use of isotopically labeled substrates to follow the reaction rate. In such cases the isotope in the substrate is transferred to product and its amount quantitated at various time intervals to estimate the *rate*. The other three categories of information obtainable from experiments using isotopically labeled substrates involve (1) the formal kinetic mechanism of a particular enzyme-catalyzed reaction, (2) non-rate-limiting steps in the formal kinetic mechanism, and (3) identification of bond-breaking steps with particular phases of the formal kinetic mechanism of a reaction.

In this chapter we examine a variety of techniques employing isotopically labeled substrates which provide information that can complement that obtained from initial rate or rapid-reaction studies and, in many cases, examine ambiguities arising from initial rate kinetic studies.

ISOTOPIC COMPOUNDS AS ALTERNATIVE SUBSTRATES

In earlier chapters we considered the use of alternative substrates in initial rate studies as a method for elucidating the formal kinetic mechanism of a reaction. One problem frequently encountered is when no *good* alternative substrate is available. Although many enzymes can use alternative substrates, it is often found that they are very much poorer than the true substrate, and as a result one is left with a question as to whether the alternative follows the same formal kinetic mechanism. One way out of this dilemma is to use a radioactive label in the substrate and to mix labeled and nonlabeled substrates and follow the rate of reaction with one or the other. It is important to be able to follow the rate with only one of the alternate substrates. So that the isotopically labeled substrate does not alter the formal kinetic mechanism, care should be taken that the isotope does not involve any bond that may be broken during the course of the reaction; otherwise, an isotope effect may alter rate-limiting steps in the overall mechanism.

If we examine two situations, a compulsory-order two-substrate mechanism and a random-order, equilibrium two-substrate mechanism, Eqs. (19-1) to (19-3) describe the use of isotopically labeled alternative substrates.

Compulsory-Order Mechanism

The mechanism, in the presence of the normal substrate, S, and the isotopically labeled substrate, S_1^*, is given in Fig. 19-1. The rate equation when S_1, S_1^*, and S_2 are present is given by

$$\frac{e}{V_0} = \phi_0 + \frac{\phi_1}{S_1}(1 + X + Y + Z) + \frac{\phi_2}{S_2} + \frac{\phi_{12}}{S_1 S_2}(1 + X + Y + Z) \qquad (19\text{-}1)$$

where

$$X = \frac{k_1' S_1^*(k_4' + k_5')}{k_2'(k_4' + k_5') + k_3' k_5'(S_2)}$$

$$Y = \frac{k_3'[S_2]X}{k_4' + k_5'}$$

$$Z = \frac{k_3' k_5'[S_2]X}{k_7'(k_4' + k_5')}$$

and where k' and so on indicate steps involving the isotope. Thus S_1^* is a competitive inhibitor versus S_1, *but* relative to S_2, the inhibition is parabolic.

If an isotope of S_2 is available (S_2^*) such that the steps

$$ES_1 + S_2^* \longleftrightarrow EP_1 P_2^* \longleftrightarrow EP_1 + P_2^*$$

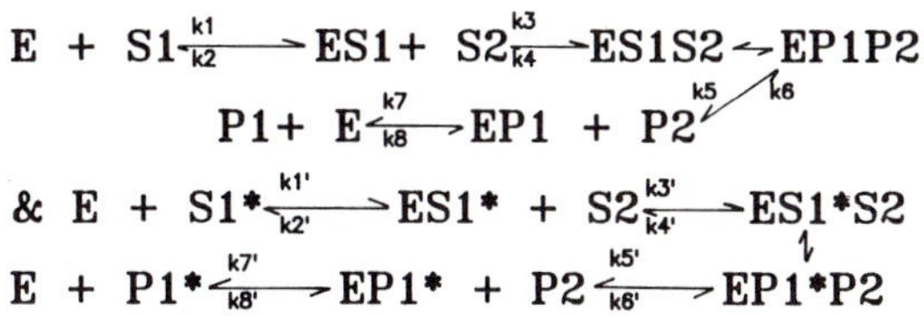

$$E + S1 \underset{k2}{\overset{k1}{\rightleftharpoons}} ES1 + S2 \underset{k4}{\overset{k3}{\rightleftharpoons}} ES1S2 \rightleftharpoons EP1P2$$
$$P1 + E \underset{k8}{\overset{k7}{\rightleftharpoons}} EP1 + P2 \overset{k5}{\underset{k6}{\nearrow}}$$

$$\&\ E + S1* \underset{k2'}{\overset{k1'}{\rightleftharpoons}} ES1* + S2 \underset{k4'}{\overset{k3'}{\rightleftharpoons}} ES1*S2$$
$$E + P1* \underset{k8'}{\overset{k7'}{\rightleftharpoons}} EP1* + P2 \underset{k6'}{\overset{k5'}{\rightleftharpoons}} EP1*P2$$

Figure 19-1 Compulsory-order mechanism in the presence of an isotopically labeled S_1^* alternative substrate.

must be included, the rate equation has to be modified:

$$\frac{e}{V_0} = \phi_0 + \frac{\phi_1}{S_1} + \frac{\phi_2}{S_2}\left[1 + \frac{k_3'S_2^*}{k_4' + k_5'}\left(1 + \frac{k_5'}{k_7'}\right)\right] + \frac{\phi_{12}}{S_1 S_2}\left(1 + \frac{k_3' k_5' S_2^*}{k_2(k_4' + k_5')}\right) \quad (19\text{-}2)$$

and now the inhibitor is competitive versus S_2 but noncompetitive versus S_1.

Random-Order, Rapid-Equilibrium Mechanism

As with the equations for a simple random-order, rapid-equilibrium (RORE) mechanism, those in the presence of isotopically labeled alternative substrate are easy to derive and are symmetrically arranged.

$$\frac{e}{V_0} = \phi_0 + \frac{\phi_1}{S_1}\left(1 + \frac{S_1^*}{K_4'}\right) + \frac{\phi_2}{S_2} + \frac{\phi_{12}}{S_1 S_2}\left(1 + \frac{S_1^*}{K_1'}\right) \quad (19\text{-}3)$$

where K_4' is

$$ES_2 + S_1^* \xrightarrow{K_4'} ES_2 S_1^*$$

and K_1' is

$$E + S_1^* \xrightarrow{K_1'} ES_1^*$$

Consequently, inhibition by S_1^* is *competitive* versus S_1 and noncompetitive versus S_2. With an isotopically labeled analog of S_2, S_2^* is competitive versus S_2 and noncompetitive versus S_1. Two examples illustrate the uses of such isotopically labeled alternative substrates.

Alcohol Dehydrogenase. In this system NAD and [^{14}C]-NAD are used as alternative substrates and Lineweaver–Burk plots drawn with either NAD or ethanol as the varied substrate. As shown in Fig. 19-2A, when [^{14}C]-NAD is examined as an inhibitor with respect to NAD, competitive inhibition results. When, however, [^{14}C]NAD is used as an inhibitor with respect to ethanol (Fig. 19-2B), the inhibition is parabolic. These results are consistent with a compulsory-order mechanism having NAD as the first substrate. Incidentally, this experiment can also be done using the thionicotinamide analog of NAD, which has an S-NAD $\rightleftharpoons$ S-NADH *isosbestic* point at 342 nm—thus measuring at 342 nm measures *only* NADH production as required.

Hexokinase. This enzyme catalyzes the phosphorylation of glucose by ATP in a magnesium-dependent reaction. [^{14}C]- and [^{12}C]glucose are convenient alternative substrates, and the following experiments explore the kinetic mechanism of the reaction.

As shown in Fig. 19-3 (left), [^{12}C]glucose is a competitive inhibitor with respect to [^{14}C]glucose when the assay is conducted by following ^{14}C transfer into product. When [^{12}C]glucose is used as an inhibitor versus the other substrate, MgATP, a linear, noncompetitive inhibition is observed. The experimental data are *not* consis-

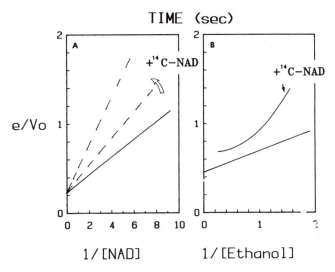

Figure 19-2 Experiments with alcohol dehydrogenase: initial rate of NADH production followed in the absence or presence (––––) of [^{14}C]NAD. Varied substrate is [^{12}C]NAD in part (A) or ethanol in part (B). Only reduction of [^{12}C]NAD is followed.

tent with a simple compulsory order of substrate addition with glucose as the first substrate, but are consistent with either a rapid-equilibrium, random-order mechanism or a compulsory-order mechanism with MgATP as the first substrate. Experiments with isotopically labeled ATP as an alternative substrate versus ATP would, of course, distinguish between these possibilities.

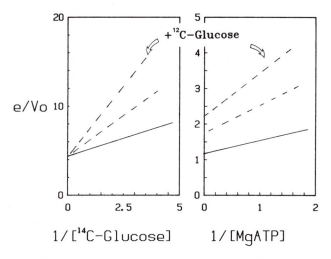

Figure 19-3 Experiments with hexokinase: initial rate of [^{14}C]glucose transfer into product is followed in the absence or presence (––––) of [^{12}C]glucose. Varied substrate is [^{14}C]glucose in part (A) of MgATP in part (B).

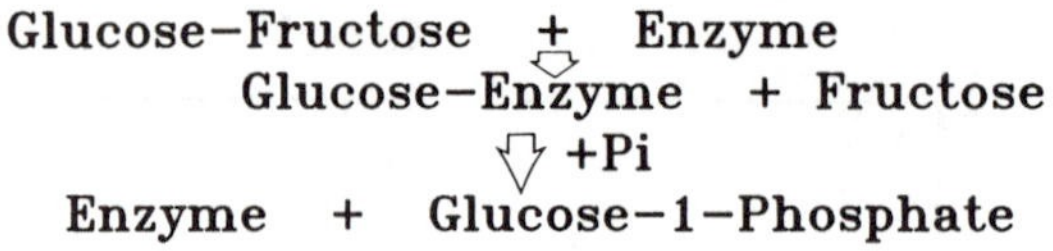

$$\text{Glucose–Fructose} \quad + \quad \text{Enzyme}$$
$$\text{Glucose–Enzyme} \quad + \quad \text{Fructose}$$
$$+\text{Pi}$$
$$\text{Enzyme} \quad + \quad \text{Glucose–1–Phosphate}$$

Figure 19-4   Enzyme-substituted mechanism for sucrose phosphorylation.

ISOTOPE EXCHANGE AT EQUILIBRIUM

Isotope-exchange studies have long been used to examine formal kinetic mechanisms of enzyme-catalyzed reactions. Early studies of isotope exchange examined if an enzyme exhibited an enzyme-substituted (ping-pong) mechanism. In such a mechanism, represented as

$$E + XG \rightleftharpoons EXG \rightleftharpoons EG + X$$

$$EG + Y \rightleftharpoons EYG \rightleftharpoons E + YG$$

the use of radioactively labeled X and nonlabeled XG allows an exchange of isotope into XG *in the absence* of Y. Such an isotope exchange is not possible in a ternary complex mechanism. Several examples illustrate the sort of observations that have been made.

In the reaction catalyzed by sucrose phosphorylase,

$$\text{glucose-fructose} + P_i \rightleftharpoons \text{G-1-P} + \text{fructose}$$

an isotope exchange between $^{32}P_i$ and glucose-1-phosphate is observed in the absence of fructose, suggesting that the enzyme operates via the enzyme-substituted mechanism illustrated in Fig. 19-4.

The biotin-dependent carboxylases, which catalyze the overall reaction,

$$\text{ATP} + \text{HCO}_3{}^- + \text{proprionyl-CoA} \xrightarrow{\text{Mg}^{2+}} \text{ADP} + P_i + \text{meth. malonyl-CoA}$$

catalyze an exchange between ATP and ^{32}Pi, but it occurs only in the presence of $\text{HCO}_3{}^-$ and Mg^{2+}. This suggests that an enzyme-substituted mechanism is in fact operating and that it involves the stages shown in Fig. 19-5.

One of the problems with this type of experiment is that the presence of a small contaminant of the second substrate allows the "exchange" to occur in the absence of *added* second substrate. In some cases small amounts of *tightly bound* second substrate on the enzyme are sufficient to allow the "exchange" to take place, when in

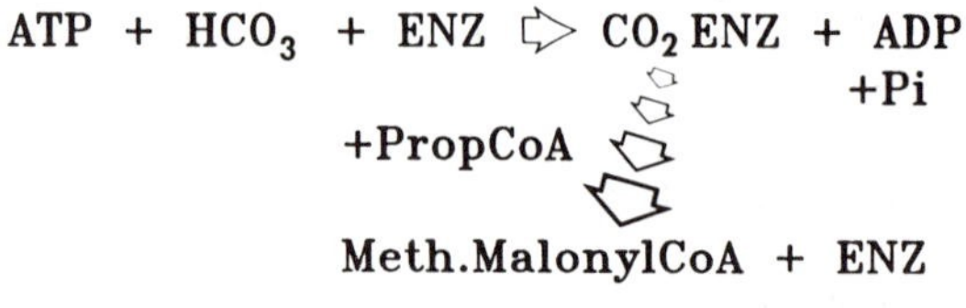

$$\text{ATP} + \text{HCO}_3 + \text{ENZ} \Rightarrow \text{CO}_2 \, \text{ENZ} + \text{ADP}$$
$$+\text{Pi}$$
$$+\text{PropCoA}$$
$$\text{Meth.MalonylCoA} + \text{ENZ}$$

Figure 19-5   Enzyme-substituted mechanism for a biotin-dependent carboxylase.

reality it results from the simple bulk transfer of substrate to product via a ternary complex mechanism rather than from the existence of an isotope exchange resulting from an enzyme-substituted mechanism.

Isotope-exchange-at-equilibrium studies are usually directed at a variety of questions asked about ternary complex mechanisms. This approach is a powerful (but as we will see, somewhat tedious) tool for distinguishing among ternary complex mechanisms, for examining the existence of abortive complexes in mechanisms, for obtaining information on non-rate-limiting steps in reaction mechanisms, and for ascertaining the true equilibrium nature of a rapid-equilibrium, random-order mechanism.

The principle of the approach is quite simple. An enzyme is mixed with substrates (or products, or a mixture of substrates and products) at a particular concentration and the reaction is allowed to proceed until chemical equilibrium is reached. At this point a small amount of isotopically labeled substrate (or product) is added and the *rate of isotope exchange* into the appropriate product measured. Because the small amount of isotopically labeled substrate (or product) is chosen such that it does not significantly displace the equilibrium position of the reactants, the rate of isotope exchange is measured *at equilibrium*. The exchange rate under such conditions can be regarded as a function of the concentration of the enzyme complex containing the substrate whose exchange rate is being monitored. As with other rate measurements, the dependence of the rate on the saturation of the enzyme with reactants is followed, bearing in mind that individual reactant concentrations cannot be altered in isolation; otherwise, a perturbation of the equilibrium results. When the concentration of a particular substrate is changed, the concentration of a product must also be altered so that the equilibrium position is not changed.

In a typical two-substrate system, various exchanges are possible, depending on the chemical nature of the reactants; thus a number of different exchanges can be followed. For the reaction $A + B \rightleftharpoons P + Q$, the following exchanges are possible: $A \rightarrow P$, $A \rightarrow Q$, $B \rightarrow P$, and $B \rightarrow Q$.

Let us consider first the case of a compulsory-order reaction, represented schematically in Fig. 19-6. For such a mechanism we can consider the effects of raising various reactant concentrations on the possible exchange rates.

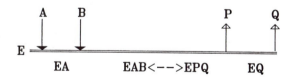

Figure 19-6   Schematic representation for a compulsory-order mechanism.

Exchanges Involving A

If an exchange of isotope *from* A is considered, the only way that the isotope of A (A*) can enter the system to exchange into P or Q involves exchange of A* with A in the complex EA. Any action that *decreases* the concentration of the complex EA decreases the rate of exchanges involving A* (i.e., A* → P or A* → Q). If the concentration of A is raised at a fixed concentration of B, the concentration of EA will increase to a maximum and exchanges involving A* will increase to a maximum. If, however, the concentration of B is raised at a fixed concentration of A, the concentration of EA will decrease eventually to zero, as all of the EA complex is bound by B and pushed into the central EAB ⇌ EPQ complexes. At this point *no* exchange involving A* is possible and the *rate* of exchanges involving A* falls to zero. Since the reaction scheme is symmetrical, similar effects of raising the concentration of P on exchanges involving Q* is observed.

In this example, not only the *dependence* of exchange rate on reactant concentration but also the *maximum rate* of exchange may differ, depending on the exchanges followed. In compulsory-order mechanisms the overall rate-limiting step is often a product release step. Consider what one expects to observe if the release of Q is rate limiting. Exchanges that do not involve Q occur at a faster maximum rate than those involving Q. Alternatively, if the overall rate-limiting step is release of P, all of the exchanges have the same maximum rate since all exchange routes depend on the release of P.

In a rapid-equilibrium, random-order mechanism, the pattern of dependence of the exchange rate on reactant concentration is easily predictable. Because all complexes in the substrate addition side of the reaction (or for that matter on the product release side) are in equilibrium, and because either substrate (A* or B*) has two routes of entry into an exchange (by combination with E or by combination with the appropriate binary complex), all possible exchange rates rise to a maximum as any

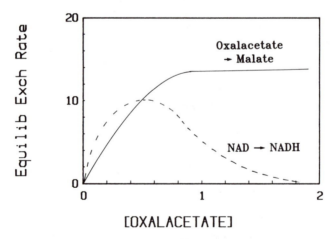

Figure 19-7 Effects of oxaloacetate/NAD concentration on the rate of isotope exchange at equilibrium for the oxaloacetate ↔ malate and the NAD ↔ NADH exchanges. The data for the NAD ↔ NADH exchange are shown at 10 times their appropriate scaled rate. The oxaloacetate/NAD ratio is kept constant as the oxaloacetate concentration is varied.

reactant concentration in increased. In addition, it is self-evident that the maximum rate of all exchanges is the same, since by definition in a rapid-equilibrium, random-order mechanism the interconversion of EAB and EPQ is rate limiting, and all exchanges must proceed through this step.

If we consider the example of malate dehydrogenase, a further useful facet of isotope-exchange studies at equilibrium is illustrated. This enzyme catalyzes the reaction

$$\text{malate} + \text{NAD} \rightleftharpoons \text{oxaloacetate} + \text{NADH}$$

If we examine the effects of raising the concentrations of NAD and oxaloacetate on the rate of NAD $\rightleftharpoons$ NADH exchange, or the rate of oxaloacetate $\rightleftharpoons$ malate exchange (Fig. 19-7), we find that the NAD $\rightleftharpoons$ NADH exchange is inhibited, while the oxaloacetate $\rightleftharpoons$ malate exchange rate reaches a maximum and shows no inhibition.

In addition, the maximum rate of the NAD $\rightleftharpoons$ NADH exchange is *slower* than that of oxaloacetate-malate. These observations are consistent with the ordered addition of NAD and malate, and the ordered release of oxaloacetate and NADH, and furthermore, suggest that NADH release may be the overall rate-limiting step in the reaction. When the concentrations of NADH and malate are raised together, it is found that the rates of *both* exchanges decrease. This suggests that an E-NADH–malate abortive complex is formed: When this occurs the concentrations of *all* the enzyme complexes on the exchange pathways are decreased, leading to an inhibition of *all* exchange rates under conditions that favor formation of the abortive complex (Fig. 19-8).

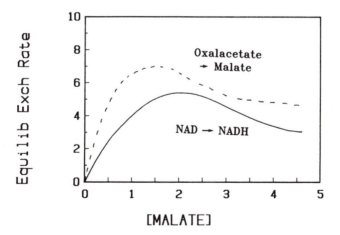

Figure 19-8 Effects of malate/NADH concentration on the rate of isotope exchange at equilibrium for the oxaloacetate ↔ malate and the NAD ↔ NADH exchange. The malate/NADH ratio is kept constant as the malate concentration is raised. Both exchange rates show inhibition as the result of abortive complex formation.

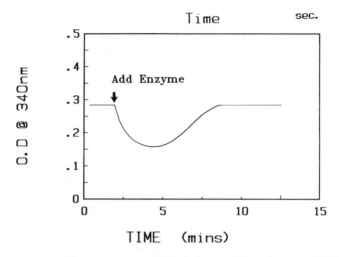

Figure 19-9   Equilibrium mixture established prior to addition of enzyme. NADP + malate-2-D (in place of malate-2-H) + CO_2 + pyruvate + NADPH. Malic enzyme as indicated.

EQUILIBRIUM PERTURBATION STUDIES

Consider an experiment involving the malic enzyme, which catalyzes the reaction

$$NADP + malate \rightleftharpoons NADPH + pyruvate + CO_2$$

The overall chemical reaction has an equilibrium constant that is first determined. From this the equilibrium concentrations of each reactant can be calculated, and an equilibrium mixture, *in the absence of the enzyme*, established. If this mixture is set up with enzyme added and the optical density at 340 nm (due to the NADPH) monitored, one expects to see no change in OD at 340 nm. If, however, one set up the same equilibrium mixture but replaced the malate with malate deuterated at the 2-position and followed the absorbance at 340 nm after the enzyme had been added; then a distinct change in absorbance is observed, as illustrated in Fig. 19-9.

This perturbation from equilibrium is caused by the more rapid reaction of NADPH + CO_2 + pyruvate to form malate + NADP than the reaction of NADP + malate-2-D to form NADPD + pyruvate + CO_2. In the first part of this discussion the rate of NADP + malate → NADPH + pyruvate + CO_2 (k_1) equals the rate of CO_2 + pyruvate + NADPH → NADP + malate (k_2), as required by the equilibrium condition. In the latter case, with malate-2-D, however, k_1 is less than k_2 because of an isotope effect. As the equilibrium process proceeds, isotopic equilibrium is reached and the system returns to chemical equilibrium. In effect, one is measuring a difference spectrum between two approximately first-order approaches to equilibrium. Apart from providing a simple way of demonstrating an isotope effect, this equilibrium perturbation method has been used to distinguish between compulsory-order and random-order mechanisms.

$$E+A \underset{k2}{\overset{k1}{\rightleftharpoons}} EA+B \underset{k4}{\overset{k3}{\rightleftharpoons}} EAB \underset{k6}{\overset{k5}{\rightleftharpoons}} EAB^* \underset{k8}{\overset{k7}{\rightleftharpoons}} EPQ^* \underset{k10}{\overset{k9}{\rightleftharpoons}} EPQ \underset{k12}{\overset{k11}{\rightleftharpoons}} EQ+P \underset{k14}{\overset{k13}{\rightleftharpoons}} E+Q$$

Figure 19-10 Compulsory-order mechanism.

For particular mechanisms, rate equations for the conversion of, for example, labeled A to P and unlabeled P to A can be derived. In terms of the mechanism shown in Fig. 19-10, the following equations describe these rates.

$$\frac{-dA_D}{dt} = \left(\frac{k_1 k_3 k_5 k_{7D}(E)}{k_2 k_4 k_6} A_D B - \frac{k_{8D} k_{10} k_{12}(EQ)}{k_9 k_{11}} P_D \right) \Big/ \left\{ 1 + \frac{k_{7D}}{k_6} \left[1 + \frac{k_5}{k_4} \left(1 + \frac{k_3 B}{k_2} \right) \right] \right.$$

$$\left. - \frac{k_{9D}}{k_9} \left(1 + \frac{k_{10}}{k_{11}} \right) \right\} \tag{19-4}$$

$$\frac{dA_H}{dt} = \left(\frac{k_{8H} k_{10} k_{12}(EQ)}{k_9 k_{11}} P_H - \frac{k_1 k_3 k_5 k_{7H}(E)}{k_2 k_4 k_6} A_H B \right) \Big/ \left\{ 1 + \frac{k_{7H}}{k_6} \left[1 - \frac{k_5}{k_4} \left(1 - \frac{k_3 B}{k_2} \right) \right] \right.$$

$$\left. - \frac{k_{9H}}{k_9} \left(1 + \frac{k_{10}}{k_{11}} \right) \right\} \tag{19-5}$$

The subscripts H and D refer to either concentrations or rate constants for hydrogen- or deuterium-substituted A or P. Under conditions where the equilibrium levels of B and Q do not change during the perturbation, we can write

$$EQ = E \left(\frac{k_{14} Q}{k_{13}} \right) \tag{19-6}$$

Furthermore, we can define an equilibrium isotope effect

$$\frac{K_{eqH}}{K_{eqD}} = \frac{k_{7H} k_{8D}}{k_{8H} k_{7D}} = \frac{1}{\beta} \tag{19-7}$$

The equation for $-dA/dt$ after rearrangement and substitution for EQ becomes

$$\frac{-dA_D}{dt} = \left(\frac{k_1 k_3 k_5 k_{7H} A_D B}{k_2 k_4 k_6} - \frac{k_{8H} k_{10} k_{12} k_{14} P_D Q}{\beta k_9 k_{11} k_{13}} \right)(E) \Big/ \left(\frac{k_{7H}}{k_{7D}} + C_f + \frac{C_r}{\beta} \right) \tag{19-8}$$

and the two phenomenological parameters C_f and C_r are defined by

$$C_f = \frac{k_{7H}}{k_6} \left[1 + \frac{k_5}{k_4} \left(1 + \frac{k_3 B}{k_2} \right) \right] \tag{19-9}$$

$$C_r = \frac{k_{9H}}{k_9} \left(1 + \frac{k_{10}}{k_{11}} \right) \tag{19-10}$$

In the context of distinguishing ordered from random mechanisms, we need only consider the parameter C_f, which, for the compulsory ordered mechanism shown earlier, has a linear dependence on the concentration of the second substrate, B. C_f

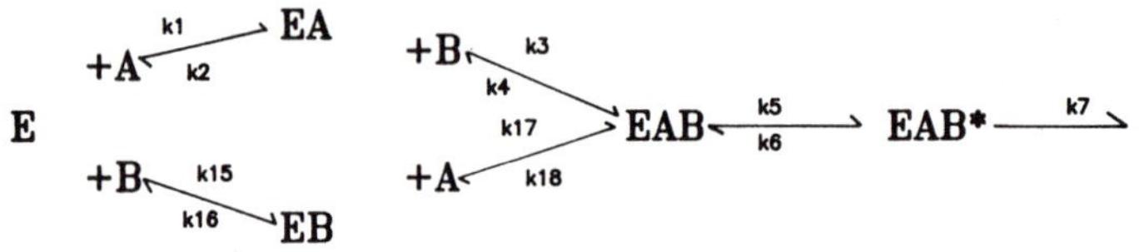

Figure 19-11   Random-order mechanism.

reflects the commitment of the ternary EAB complex to catalysis. In a compulsory-order mechanism with A as the leading substrate, C_f is infinite at an infinite concentration of B. If, however, A can appreciably dissociate from EAB (i.e., the mechanism has some component of randomness). C_f is not infinite at an infinite concentration of B. For a random addition of A and B, as shown in Fig. 19-11, the expression for C_f is given by

$$C_f = \frac{k_{7H}}{k_6}\left(1 + \frac{k_5}{k_{10}}\right) \tag{19-11}$$

which, at infinite concentrations of B, becomes

$$C_f = \frac{k_{7H}}{k_6}\left[1 + \frac{k_5}{k_4/(1 + k_3 B/k_2) + k_{10}}\right] \tag{19-12}$$

Values of C_f (and C_r) are obtained from perturbation experiments by solving these (and other derived) equations, usually by computer, but this requires a knowledge of the equilibrium isotope effect. Equilibrium isotope effects are determined by careful determination of the equilibrium constant with protonated or deuterated compounds.

MEASUREMENT OF FLUX RATIOS

One of the problems that many approaches available for determining the formal kinetic mechanism of an enzyme-catalyzed reaction have is the fact that any complexity, such as abortive complex formation or allosteric interactions and enzyme isomerizations, can interfere with the interpretation of initial rate studies, inhibition studies, or even isotope exchange at equilibrium studies. One approach circumvents most of these difficulties by isolating the *order* of substrate binding from other aspects of the mechanism. This is the so-called *flux ratio* method. In many respects it resembles a product inhibition technique, but rather than examining effects on initial rates, it examines the fate of individual product molecules participating in inhibitory reactions. The approach is best understood by examining what one would expect in an experiment involving a two-substrate reaction:

$$A + B \rightleftharpoons P + Q$$

Using a suitably labeled product, P, one can measure *two* fluxes, one involving P → A and the other P → B. If we consider the mechanism shown in Fig. 19-12, equations for these fluxes can be derived, but the expressions would be extremely

Figure 19-12 Mechanism involving a compulsory order of substrate addition and a random order of product release.

complex. However, as we will see, the useful parameter is the *ratio* between the fluxes for P → A and P → B.

For this mechanism, consider the *fate* of a molecule of P that traps a molecule of EQ produced as an intermediate in the breakdown of EAB. Once P has bound to EQ, we get formation of EPQ, which can do one of many things, as outlined in Fig. 19-13.

Figure 19-13 Fates of EPQ molecule possible in the mechanism shown in Fig. 19-12.

Obviously only EPQ molecules following the route to EAB can contribute to the flux of P → A. Once EAB is formed various routes can be taken. EAB can be converted to EA + B or back to EPQ and, as before, only EA can contribute to the flux of P → A. EA, once formed, can dissociate to E and A or bind B and return to EAB. Similar considerations contribute to the flux of P → B, and it is quite easy to see why expressions for these two fluxes are complex.

However, if one examines the pathways for P → A and for P → B, they are essentially similar; the formation of A requires the extra step EA → E + A. As a result, the flux ratio [Eq. (19-13)] is determined essentially by the immediate fate of EA formed by back reaction from P′.

$$\text{flux ratio} = \frac{F_{P \to B}}{F_{P \to A}} \tag{19-13}$$

The *fate* of this molecule of EA depends on the *concentration* of B but not A. At low concentrations of B, EA appreciably dissociates to give E + A (i.e., flux of P → A occurs). At high concentrations of B, EA is trapped back into EAB and flux does not take place. Thus we can write

$$\frac{F_{P \to B}}{F_{P \to A}} = 1 + \alpha[B] \tag{19-14}$$

where α is a constant. Figure 19-14 illustrates the dependence of the flux ratio for this compulsory order of addition of A and B on the concentrations of A and B.

If there is a *random* order of addition of A and B to give a ternary complex, a more complex situation holds, and nonlinear dependence of the flux ratio on the

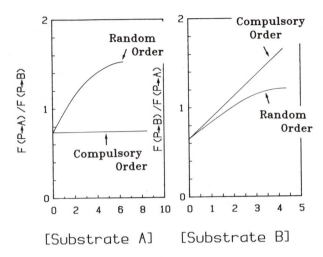

Figure 19-14 Expected patterns for flux-ratio measurements for random-order mechanisms (– – – –) and for compulsory-order mechanisms (——).

concentration of either A or B results, as shown by the dashed lines in Fig. 19-14.

To examine the order of product release (i.e., the order of substrate addition in the reverse reaction), flux ratios for A → P and A → Q must be examined as a function of the concentrations of P and Q.

SUBSTRATE "STICKINESS"

In considering *random-order* mechanisms it is possible to determine when a mechanism is rapid equilibrium as opposed to steady state. In rapid-equilibrium order, one is basically saying that the interconversion of EAB → EPQ is a much slower step than EA + B ⇌ EAB or EB + A ⇌ EAB. This assumption can be directly tested experimentally using isotopically labeled substrates.

Consider the following experiment. Enzyme is premixed with isotopically labeled A (A*) to give EA*. This EA* is then added to a large excess of A and B and the amount of A* transferred into product is measured.

If the mechanism is true rapid equilibrium, random order, the *specific activity* of the product should approach that of A* *after* the addition of the excess A. This is so because each preformed molecule of EA* dissociates far more rapidly than it proceeds to product, allowing isotopic equilibration before significant product formation occurs; therefore, product and A* (after mixing) have similar specific activities. If, on the other hand, one approaches the situation where $k_{cat} = k_8$ in the scheme in Fig. 19-15, the specific activity of the product formed in the experiment above is much higher than in the true rapid-equilibrium, random-order case. Similarly, if one premixes E with B* to give EB* and follows a similar procedure, the relative magnitudes of k_{cat} and k_6 in the Fig. 19-15 scheme can be determined.

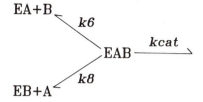

Figure 19-15 Scheme of a random-order mechanism.

This type of experiment is extremely useful in determining the "degree of randomness" of a random-order mechanism, or in other words, how closely the mechanism approaches the true rapid-equilibrium, random-order situation. If it does not obey rapid equilibrium kinetics, relationships giving dissociation constants for binary EB complexes and the equality of Michaelis constants with dissociation constants of the appropriate ternary complexes do not hold, and the mechanism can exist as a kinetically preferred pathway of substrate addition.

ELUCIDATION OF RATE DETERMINING STEPS

For an enzyme having a random-order addition of substrates that does obey the rapid-equilibrium condition, k_{cat} is the overall rate-limiting step. It is usually written as EAB → EPQ, implying that this rate constant is that for the chemical reaction. Although this may well be the case, the rapid-equilibrium condition requires only that some step (other than substrate addition or product release) involving the central complexes be rate limiting. This step could equally well be a rate-limiting conformational change involving EAB or EPQ required to allow catalysis or product release, as summarized in Fig. 19-16.

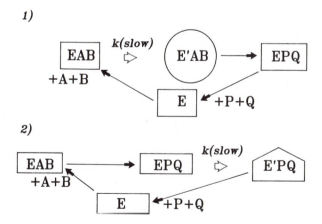

Figure 19-16 Schemes of ternary complex mechanisms involving rate-limiting conformational changes.

In such a case, studies of isotope effects on the V_{max} parameter (θ_0) allows a distinction to be made. For a rapid-equilibrium, random-order mechanism, $\theta_0 = 1/k_{cat}$, and if k_{cat} is the *chemical* step in the interconversion of EAB → EPQ, use of a substrate with an isotope at the bond-breaking atom gives a primary isotope effect, detected by initial rate studies as a change in θ_0. If, on the other hand, a rate-determining conformational change is occurring, no isotope effect on θ_0 is expected. In the case of a mechanism involving a hydride transfer step (e.g., a dehydrogenase), substitution with tritium results in a large primary isotope effect (in the range 3- to 10-fold) if the hydride transfer step is, in fact, rate limiting. The magnitude of the effect is important, as the isotopic substitution could make the hydride transfer step become rate limiting even though some other step is, with the normal substrate, slower. Under this circumstance, the magnitude of the effect is less than that expected for a primary isotope effect.

For non-rapid-equilibrium, random-order mechanisms, isotope effects on individual θ parameters are usually smaller than the effects observed on θ_0 in a rapid-equilibrium, random-order mechanism. This is so because in general the k_{cat} term appears, together with other rate constants, in a number of the θ parameters. For a compulsory-order mechanism only the θ_1 parameter does *not* contain k_{cat} and as a result is immune to an isotope effect. The other parameters show an isotope effect, but because of a "buffering effect" of rate constants not involved in the chemical interconversion, they have a less-than-maximal effect.

Stopped-flow studies in these cases allow examination for the existence of a primary isotope effect. In a compulsory-order mechanism, with rate-limiting product release, a pre-steady-state "burst" is observed and its rate includes the chemical interconversion step (strictly speaking, the k_{cat} step, as in the previous discussion for rapid-equilibrium, random-order mechanisms). This allows one to look for an isotope effect in the "burst" phase of the reaction.

We have considered isotope effects here as involving substitution of a proton by tritium or deuterium, cases that clearly optimize the magnitude of an expected isotope effect. In steady-state studies it is not unreasonable to expect to be able to experimentally determine *primary* isotope effects associated with bond-breaking steps involving carbon or nitrogen atoms, although such effects would be difficult to establish in pre-steady-state studies.

Isotopically labeled substrates provide a variety of means for obtaining information on the kinetic mechanism of an enzyme-catalyzed reaction. Apart from experiments involving substrate "stickiness" and isotope effects, much knowledge can also be gained in principle from the kinetic techniques discussed in Chaps. 13 to 15. Isotopically labeled substrates are invaluable where the approaches described in those chapters cannot readily be employed. As stated in the last two sections of this chapter, the use of isotopically labeled substrates can provide a wealth of information that is not available from initial rate kinetic studies.

20

Effects of pH, Temperature, and Isotopically Labeled Substrates on Enzyme Activity

INTRODUCTION

In Chaps. 13 to 19 we examined various kinetic aspects of enzyme mechanisms. Several of the examples discussed throughout this book refer to particular amino acid side chains that may be involved in various aspects of a mechanism. Previously, the rate-limiting steps were discussed in terms of whether a particular bond-breaking event is involved. In both of these areas much useful information can be obtained from kinetic studies. In some instances this is preliminary to other studies, while in other cases the information can directly provide the answer to a question regarding the enzyme mechanism. In this chapter we examine the usefulness of pH studies, isotope effect studies, and temperature studies of various kinetic parameters.

pH-DEPENDENCE STUDIES

There are many ionizable side chains in a protein molecule. In addition, the α-amino and carboxyl terminal residues must be considered. These ionizable groups have a variety of potential roles in a protein's structure and function. Ionic interactions can clearly be important in secondary, tertiary, and quaternary structure. Ionizable groups also often play important roles in catalysis. In Chap. 7 we looked at a variety of ways to chemically modify amino acid side chains, with the intention of identifying which types of residues may be important in the function of the protein. A particularly useful guide to such chemical modification experiments can be pH studies of various kinetic parameters for the enzyme; they can yield information about the pK_a values of groups involved in a variety of functional properties of the protein.

Basic Theory

The state of protonation of a group that can undergo a reversible protonation ($B + H \rightleftharpoons BH$) is given by the Henderson–Hasselbalch equation,

$$pH = pK_a + \log \frac{[B]}{[BH^+]} \tag{20-1}$$

where pK_a is given by

$$pK_a = -\log K_a \tag{20-2}$$

and K_a is the ionization constant, defined by

$$K_a = \frac{[B][H^+]}{[BH^+]} \tag{20-3}$$

Equation (20-1) is derived by rearranging Eq. (20-3) to the form

$$[H^+] = \frac{K_a[BH^+]}{[B]} \tag{20-4}$$

Taking the negative logarithm of Eq. (20-4) leads to Eq. (20-1). In terms of a protonatable group which upon association gives a neutral species ($HA \rightleftharpoons H^+ + A^-$), the Henderson–Hasselbalch equation becomes

$$pH = pK_a + \log \frac{[A^-]}{[HA]} \tag{20-5}$$

As discussed in Chap. 7, the pK_a of a particular side chain is characteristic of that group, with the reservation (stressed in Chap. 7) that the pK_a of a particular chemical group may, in the three-dimensional structure of the protein, be quite perturbed from its "normal" value.

pH studies of enzyme kinetics offer a means of obtaining information concerning pK_a values of groups involved in enzyme function. Before discussing possible interpretation of the pH parameters, we consider the way such data are usually presented.

Consider a parameter, K_{obs}, that is dependent on the state of ionization of a side chain in the protein. First, the side chain may need to be protonated or unprotonated (but not both for the value of K_{obs} to be maximal. Figure 20-1 illustrates two ways of plotting the dependence of K_{obs} on pH. K_{obs} may be plotted directly versus pH or log K_{obs} versus pH may be plotted. Figure 20-1 illustrates how the pK_a is obtained from either of these plots. In the example, K_{obs} is maximal when the ionizable group is in its unprotonated state. If K_{obs} required the ionizable group to be protonated, the two plots would be the inverse of those shown.

It is quite conceivable that K_{obs} depends on the state of ionization of two groups on the protein surface. As illustrated in Fig. 20-2, this results in a bell-shaped curve.

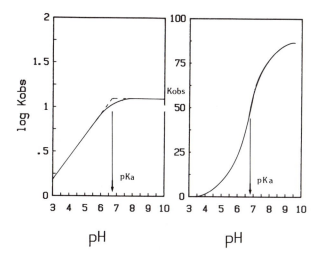

Figure 20-1   Methods of plotting the pH dependence of a parameter, K_{obs}, and the determination of pK_a.

Effects on K_m

As discussed extensively in Chaps. 13 to 15, the K_m of an enzyme-catalyzed reaction can, depending on the formal kinetic mechanism, have a variety of physical significances. In an equilibrium mechanism the Michaelis constant for a particular substrate can be equated with the dissociation constant of the substrate from the ternary or quaternary enzyme–substrate complexes for a two or three-substrate

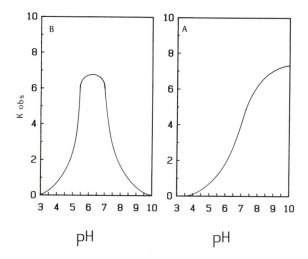

Figure 20-2   Plots of K_{obs} versus pH for a system dependent on the ionization of a single side chain (A), or one dependent on two side chains (B).

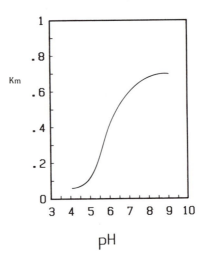

Figure 20-3   Typical plot of K_m versus pH. From the plot shown, a pK_a of approximately 6 is obtained.

reaction, respectively. Clearly, in such an instance a pK_a from a plot of K_m versus pH (as in Fig. 20-3) gives information concerning a group in the enzyme–substrate complex involved in the stability of the complex. This group can be on either the substrate or the enzyme. If the pK_a values (if any) of ionizable groups on the substrate are known (or independently determined), it is possible that the pK (or pK's) obtained from a plot of K_m versus pH can be assigned to ionizable groups on the enzyme.

Where a steady-state mechanism is involved, the K_m for a substrate has no simple physical significance. Any pK values obtained in such a case may reflect groups on the enzyme (assuming that substrate pK values have been eliminated from consideration) that are involved with any of the rate constants in the K_m term.

Effects on V_{max}

As with the interpretation of the pH dependence of K_m, the significance of any pK_a values determined in plots of V_{max} versus pH (Fig. 20-4) depends on knowledge of the formal kinetic mechanism. In equilibrium mechanisms the V_{max} is directly proportional to the rate constant for the interconversion of central ternary or quaternary complexes. In such an instance the pK (or pK's) obtained reflect a group (or groups) involved in the catalytic mechanism of the enzyme. The nature of such involvement can range from a group's participating in acid–base catalysis or some other aspect of the chemical mechanism to one whose ionization is important in maintaining the correct conformation of the enzyme's active site, thus allowing efficient catalysis to occur. In addition, the interconversion of central complexes may involve a rate-limiting conformational change that is dependent on the ionization state of one or more amino acid side chains.

When the formal kinetic analysis of the reaction indicates a steady-state mechanism, the V_{max} reflects the rate-limiting step. In steady-state mechanisms this can

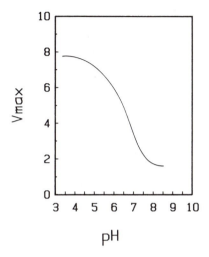

Figure 20-4 Typical plot of V_{max} versus pH. From the plot shown, a pK_a of approximately 5.8 is estimated.

be a substrate binding step, a product release step, a component of the catalytic reaction, or a combination of several of these stages. If the overall rate-limiting step in the reaction is known from other experiments, the interpretation of the pH dependence of V_{max} is straightforward. In the absence of this it is sometimes possible to obtain information by examining the pH dependence of K_m for substrates in the reverse reaction. Such substrates are of course the products of the reaction whose V_{max} is being studied. In the case where product release is rate limiting, any pK associated with a group involved in product release is also associated with the reverse reaction. A pK reflected in V_{max} in this case also appears in K_m (for the product as a substrate in the reverse reaction).

Clearly, knowledge of the formal kinetic mechanism is of considerable help in the interpretation of the pH dependence of an enzyme reaction. In place of plotting K_m versus pH it is often advantageous to plot a particular θ parameter, or ratio of θ parameters versus pH, as in Fig. 20-5.

As discussed in Chaps. 14 and 15, depending on the formal kinetic mechanism, these parameters (or ratios) can be related uniquely to particular steps in the overall mechanism. For example, in compulsory-order steady-state mechanisms $\theta_1 = 1/k_1$, where k_1 is the rate constant for the first substrate binding. Other such identities were given in Tables 14-1, 14-3, and 15-1.

A particularly interesting example that may be considered is the enzyme dihydrofolate reductase. This enzyme uses NADPH to reduce dihydrofolate in an overall reaction that includes the hydride transfer from the reduced nicotinamide ring of NADPH and a protonation (using a solvent proton) of an imine bond in dihydrofolate. The reaction, which in terms of the involved atoms is shown by

$$-N{=}C{\diagdown}^{\diagup} \longrightarrow -NH-CH{\diagdown}^{\diagup}$$

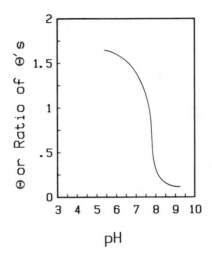

Figure 20-5   Plot illustrating the dependence of either an individual rate parameter or some ratio of rate parameters on pH. From the curve shown, a pK_a of approximately 7.5 is estimated.

is presumed to involve the protonation of the nitrogen prior to the hydride transfer to the carbon since this would enhance the carbonium-ion character of the carbon. Two important questions that can be raised involve the protonation pathway and stabilization of the protonated transition state, which will be positively charged.

Many genetic varients of dihydrofolate reductase have been described and a particular residue, Asp-27, appears from the crystal structure to be in a position to play a role in either the protonation or the transition-state stabilization or both.

Elegant studies using site-directed mutagenesis have provided an answer to the role of Asp-27 in the mechanism and shed further insight into the delicate balance of factors contributing to the overall catalytic efficiency of dihydrofolate reductase. Oligonucleotide-directed mutagenesis (see Chap. 6) was used to generate a mutant gene with asparagine in position 27. In addition, a primary site revertant of the Asn-27 gene, where the spontaneous transition AAC (Asn) to AGC (Ser) had occurred, was obtained. Native enzyme, together with both mutants, gave binary complexes with the inhibitor methotrexate, which gave isomorphous crystal structures. The major difference between the three structures (obtained to 1.9-Å resolution) involved a water molecule which in the wild-type enzyme is hydrogen bonded to the carboxyl of Asp-27.

The native enzyme shows initial rate kinetics consistent with a rapid-equilibrium, random-order mechanism. In such a mechanism either hydride transfer or proton transfer could be rate limiting, although the lack of an effect on V_{max} of deuterated coenzyme suggests that the hydride transfer step is not rate limiting. Detailed kinetic analysis of all three enzymes indicates that in both mutants k_{cat} is decreased while K_m is increased, indicating a role for Asp-27 in both binding and catalysis. The pH dependence of these parameters, however, gives a clue to the detailed role of Asp-27. With both mutant enzymes k_{cat} increases as the pH decreases, unlike the wild-type enzyme, which shows pK values of approximately 5 and 8. This suggests that when the substrate is pre-protonated (as it will be increasingly at low pH), it can function independently of the presence of Asp-27, indicating that the primary role of this carboxyl group involves substrate protonation, presumably via a concerted mechanism involving the hydrogen-bonded water molecule rather than being involved in the hydride transfer step in any way.

Interestingly, with the mutant enzymes at low pH (where, as indicated above, the substrate is pre-protonated) a kinetic isotope effect with deuterated NADPH is now observed, showing that in the mutant enzymes hydride transfer is now rate limiting.

Before leaving the topic of pH dependence of kinetic parameters, it must be reiterated that any pK assigned to a group on the enzyme, on whatever basis, need not necessarily reflect the pK of a group involved directly with the substrate or the catalytic reaction. It is possible that such a group does reflect the pK of a group involved directly with the substrate or the catalytic reaction; it is also possible that such a group can affect the overall stability of the protein (and hence have an effect on V_{max} by decreasing the effective enzyme concentration) or the K_m of a substrate via a conformational change triggered by the ionization of a group located spatially far from the binding site.

Effects on Protein Structure

The protonatable groups of a protein can interact to form noncovalent ion pairs that act to stabilize the protein's structure. At neutral pH there are a number of both positive and negative charges in a protein. The positive charges come from the side chains of lysine, arginine, and histidine (which, depending on pK_a, are positive at pH 7). The N-terminal amino acid, in addition, has an α-amino group. Negative charges arise from glutamate and aspartate carboxyl side chains as well as the C-terminal residue. The crystal structures of a variety of proteins have shown the existence of ion pairs involving various partners. In hemoglobin (deoxy) they appear to be involved in both intra-subunit and inter-subunit interactions, and both side chain moieties and terminal amino acid residues are involved. (The making and breaking of these ion pairs in terms of the allosteric properties of hemoglobin are discussed in more detail in Chap. 21). Such ion-pair formation is an entropy-driven process since it results in the disordering of the solvating water molecules of the individual partners. The disruption of ion–dipole interactions with the solvating water molecules results in a positive ΔH on complex formation.

As the pH of the protein solution is varied, groups involved in such ion-pair formation are titrated and the stabilization energy provided by their interaction is lost. On a local scale, such effects might be expected to produce conformational changes affecting the activity by influencing K_m or V_{max}. On a larger scale, if enough ion pairs are involved in the overall structure and stability of the protein, it will

TABLE 20-1   *H* ion for titratable groups in proteins

Group	ΔH ion (kcal/mol)
Carboxyl	±1.5
Imidazole	6.9–7.5
Amino	10–13
Sulfhydryl	6.5–7.0
Guanidine	12–13
Tyrosyl hydroxyl	6

unfold at extremes of pH. In general, proteins do denature at either low or high pH, and disruption of ion pairs contributes to this instability.

Temperature Dependence of pK$_a$ Values

As indicated, local environmental effects can have significant influence on the pK$_a$ of a group, which can complicate the interpretation of an experimentally determined pK$_a$. The temperature dependence of pK$_a$, however, can give a more reliable indication of which particular side chain is being titrated. As shown in Table 20-1, each type of side chain has a characteristic ΔH ion.

Since the standard free change (ΔG°) of a reaction is given by

$$\Delta G^\circ = -2.3RT \log K = \Delta H^\circ - T\,\Delta S^\circ \tag{20-6}$$

where K is the equilibrium constant, ΔH° the standard enthalpy change, and ΔS° the standard entropy change, one can write

$$-\log K = \frac{H^\circ}{2.3R}\frac{1}{T} - \frac{\Delta S^\circ}{2.3R} \tag{20-7}$$

The effects of temperature (T) on K are given by the van't Hoff equation,

$$\frac{d \ln K}{dt} = \frac{\Delta H^\circ}{RT} \tag{20-8}$$

which can be rewritten as

$$\frac{d \ln K}{d(1/T)} = -\frac{\Delta H^\circ}{R} \tag{20-9}$$

Equation (20-9) can be integrated between the limits of K_1 and K_2 at T_1 and T_2 to give

$$\log \frac{K_2}{K_1} = \frac{\Delta H^\circ}{2.3R}\frac{T_2 - T_1}{T_2 T_1} \tag{20-10}$$

and ΔH° can be estimated by determining the equilibrium constant at two different temperatures. More reliably, ΔH° is determined using Eq. (20-7), which indicates that a plot of $-\log K$ versus $1/T$ is linear, with a slope $= \Delta H°/2.3R$. In the context of pK_a values of groups, $-\log K$ is of course pK_a and all that is necessary is to measure the temperature dependence of pK_a and $\Delta H°$ ion is readily determined.

ISOTOPE EFFECTS

Although isotopes have been discussed in many chapters, their use in determining potential rate-limiting steps has not been formally examined. If we consider an example, the hydride transfer step involved in many dehydrogenase reactions, possible uses of isotopically labeled substrates in establishing a reaction's rate-limiting step become obvious. If a substrate is deuterium labeled in the position to be transferred, various experimental observations are possible. The deuterated substrate may show essentially identical steady-state kinetics to the normal substrate: Since an isotope effect is not observed, the inescapable conclusion is that the hydride transfer step is not the rate-limiting step. In such a case, an effect of the deuterated substrate on the pre-steady-state phase of the reaction may well be seen. As detailed in Chap. 18, if the reaction does not involve a rapid-equilibrium mechanism, it is likely that a pre-steady-state "burst" of product accumulation occurs. If the rate-limiting step of this pre-steady-state phase is the hydride transfer step, an isotope effect is expected. As illustrated in Fig. 20-6, an effect on this pre-steady-state phase has a minimal effect on the steady-state rate.

In a rapid equilibrium mechanism no pre-steady-state phase is observed since k_{cat} is the overall rate-limiting step in the reaction. Under these circumstances an isotope effect is seen but only in steady-state kinetic studies.

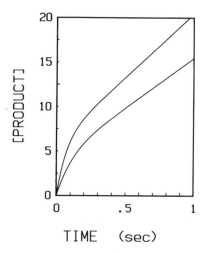

TIME   (sec)

Figure 20-6 Simulated pre-steady-state kinetic results for a normal and an isotopically labeled substrate where the isotope effect is seen in the pre-steady-state phase of the reaction.

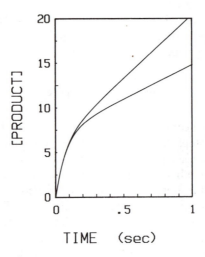

Figure 20-7 Simulated pre-steady-state kinetic results where the isotope effect is seen only in the steady-state rate.

In both cases we have assumed that the rate-limiting step contributing to k_{cat} or the pre-steady-state burst is the hydride transfer itself. It is, of course, possible that a rate-limiting conformational change in the central complex governs the rate of hydride transfer; in such an instance no isotope effect is observed on any phase of the reaction.

In cases where an isotope effect is seen in a step involving bond making or breaking, the magnitude of the effect depends on the atom involved. With hydrogen–deuterium or hydrogen–tritium substitutions, isotope effects up to two- or threefold are often observed.

In some circumstances, as examined in Chap. 15, the formal kinetic mechanism may be random order but not obey the rapid-equilibrium conditions. Under such conditions the hydride transfer step may not be rate limiting and a burst is seen. If the steady-state rate is affected but not the burst rate, as illustrated in Fig. 20-7, it indicates that the isotopically labeled product is released more slowly than the normal product, and that this is the rate-limiting step of the overall mechanism.

Although a large isotope effect in such an instance is unlikely unless the isotopically labeled product is a small molecule, small isotope effects may well be observed if the isotope atom is involved directly in some interaction with the protein.

TEMPERATURE EFFECTS

The effects of temperature on an enzyme-catalyzed reaction are diverse. As with the pH studies already examined, a detailed knowledge of the formal kinetic mechanism of the reaction considerably increases the merit of such experiments. Before considering some of the enzyme parameters whose temperature dependence might be examined, it should be noted that the stability of the protein itself has a temperature dependence that must be taken into account when V_{max} effects are studied.

A variety of thermodynamic parameters can be obtained from temperature studies. As discussed in Chap. 13, the Arrhenius equation [Eq. (20-11)] relates the activation energy, E_a, to the rate constant, k, of a process;

$$k = Ae^{-E_a/RT} \qquad (20\text{-}11)$$

where A is a constant related to the probability of the reaction occurring. As with the van't Hoff equation [Eq. (20-8)], Eq. (20-11) can be written in a linear form [Eq. (20-12)] or an integrated form [Eq. (20-13)].

$$\log k = -\frac{E_a}{2.3R}\frac{1}{T} + \log A \qquad (20\text{-}12)$$

$$\log\frac{k_2}{k_1} = \frac{E_a}{2.3R}\frac{T_2 - T_1}{T_2 T_1} \qquad (20\text{-}13)$$

The activation energy can be obtained by determining values for k at several temperatures. Enthalpies can be calculated from the temperature dependence of equilibrium constants in much the same way as described earlier for the temperature dependence of pK_a. Initial rate kinetic studies can yield much information concerning the rates of substrate binding or product release (in certain steady-state mechanisms), allowing the calculation of activation energies for these steps. The formal kinetic mechanism must be established, but once this is done the temperature dependence of the various θ parameters or ratios of θ parameters can yield a wealth of thermodynamic information.

In equilibrium mechanisms k_{cat} is the rate-limiting step and its temperature dependence allows the calculation of the activation energy for the chemical reaction (under conditions where it can be established that k_{cat} is a direct measure of the chemical reaction rather than a rate-limiting conformational change). Various relationships between initial rate parameters, as detailed in Chaps. 14 and 15, give a variety of equilibrium constants for substrate binding in various enzyme–substrate complexes. The temperature dependence of these equilibria permit the calculation of thermodynamic parameters.

Although the discussion so far has involved initial rate studies, similar experiments can be accomplished using rapid kinetic approaches (described in Chap. 18) where direct estimates of rate constants can be made. Similarly, the temperature dependence of conformational parameters can be employed to give thermodynamic information concerning interactions that may be involved in conformational changes. The temperature dependence of hydrophobic interactions or ion-pair formation are readily distinguishable from other types of interactions since they each have positive ΔH values.

The major problem with temperature studies of the types examined here is that it is possible that the formal kinetic mechanism may change in different steps in the reaction contributing to particular kinetic parameters, which complicates the interpretation. Such changes may result from the temperature dependence of the various steps in the mechanism themselves; as the temperature is changed, new steps may become rate limiting. Alternatively, protein conformational effects may alter the formal kinetic mechanism: These are usually manifested as nonlinear van't Hoff of Arrhenius plots, emphasizing the need to use the linearized forms of these equations rather than calculating parameters from the integrated forms and experimental data at only two temperatures.

CRYOENZYMOLOGY

In the preceding section of this chapter we discussed the temperature dependence of kinetic parameters. Such effects are taken to extremes in the concept of cryoenzymology, the study of enzyme reactions at low temperatures. Under such conditions it should be (in theory) possible to measure steps in a reaction pathway which, at room temperature, are too rapid for techniques such as stopped flow. Since temperatures below $0°$ are often used, it is necessary to use antifreeze solvent systems, and this presents the first problem—it is essential to demonstrate that the solvent system does not affect the process being studied. The major problem, however, is the fact that very different rate-limiting steps may exist at low temperatures, meaning that simply determining kinetic parameters and interpreting their temperature dependence is no easy task. Despite these problems, cryoenzymology is being used with some systems.

Each of the protocols in this chapter (i.e., the study of pH effects, isotope effects, or temperature effects, etc.) has been examined from the standpoint of kinetic studies. The detailed interpretation of such studies usually requires a knowledge of the formal kinetic mechanism of the enzyme. In some instances similar information can be obtained from direct ligand binding studies. It is often easier, however, to use initial rate studies.

Although it might seem that the topics covered in this chapter are rather diverse, the *combination* of approaches is essential in the detailed interpretation of pH or temperature dependence of kinetic parameters. Each technique offers different insights into the makeup of the various kinetic constants that constitute the initial rate equation. Their application, in conjunction with thorough initial rate or stopped-flow studies, allows the enzymologist to dissect conformational effects from chemical steps and ligand binding or release steps in a kinetic mechanism. This in turn permits a far more detailed interpretation of the pH dependence of kinetic parameters in terms of which amino acid side chains may be involved in the actual chemical catalysis steps of the reaction.

21

Allosteric Models
of Enzyme Regulation

INTRODUCTION

So far we have looked at a number of theoretical aspects of protein conformation and biological activity, especially in the context of enzyme kinetics. In Chaps. 9 to 11 we examined various features of protein conformation, and in Chaps. 13 to 15 we discussed the basics of enzyme kinetics. In the later chapters no assumptions were made about the molecular structure of the enzyme, and thus it was tacitly assumed that if the molecule contains more than one active site (i.e., is oligomeric), they all behave independently and identically. In Chap. 16 we considered the consequences when the active sites within an oligomer are not identical, and first brought up the idea of an oligomeric protein exhibiting allosteric interactions.

The concept of allosteric interactions implies ligand conformational changes. In Chap. 12 we discussed a variety of experimental techniques for following conformational changes in proteins. In this chapter we consider whether such approaches can give useful information about which of the various formal allosteric models examined might be more appropriate.

LIGAND-INDUCED COOPERATIVITY

Cooperativity in ligand binding to proteins, whereby the first molecule of ligand binding to an oligomer facilitates the binding of subsequent ligand molecules, was first detected in hemoglobin by Bohr and co-workers in 1904, and concepts relating to this cooperativity were later formalized by Adair, who in 1925 proposed that if the four heme groups in the hemoglobin molecule are equivalent, oxygenation of one heme somehow increases the oxygen affinity of some or all of the remaining hemes, resulting in a sigmoidal saturation curve, as was found experimentally. This site-site interaction hypothesis is symmetrical in that it allows both positive and negative interactions to occur, but Adair did not suggest a mechanism for it. Following the x-ray crystallographic studies of hemoglobin by Perutz in 1960, which showed that the heme groups are far apart, it was established that the molecular mechanism for their interaction involved protein conformational changes rather than direct interaction, as had previously been assumed.

The observations of rate cooperativity with enzymes, especially aspartate transcarbamoylase, led to the concept of allosteric transitions of regulatory proteins and to the subsequent proposal of the concerted symmetry model by Monod, Wyman, and Changeux in 1965. In this, two conformations of the unliganded oligomer exist in equilibrium, one of which preferentially binds ligand. The conformation and symmetry of the oligomer are maintained on ligand binding. An alternative model for allosteric proteins proposed by Koshland, Nemethy, and Filmer in 1966 depends on ligand-induced conformational changes to affect the affinity of unliganded sites. It is assumed in both models that all the subunits in the unliganded oligomers are equivalent, that is, that they have the same intrinsic affinity for ligand. The essential difference between the two models is that in the Monod, Wyman, and Changeux model there is an equilibrium between unliganded forms with different intrinsic affinities while in the Koshland, Nemethy, and Filmer model symmetry of the oligomer need not be conserved.

Although a number of enzymes have been shown to exhibit positive cooperativity, evidence for negative cooperativity was first obtained in kinetic and ligand binding studies with the oligomeric enzymes glutamate dehydrogenase and glyceraldehyde-3-phosphate dehydrogenase, respectively.

In 1968, Dalziel and Engel observed several apparently linear regions in Lineweaver–Burk plots for glutamate dehydrogenase, with NAD or NADP as the varied substrate, that could be described in terms of increasing K_m and V_{max} values as the coenzyme concentration was increased. They suggested that these observations could be explained in terms of negative homotropic interactions, whereby the binding of coenzyme at one active center in the hexamer weakens binding at subsequent sites. Independently, it was suggested that the binding of NAD to rabbit muscle glyceraldehyde-3-phosphate dehydrogenase, which appeared to become weaker as saturation of the tetramer was approached, could be explained in the same terms. Although other interpretations of the data are possible, a model involving protein conformational changes was proposed. For kinetic data giving Lineweaver–Burk plots that are concave downwards, as is the case with glutamate dehydrogenase, several purely kinetic explanations exist. If some form of activation at high substrate concentration occurs, Lineweaver–Burk plots that are concave downward may be observed, as discussed in Chap. 16. Such effects might be due to substrate binding to an additional site on the protein resulting in activation at the active site, or to substrate binding to an enzyme–product complex, facilitating product release where the latter is the rate-limiting step. It has also been shown that for a two-substrate random-order mechanism, unless the rapid-equilibrium assumption can be made, the rate equation does not predict linear Lineweaver–Burk plots, and that deviations either concave downward or upward may be obtained depending on the values of various rate constants for the mechanism. The demonstration that ligand binding data cannot be described by a single dissociation constant in such cases can be used to rule out these purely kinetic explanations. For glutamate dehydrogenase, evidence of apparent negative cooperativity in the equilibrium binding of NAD and NADP has been obtained, showing that such effects in this system are not purely kinetic.

There are a number of plausible explanations of kinetic and ligand binding data of the type observed by Dalziel and Engel and by Conway and Koshland. The oligomeric proteins may simply be made up of subunits with different intrinsic binding constants and possibly different catalytic rate constants. This could be the result of the oligomer containing chemically different subunits, or chemically identical subunits so arranged geometrically that the active centers are not equivalent.

If the subunits are spatially arranged such that each is in an identical environment, negative cooperativity can be explained either in terms of direct ligand–ligand interactions or protein conformational changes. Direct ligand-ligand interactions may be electrostatic or steric, and the ligand binding sites on separate subunits in the oligomer would be expected to be close to one another, as electrostatic interactions do not occur to a significant extent over more than a few angstroms. Steric interaction also depends on the closeness of the ligand binding sites and the size of the ligand molecule. Negative cooperativity can be explained by ligand-induced conformational changes of the oligomer if these are such as to decrease the affinity of unliganded sites for ligand. However, one of the models involving these changes, that of Monod, Wyman, and Changeux, cannot explain negative cooperativity; in any model involving preexistent equilibria between unliganded forms of the oligomer, only positively cooperative effects can be accounted for.

Equilibrium binding studies do not allow distinctions to be made between intrinsically nonidentical binding sites and negative cooperativity arising from ligand-induced conformational changes. However, it is in theory possible to establish whether observed interactions are due solely to electrostatic effects, provided that an estimate can be made of the interaction between the occupied sites. It has been shown that for the binding of anions to bovine serum albumin, simple electrostatic interactions do not account for the saturation curves, suggesting that either multiple classes of binding sites exist or that negative cooperativity is involved.

In studies of apparent negative cooperativity, these considerations put a great deal of emphasis on the detection of ligand-induced conformational transitions for distinguishing among the possible models involved. In this context it must be pointed out that ligand-induced conformational changes affecting the affinity of unoccupied subunits must be separated from those that might be expected to occur as a result of ligand binding, even with an oligomer that shows no cooperative behavior. Moreover, if one considers the free-energy changes involved in ligand binding and subunit interactions, then, even in an oligomer showing no cooperative binding effects, conformational changes may be induced in unliganded subunits without necessarily affecting the ligand binding site. As a result, conformational changes accompanying ligand binding need not be proportional to fractional saturation. The demonstration that conformational changes accompanying ligand binding are not proportional to fractional saturation does not, therefore, constitute evidence for any particular allosteric model.

Thus in any study of negative cooperativity the existence of ligand-induced conformational changes that affect the affinity of unliganded binding must be demonstrated directly, to rule out possible models that explain kinetic and binding data without invoking such changes.

To date, three different types of approaches have been employed to obtain evidence for ligand-induced conformational changes occurring across subunit interfaces. In the first (discussed in Chap. 12), the fluorescence of the natural coenzyme NAD(P)H was used to monitor conformational changes in the active sites of bovine glutamate dehydrogenase and rabbit muscle glyceraldehyde-3-phosphate dehydrogenase induced by the oxidized coenzyme. Since oxidized and reduced coenzymes cannot bind to the same active site at the same time, such changes result from subunit–subunit interactions. The second approach can be illustrated by the elegant experiments of Yang and Schachman using a combination of chemical modifications and hybridization experiments with aspartate transcarbamoylase to produce a form of the enzyme containing one catalytic trimer in its active state and one inactivated but spectroscopically labeled trimer. Such hybrids allow one to look at the effects that ligands added to the catalytically active trimer may have on the inactive but spectroscopically labeled trimer, yielding information about conformational changes induced across subunit–subunit interfaces. The third approach, which does not detect conformational changes by means of direct spectroscopic observation, has been developed to examine the effects of induced conformational changes in the catalytically active form of the enzyme. With glutamate dehydrogenase use has been made of active coenzyme analogs, whose products can be separated spectrally, to show that the addition of one analog to an active site in an oligomeric enzyme has an effect on the rate of utilization of another analog. Since both cannot reside in the same active site at the same time, such effects can only be attributed to conformational changes induced between subunits within an oligomer.

Homotropic Versus Heterotropic Effects

So far we have considered models that give rise to homotropic effects. In many systems, however, ligand-induced conformational changes result in heterotropic effects, where a ligand binding to a regulatory site on the protein induces a conformational change affecting the properties of the active site. Although such effects undoubtedly constitute the major number of cases of so-called allosteric regulation, it is not necessary to consider heterotropic effects in great detail since mechanistically they are quite straightforward—the ligand binds and induces a conformational change affecting activity. It is perhaps useful to consider heterotropic effects as involving not just different types of ligand but also chemically or conformationally distinct sites for the same type of ligand. Using this definition, it is possible to include models that may have a second, nonactive site for a ligand that also acts as a substrate. The formal models for homotropic effects considered allow for the possibility of heterotropic in addition to homotropic effects, but differ radically in the physical mechanism by which such effects are elicited.

THE FORMAL MODELS

In keeping with the historical development of ideas concerning cooperative interactions in proteins, it is easiest to consider the formal models that have been proposed to account for allosteric interactions in terms of ligand binding and the effects that the degree of saturation may have on the affinity of a binding site for its ligand.

Consider the binding of a ligand, L, to a protein, P; we get the equilibrium $P + L \rightleftharpoons PL$, and a dissociation constant, K_d, can be written, as in

$$K_d = \frac{[P][L]}{[PL]} \tag{21-1}$$

Thus

$$[PL] = \frac{[P][L]}{K_d} \tag{21-2}$$

The fractional saturation, Y, of the protein, is given by

$$Y = \frac{[PL]}{[P] + [PL]} \tag{21-3}$$

and represents the concentration of binding sites in the protein that are actually bound with ligand divided by the total concentration of ligand binding sites.

Substituting into Eq. (21-3) the concentration of PL from Eq. (21-2), we get

$$Y = \frac{[L]}{K_d + [L]} \tag{21-4}$$

which yields a "normal" saturation curve when Y is plotted as a function of $[L]$, as shown in Fig. 21-1.

The Hill Equation

An equation similar to Eq. (21-3) was derived by Hill in 1910 to account for the sigmoidal Y versus $[L]$ plots obtained with hemoglobin. Hill assumed that a protein, P, has n binding sites for ligand, L, and that as soon as one molecule of L binds, the remaining $(n - 1)$ sites are immediately occupied; that is, there is extreme cooperativity in the binding of L to the protein. As a result, one can consider an equilibrium with no significant contribution from protein molecules with fewer than n moles of ligand bound per mole, and the system may be represented as $P + nL \rightleftharpoons$

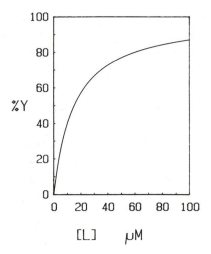

Figure 21-1 Normal saturation curve obtained in a plot of the fractional saturation, Y, versus the free ligand concentration, [L], with a dissociation constant of 15 μM.

PL_n. The dissociation constant for L may be written as in

$$K_d = \frac{[P][L]^n}{[PL_n]} \tag{21-5}$$

The resultant fractional saturation becomes

$$Y = \frac{[PL_n]}{[P] + [PL_n]} \tag{21-6}$$

which, using Eq. (21-5) for the dissociation constant, becomes

$$Y = \frac{[L]^n}{K_d + [L]^n} \tag{21-7}$$

where n is the Hill coefficient. Equation (21-7) can be rearranged to give

$$\frac{Y}{1 - Y} = \frac{[L]^n}{K_d} \tag{21-8}$$

and therefore,

$$\log \frac{Y}{1 - Y} = n \log [L] - \log K_d \tag{21-9}$$

and a plot of $\log (Y/(1 - Y))$ versus $\log [L]$ is linear with a slope of n.

This derivative depends on assumed extreme cooperativity among ligand molecules binding to protein. Under such conditions n should equal the number of subunits within the oligomer. If n is equal to 1 (i.e., there is only one binding site per molecule) or if the subunits contain identical and independent binding sites, Eq. (21-7) reduces to Eq. (21-4), and a normal saturation curve, as in Fig. 21-1, is observed; the slope of the Hill plot is equal to 1.

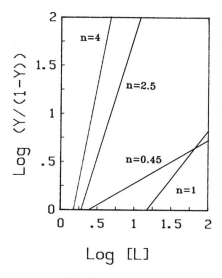

Figure 21-2 Hill plot obtained with (1) $n = 4$ for a tetramer, (2) $n = 2.5$, (3) $n = 1$, and (4) $n = 0.45$.

In many cases the assumption of extreme cooperativity in ligand binding is not justified and n, calculated from a plot of $\log[Y/(1 - Y)]$ versus $\log[Y]$, is less than the number of subunits but greater than 1 if cooperative binding occurs. The Hill plot, and calculation of n, are given in Fig. 21-2. Also shown is a Hill plot for a case where n is less than 1, a situation arising when ligand binding displays negative homotropic interactions. The saturation plots, Y versus [L], for these are shown in Fig. 21-3.

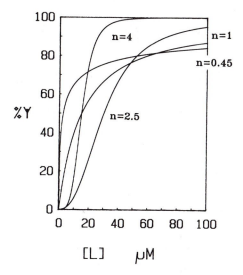

Figure 21-3 Plots of Y versus [L] for the four situations illustrated in Fig. 23-2. For line 1, $K_d = 50,000\ \mu M$; for line 2, $K_d = 5000\ \mu M$; for line 3, $K_d = 15\ \mu M$; and for line 4, $K_d = 1.5\ \mu M$.

The Adair Equation

Adair considered that the elimination of the intermediate liganded forms of the protein as proposed by Hill was too simplistic, and derived an equation that took into account all possible intermediate forms.

Dalziel and Koshland independently recognized that the ideas of Adair could be fit into the concept of allosteric proteins, and proposed models giving a mechanistic explanation to the proposed equations of Adair. Before considering such models it is informative to examine the Adair equation in some detail.

In a system in which one molecule of protein, P, can bind up to n molecules of ligand, L, a number of intermediate stages have to be considered, as shown in Table 21-1. For each step in Table 21-1 a dissociation constant can be defined by a series of equations, as represented by

$$K_{d_1} = \frac{[P][L]}{[PL]} \tag{21-10}$$

$$K_{d_n} = \frac{[PL_{n-1}][L]}{[PL_n]} \tag{21-11}$$

Defining a fractional saturation, Y, by

$$Y = \frac{\text{total concentration of L bound}}{\text{total concentration of sites for L}} \tag{21-12}$$

we get

$$
\begin{aligned}
Y &= \frac{[PL] + 2[PL_2] + 3[PL_3] + \cdots + n[PL_n]}{n([P] + [PL] + [PL_2] + \cdots + [PL_n])} \\
&= \frac{[P][L]/K_{d_1} + 2[P][L]^2/K_{d_1}K_{d_2} + \cdots + n[P][L]^n/K_1K_2\cdots K_n}{n([P] + [P][L])/K_{d_1} + [P][L]^2/K_{d_1}K_{d_2} + \cdots + [P][L]^n/K_{d_1}K_{d_2}\cdots K_{d_n}}
\end{aligned}
\tag{21-13}
$$

TABLE 21-1   Intermediate stages in the saturation of a protein with n binding sites for a ligand, L

Step	Sites occupied	Sites empty
$P + L \quad\quad PL$	1	$n-1$
$PL + L \rightleftharpoons PL_2$	2	$n-2$
$PL_2 + L \rightleftharpoons PL_3$	3	$n-3$
$PL_{n-1} + L \rightleftharpoons PL_n$	n	0

Dividing both the numerator and the denominator by [P] reduces Eq. (21-13) to

$$Y = \frac{[L]/K_{d_1} + 2[L]^2/K_{d_1}K_{d_2} + \cdots + n[L]^n/K_{d_1}K_{d_2}\cdots K_{d_n}}{n(1 + [L]/K_{d_1} + [L]^2/K_{d_1}K_{d_2} + \cdots + [L]^n/K_{d_1}K_{d_2}\cdots K_{d_n})} \qquad (21\text{-}14)$$

which is the Adair equation.

For a tetrameric protein with a single ligand binding site per subunit, Eq. (21-14) becomes

$$Y = \frac{[L]/K_{d_1} + 2[L]^2/K_{d_1}K_{d_2} + 3[L]^3/K_{d_1}K_{d_2}K_{d_3} + 4[L]^4/K_{d_1}K_{d_2}K_{d_3}K_{d_4}}{4(1 + [L]/K_{d_1} + [L]^2/K_{d_1}K_{d_2} + [L]^3/K_{d_1}K_{d_2}K_{d_3} + [L]^4/K_{d_1}K_{d_2}K_{d_3}K_{d_4})}$$

$$(21\text{-}15)$$

The dissociation constants in Eq. (21-15) K_{d_1}, K_{d_2}, K_{d_3}, and K_{d_4}, are related to the intrinsic dissociation constants (K') by statistical factors. For a protein containing four sites these relationships are given by

$$K_{d_1} = \frac{K'_1}{4} \qquad (21\text{-}16)$$

$$K_{d_2} = \frac{2K'_2}{3} \qquad (21\text{-}17)$$

$$K_{d_3} = \frac{3K'_3}{2} \qquad (21\text{-}18)$$

$$K_{d_4} = 4K'_4 \qquad (21\text{-}19)$$

If $K'_1 = K'_2 = K'_3 = K'_4$, the binding has no interactions. If $K'_1 > K'_2 > K'_3 > K'_4$, the binding shows cooperativity at each stage, and if $K'_1 < K'_2 < K'_3 < K'_4$, negative interactions are occurring at each binding step. Figure 21-4 gives plots of the fractional saturation, Y, versus [L] for each of these cases.

Examination of the plots from Figs. 21-3 and 21-4 indicates that both the Hill and Adair equations can generate two types of curves that differ from the normal saturation behavior in Fig. 21-1. These deviations result in a sigmoidal saturation function or a saturation function that is steeper than expected at low degrees of saturation but less steep than expected at higher saturation levels. Again it must be emphasized that these equations do not imply a mechanism for the differences in intrinsic dissociation constants, although the sigmoidal curve is explicable only by cooperative interactions.

If we assume that cooperative interactions are the cause of the changes in shape of the saturation functions in Fig. 21-4, we can calculate the interaction energy associated with the differences in intrinsic dissociation constants for the sites. The standard free energy of *dissociation* (ΔG_d°) of a particular ligand is given by

$$\Delta G_d^\circ = RT \ln K_d \qquad (21\text{-}20)$$

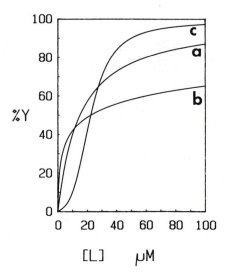

Figure 21-4   Plots of Y versus [L] for a tetrameric protein: (A) $K'_1 = K'_2 = K'_3 = K'_4 = 15\ \mu M$; (B) $K'_1 = 2\ \mu M$, $K'_2 = 10\ \mu M$, $K'_3 = 50\ \mu M$, $K'_4 = 250\ \mu M$; (C) $K'_1 = 250\ \mu M$, $K'_2 = 50\ \mu M$, $K'_3 = 10\ \mu M$, $K'_4 = 2\ \mu M$.

which contains a purely statistical component given by $RT \ln (\Omega n,\, i - 1/\Omega n,\, i)$. When this is taken into account we obtain an expression for the intrinsic standard free-energy change of dissociation, $\Delta \bar{G}_d^\circ$:

$$\Delta \bar{G}_d^\circ = RT \ln K_d - RT \ln \left(\frac{\Omega n,\, i - 1}{\Omega n,\, i} \right) \tag{21-21}$$

The interaction free energy, ΔG_{int}, is the difference in intrinsic free energy for *association* of, for example, the first and second ligands bound. If we consider the example in Fig. 21-4, where $K'_1 = 2\ \mu M$ and $K'_2 = 10\ \mu M$, the interaction energy is given by

$$\Delta G_{\text{int}} = -RT \ln \left(\frac{K'_1}{K'_2} \right) \tag{21-22}$$

and ΔG_{int} is positive. When $K'_1 = 250\ \mu M$ and $K'_2 = 50\ \mu M$, ΔG_{int} is negative. These two examples correspond to negative homotropic interactions and positive cooperativity, respectively. Depending on the ambient temperature, the interaction energies are of the order of 1 *kcal*/mol, either positive or negative. In the case of cooperative interactions this indicates that site–site interactions stabilize the second molecule of ligand bound by about 1 *kcal*/mol compared to the first molecule of ligand bound.

Sequential Models

The Adair equation is the basis of a model for allosteric interactions known as the *sequential model*. In it, proposed independently by Koshland and Dalziel, homotropic cooperativity is explained by conformational changes induced as each ligand molecule binds. Positive cooperativity is the result when the first molecule of ligand binding induces a conformational change that increases the affinity for subsequent ligand molecules. Negative cooperativity is produced by the affinity of subsequent ligand molecules being decreased by the first ligand molecule binding. The model requires that the protein be oligomeric and that conformational changes be transmitted across subunit interfaces in the oligomer.

Although the model in its simplest form implies that each ligand molecule to bind induces a conformational change that results in altered affinity for subsequent ligand molecules, the Adair equation is quite consistent with alternative "sequential" models where, for example in a tetramer, the conformational change induced by ligand binding does not occur until two of the four sites in the oligomer are liganded.

The Monod–Wyman–Changeux Model

The Monod–Changeux (MWC) model was designed primarily to account for a certain class of cooperative interactions observed for the binding of ligands to regulatory enzymes. An allosteric protein is postulated to comprise a small number of identical subunits called protomers, and to equilibrate among a small number of conformational states that differ in their ligand affinity. The most stable, and consequently the predominant states, are those in which all protomers of the protein molecule have the same conformation (i.e., symmetrical states). Cooperativity in ligand binding arises from a coordinated transition of all protomers of a protein molecule to the conformational state for which the ligand has greater affinity. As a result, there does not exist a strict linear relationship between the fraction of sites occupied (the saturation function, Y) and the fraction of molecules in a given conformation (the conformational state function, R).

This prediction, and the underlying assumption that the conformational equilibrium preexists the binding of effectors, distinguishes this model from others in which the conformational changes are assumed to be induced by, and thus coincident with, ligand binding.

The detailed predictions of the model are embodied in two analytical functions: the binding function, Y, which represents the fraction of specific sites in the total protein population occupied by the considered ligand, and the state function, R, representing the fraction of molecules in the R conformation. These two functions are given by

$$Y = \frac{\alpha(1+\alpha)^{n-1} + L\alpha C(1+\alpha C)^n}{(1+\alpha)^n + L(1+\alpha C)^n} \qquad (21\text{-}23)$$

$$R = 1 - T = \frac{(1+\alpha)^n}{(1+\alpha)^n + L(1+\alpha C)^n} \qquad (21\text{-}24)$$

L is the apparent conformational equilibrium constant in the absence of substrate and is defined by

$$L = \frac{T}{R} \tag{21-25}$$

n, the number of equivalent, independent binding sites for each type of ligand, corresponds to the number of identical subunits, and α and C are constants related to the intrinsic dissociation constants of the two states (K_R and K_T) and the free ligand concentration (F) by

$$C = \frac{K_R}{K_T} \tag{21-26}$$

$$\alpha = \frac{F}{K} \tag{21-27}$$

These equations are derived in the next section.

Several important points can be made based on Eq. (21-23):

1. If $K_T \gg K_R$ and/or L is very small, the second term becomes negligible compared to the first, resulting in

$$Y = \frac{F/K_R(1 + F/K_T)^{n-1}}{(1 + F/K_R)^n} \longrightarrow \frac{F/K_R}{1 + F/K_R} \tag{21-28}$$

and a rectangular hyperbola saturation curve is obtained; similarly if $K_R \gg K_T$.

2. If $n = 1$, the equation for Y reduces to

$$Y = \frac{(1/K_R + L/K_T)F}{(1 + L) + (1/K_R + L/K_T)F} \tag{21-29}$$

and once again a rectangular hyperbola saturation results.

3. If $K_R = K_T$, the equation for Y again reduces to a rectangular hyperbola. To obtain a sigmoidal saturation curve, however, K_R and K_T must be different but not too different. If $K_T \gg K_R$ *and* L is large, the R form binds ligand more tightly than the T form but $[T] \gg [R]$ and a sigmoidal saturation curve results. If T does not bind ligand, we get

$$Y = \frac{F/K_R(1 + F/K_R)^{n-1}}{L + (1 + F/K_R)^n} = \frac{\alpha(1 + \alpha)^{n-1}}{L + (1 + \alpha)^n} \tag{21-30}$$

and again sigmoidal saturation is obtained. These effects are illustrated by the saturation curves in Fig. 21-5 for the various instances discussed.

Equation (21-23) readily accounts for normal behavior or for sigmoidal saturation curves resulting from cooperative ligand interactions, but unlike the Hill or Adair equations, cannot account for saturation plots indicative of negative inter-

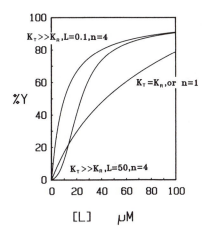

Figure 21-5 Saturation curves according to Eq. (21-23). The appropriate values of K_R, K_T, and L used in these simulated curves are indicated.

actions. This is a direct consequence of the existence of an equilibrium between two forms of the enzyme in the absence of ligand.

Derivation of Equations for the MWC Model. Consider a tetramer of four identical subunits. Each subunit can bind only one molecule of ligand and can exist in two forms, whether or not a ligand molecule is bound. These conformations and the saturation process are shown in Fig. 21-6. In each form the subunits have identical intrinsic affinities for ligand, and the equilibrium between the two forms is governed by the equilibrium constant L.

$$L = \frac{T}{R} \qquad (21\text{-}31)$$

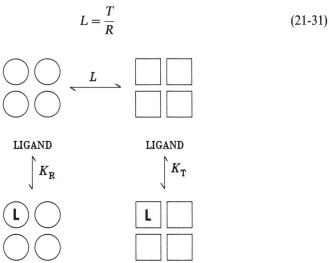

Figure 21-6 Scheme of a MWC model.

The fractional saturation, Y, is the fraction of the total number of sites with ligand and is given by

$$Y = \frac{(R_1 + 2R_2 + 3R_3 + \cdots + nR_n) + (T_1 + 2T_2 + \cdots + nTn)}{n(R_0 + R_1 + R_2 + \cdots + R_n) + (T_0 + T_1 + T_2 + \cdots + T_n)} \quad (21\text{-}32)$$

A "function of state," R (or T), can be defined as

$$R = \frac{R_0 + R_1 + R_2 + \cdots + R_n}{(R_0 + R_1 + R_2 + \cdots + R_n) + (T_0 + T_1 + T_2 + \cdots + T_n)} \quad (21\text{-}33)$$

and the ratio

$$\frac{T}{R} = \frac{T_0 + T_1 + \cdots + T_n}{R_0 + R_1 + \cdots + R_n} \quad (21\text{-}34)$$

Intrinsic dissociation constants, K_R and K_T, for ligand binding to the R form or the T form are defined as follows:

$$K_1 = \frac{R_0 F}{R_1}$$

Therefore,

$$R_1 = \frac{R_0 F}{K_1} \quad (21\text{-}35)$$

But

$$K_1 = \frac{K_R}{n}$$

Therefore,

$$R_1 = \frac{R_0 F_n}{K_R} \quad (21\text{-}35a)$$

$$K_2 = \frac{R_1 F}{R_2}$$

Therefore,

$$R_2 = \frac{R_1 F}{K_2} \quad (21\text{-}36)$$

But

$$K_2 = \frac{2K_R}{n-1}$$

Therefore,

$$R_2 = \frac{R_1 F(n-1)}{2K_R} \quad (21\text{-}36a)$$

Therefore,

$$R_2 = \frac{n(n-1)(n-2)R_0 F^3}{3 \times 2K_R^3} \tag{21-36b}$$

$$K_3 = \frac{R_2 F}{R_3}$$

Therefore,

$$R_3 = \frac{R_2 F}{K_3} \tag{21-37}$$

But

$$K_3 = \frac{3K_R}{n-2}$$

Therefore,

$$R_3 = \frac{R_2 F(n-2)}{3K_3} \tag{21-37a}$$

Therefore,

$$R_3 = \frac{n(n-1)R_0 F^2}{2K_R^2} \tag{21-37b}$$

Substituting into Eq. (21-32) for Y we get

$$Y = \frac{(R_0 nF/K_R)(1 + F/K_R)^{n-1} + (T_0 nF/K_T)(1 + F/K_T)^{n-1}}{n[R_0(1 + F/K_R)^n + T_0(1 + F/K_T)^n]} \tag{21-38}$$

Substituting for T $[= LR$ from Eq. (21-31)$]$, we get

$$Y = \frac{(R_0 F/K_R)(1 + F/K_R)^{n-1} + (LR_0 F/K_T)(1 + F/K_T)^{n-1}}{R_0(1 + F/K_R)^n + LR_0(1 + F/K_T)^n} \tag{21-39}$$

Therefore

$$R = \frac{(1 + F/K_R)^n}{(1 + F/K_R)^n + L(1 + F/K_T)^n} \tag{21-40}$$

and

$$\frac{T}{R} = \frac{L(1 + F/K_T)^n}{(1 + F/K_R)^n} \tag{21-41}$$

From Eq. (21-39) we get Eq. (21-42) by dividing all terms by R_0:

$$Y = \frac{(F/K_R)(1 + F/K_R)^{n-1} + (LF/K_T)(1 + F/K_T)^{n-1}}{(1 + F/K_R)^n + L(1 + F/K_T)^n} \tag{21-42}$$

Defining $K_R/K_T = C$ and $F/K_R = \alpha$, it is apparent that $F/k_T = C\alpha$, and, substituting into Eq. (21-42) for Y, we get

$$Y = \frac{\alpha(1 + \alpha)^{n-1} + LC\alpha(1 + \alpha)^{n-1}}{(1 + \alpha)^n + L(1 + C\alpha)^n} \tag{21-43}$$

Heterotropic Effects

Although we have considered only homotropic effects so far, each of the models presented adequately explains heterotropic effects. In the Adair–Koshland–Dalziel type, heterotropic effects are explained quite separately from homotropic effects simply by postulating the appropriate regulatory site on each subunit. The MWC model offers a more elegant (but in no way more correct) solution. Activators stabilize the so-called T form, thus displacing the equilibrium between the two forms and leading to a more or less active form of enzyme.

OTHER IDEAS ON COOPERATIVE INTERACTIONS

The models examined so far represent just a few of the many models that have been proposed to explain anomalous binding or kinetic data for certain proteins. The ideas of Hill; Adair; Monod, Wyman, and Changeux; Dalziel; and Koshland, Nemethy, and Filmer have been influential in the field of allosteric interactions, and the concepts embodied in the models discussed appear in one form or another in most others.

Two other ideas concerning allosteric proteins must be introduced since they play an important part in current thinking about enzyme regulation. These ideas are those found in (1) reciprocating subunit models, and (2) hysteresis models.

Reciprocating Subunit Models

There are two types of reciprocating subunit models. In each the central point is that only half of the subunits (in the simplest approach, although in reality the only requirement is that not all of the subunits) participate in the reaction at any given time. After an initial reaction phase has taken place on half of the subunits in the oligomer, a conformational change is induced in the remaining half, allowing them to bind substrate more readily. As substrate binds to the second half of the oligomer, a reciprocal conformational change is induced in the first half, allowing the reaction to go to completion. This reciprocal interaction can be envisaged either as allowing the chemical steps of the reaction to proceed more rapidly (i.e., by lowering the activation energy of the transition state) or by allowing product to be released more rapidly (i.e., by decreasing the negative free energy of product binding). Clearly, the mode of action depends on the formal kinetic mechanism of the enzyme involved. In an enzyme with a true rapid-equilibrium, random-order mechanism, there is no overall advantage to speeding the rate of product release since that step is never rate limiting. In other enzymes advantage may result from either or both mechanisms of reciprocation, and it is likely that the overall mechanism contains energy imput into both stages. Both forms of the reciprocating subunit model are outlined in Fig. 21-7.

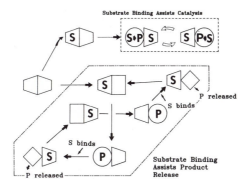

Figure 21-7   Schematic representation of reciprocating subunit models for a dimeric enzyme.

An enzyme showing such a mechanism must show negative homotropic interactions in substrate binding; otherwise, a fully saturated enzyme oligomer is formed that would prevent the reciprocal interactions form occurring. The basic concepts for reciprocating subunit models were originally developed for alkaline phosphatase. Two other enzymes have been shown to operate via reciprocating subunits: ATPase and glutamate dehydrogenase. ATPase studies with isotopically labeled substrates and with ATP analogs have shown that when one of the three active sites in the molecule has bound ATP, hydrolysis proceeds very slowly, but that when a second molecule is bound, the binding energy is used to enhance the rate of hydrolysis at its first site as well as the rate of product release from the first site.

Glutamate dehydrogenase operates as a dimer of trimers with the reciprocal interactions occurring between the trimers. Substrates bind to the first trimer (negative homotropic interactions hinder higher degrees of saturation) and react to form enzyme-bound products. This induces a conformational change that enhances substrate binding to the second trimer which then binds substrates, reciprocal interaction occurs and release of product is enhanced. Unlike ATPase, clear evidence has been presented for each phase of the reciprocating subunit mechanism that operates in both the oxidative deamination reaction and in the reductive amination reaction.

In the case of ATPase it appears that the main purpose of the reciprocating subunit mechanism is to enhance the enzyme's catalytic efficiency. In the case of glutamate dehydrogenase it appears that the mechanism may also be involved in the rather complex regulation of this enzyme.

Hysteresis Models

The concept of hysteresis (a rate-limiting conformational change) was first applied to enzyme systems by Frieden, who observed that in a number of cases reactions either auto-activate or auto-inhibit well after the "steady state" has been reached. In such instances it was postulated that a very slow conformational change induced by a ligand was occurring, leading to the changes in activity. A typical scheme representing an enzyme displaying hysteresis is illustrated in Fig. 21-8.

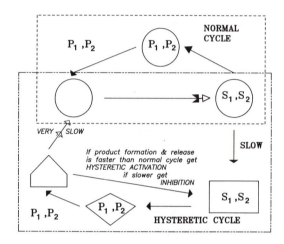

Figure 21-8   Scheme of an enzyme exhibiting a hysteretic response.

In many cases, the slow conformational change has been detected and shown to account for the auto-effect. However, in at least one instance of apparent hysteresis it has been demonstrated that the auto-activation resulted from product accumulation, where the product bound to vacant sites in an oligomer and relieved a substrate inhibition phenomenon.

COOPERATIVE EFFECTS IN HEMOGLOBIN

Hemoglobin may well be the most thoroughly studied protein in biochemistry or biophysics in terms of allosteric models as well as in terms of general protein chemistry. The molecule consists of four polypeptide chains, two α and two β, arranged in a tetramer. Although strictly a heteropolymer, hemoglobin has been used to illustrate the essential features of a variety of allosteric models, since the α and β subunits are almost chemically identical. In this section of the chapter we outline some of the essential features of hemoglobin and its allosteric properties.

Oxygenation is accompanied by structural changes in the subunits triggered by shifts of the iron atoms relative to the porphyrin, and in the β subunits, by the steric effect of oxygen itself. The oxygen-free form is constrained by salt bridges which are broken by the energy of heme–heme interaction with the release of H. 2-3-Diphosphoglycerate may add to the constraints by being stereochemically complementary to a site between the β chains; this complementarity is lost on oxygenation.

The Molecular Mechanism

Nature of Heme–Heme Interaction. If Y is the fractional saturation of hemoglobin with oxygen, p its partial pressure, and K and n constants, the saturation curve for oxygen can be represented by the Hill equation and is observed to be sigmoidal:

$$Y = \frac{Kp^n}{1 + Kp^n} \tag{21-44}$$

which corresponds to the equilibrium $Hb_n + nO_2 \rightleftharpoons Hb_n(O_2)_n$.

Two possible interpretations of the hemoglobin saturation effects are:

1. The first oxygen enters the hemoglobin molecule with more difficulty than the second, owing to the necessity of breaking up a preexisting partnership between the hemes.
2. The second oxygen enters more easily, owing to the decoying effect of the first.

The oxygen affinity of isolated α and β subunits corresponds to that of the fully combined oxyhemoglobin, indicating that it is the low oxygen affinity of deoxyhemoglobin that must be the result of constraints imposed on this form.

Movements of Iron Atoms Relative to the Porphyrin. X-ray studies of iron porphyrin compounds demonstrate that the length of the bond from the iron atom to the N atoms of the ring is 2.061 Å in high-spin ferric and 1.99 Å in low-spin ferric compounds. In metal-free porphyrin the distance to the center of the ring is 2.01 Å. In six-coordinated low-spin ferric complexes, therefore, the iron atom tends to lie within 0.05 Å of the plane of the four nitrogen atoms. In five-coordinated high-spin ferric compounds the iron atom is forced to lie outside that plane by between 0.38 and 0.47 Å.

The radius of ferrous iron in low-spin compounds is the same as in low-spin ferric compounds; therefore, the iron atom in low-spin ferrous compounds should lie within 0.05 Å of the plane of the nitrogen atoms. The radius of high-spin ferrous iron, however, exceeds that of high-spin ferric by 0.12 Å; this would place the iron atom 0.83 Å outside the plane of the four nitrogen atoms.

Of ferrous hemoglobin derivatives the deoxy form is high-spin and five-coordinated, so the iron atoms should lie approximately 0.8 Å out of plane to the nitrogens, while the oxy and carbonmonoxy forms are both low-spin and six-coordinated, so the iron should lie within that plane. All the ferric (met) derivatives are six-coordinated, the acid and fluoride met derivatives being high-spin and the remainder low-spin.

The quaternary structure of mammalian hemoglobin depends only on the coordination of the iron atom, not on its valency or spin state. The deoxy form is unique; all the other forms have the same quaternary structure. In deoxyhemoglobin the displacement is 0.75 Å, in close agreement with the 0.83 Å calculated. Both displacements are on the side of the proximal histidine (F_8).

Heme is attached to globin by a covalent bond extending from the iron atom to the imidazole nitrogen of the proximal histidine F_8. On the distal side lie histidine E_7 and valine E_{11}. The ligand comes to lie between the iron atom and these two residues. Further removed from the iron atom, 60 atoms of the globin are in van der Waals contact with the porphyrin ring. The smallest ligand that produces full heme–heme interactions is the hydroxyl ion, with a van der Waals radius of 1.5 Å. It becomes attached to the iron atom in the oxidation of deoxy- to met-hemoglobin at alkaline pH. A hydroxyl ion placed at the expected distance of 2.1 Å from the iron atom in either the α or β subunits of the deoxy form fits without making short contacts. In the β subunits of the deoxy form, the β-methyl of valine E_{11} lies too close, so there is no room for a ligand.

Direct comparison of the electron density maps of the β chains confirms that the distance between the porphyrin ring and valine E_{11} shrinks by approximately 1 Å on going from the met or oxy form to the deoxy form.

Also in difference electron density maps of deoxy- and met-BME-hemoglobin (hemoglobin cross-linked with bis-maleimido-methyl ester) there is a striking movement of valine E_{11} toward the heme group. This demonstrates that in the deoxyform of the β chains, the valine E_{11} must move relative to the porphyrin ring before the iron atoms can react, and that in going to the oxy form the distance between the porphyrin and helix E widens to make room for the ligands.

Role of the C-Terminal Residues. In oxy- or met-hemoglobin, the C-terminal residues of all four chains have complete freedom of rotation and the penultimate tyrosines have partial freedom. In the deoxy form, each of the C terminals is doubly anchored by salt bridges: the α-carboxyl of arginine $HC_3(141)$ α-1 is linked to the α-amino group of valine $Na_1(1)\alpha$-2, and its quanidinium group to aspartate $H_9(126)$ α-2. The α-carboxyl of histidine $HC_3(146)$ β-1 is linked to the ε-amino group lysine $C_5(40)$ β-2, and its imidazole to aspartate $FG_1(194)$ β-1. All the penultimate tyrosines are firmly anchored in pockets between helices F and H, partly by van der Waals contacts with the helices and partly by hydrogen bonds between their OH groups and the carbonyls of the peptide bonds involving valines FG_5. The tyrosines cannot be displaced from their pockets without also displacing the C-terminal residues and rupturing their salt bridges.

The enzymatic removal of the C-terminal residues affects heme–heme interaction. Removal of His $HC_3(146)$ β has the smallest effect; it reduces the Hill coefficient from 2.7 to 2.5, but the salt bridge is probably replaced by reformation with the α-carboxyl of tyrosine $HC_2(145)$.

N-substituted maleimides on cysteine $F_9(93)$ reduce the Hill coefficient to approximately 2.0. There is no salt bridge formation with lysine $C_5(40)$ α, due to displacement of the C-terminal histidine, but the remaining structure is undisturbed in the oxy and deoxy forms. Removal of arginine $HC_3(141)$ α reduces the Hill coefficient to 1.7, and removal of both Arg and His reduces it to 1.0 as well as inhibiting most or all of the Bohr effect.

Because these residues are free in oxyhemoglobin, they serve no function there. In deoxyhemoglobin they also have no conformational or environmental effect on the heme groups. Therefore, they must function as cross-links between the subunits in deoxyhemoglobin. This was confirmed by x-ray work. Human Des His-146 β and Des Arg-141 α deoxyhemoglobin each crystallize isomorphously with the normal deoxy form, *but* Des His-146 β plus Des Arg-141 α together crystallize in a form closely resembling the normal oxy form. Thus in the absence of constraining salt bridges the quaternary deoxy structure is unstable even if all four chains are deoxy.

Conformational Changes Within the α and β Subunits

X-ray studies of hemoglobin derivatives locked in the deoxy form have indicated changes in the tertiary structure occurring upon oxygenation. This is achieved using bis-maleimidomethyl ether, which inhibits all cooperative effects by linking Cys $F_9(93)$ to His $FG_4(97)$ in the same β chain. The reagent lies at the α_1–β_1 contact and also blocks entry of the penultimate tyrosines $HC_2(145)$ of the β chains into their pockets between helices F and H. Crystals of met- and deoxy-BME-hemoglobin are isomorphous with native met-hemoglobin.

When acid met-hemoglobin is reduced, the heme-linked water molecule is removed and produces a positive peak in a difference map of met–deoxy. Using ferrous citrate as the reductant, most β subunits but very few α subunits are reduced. With sodium thiosulfate all β subunits and most α subunits are reduced, indicating that ligands must be removed from the β chain more easily than from the α chain.

In the α chains, the positive peak representing the entry of the heme-linked H_2O is matched by a negative peak of the same magnitude representing the expulsion of Tyr $HC_2(140)$ from its pocket between helices F and H. There is a negative peak in the position of the amide group normally linking Tyr $HC_2(140)$ to Arg $HC_3(141)$, corresponding to the fixed position that this amide takes up in native deoxyhemoglobin; it is rotating freely in BME-met- and native met-hemoglobin. There are peaks (positive and negative) indicating that helix F moves inward toward the center of the molecule, narrowing the pocket between it and helix H. This movement is probably responsible for the expulsion of Tyr HC_2. There are positive and negative peaks on either side of the proprionic acid side chains of the heme group, demonstrating that its inclination is becoming less steep in the liganded form.

In the β chains, entry of Tyr $HC_2(145)$ into its pocket (between helices F and H) is blocked by the BME group, but helix F shows movement similar to that of the α-chain F helix. There is a positive peak in the position of Val $E_{11}(67)$, indicating a widening of the space between the porphyrin ring and helix E on ligand binding. There are also peaks near the heme group, suggesting that its tilt becomes less steep in the liganded form; its iron has moved away from Val E_{11}.

Conformational Changes at the Subunit Interfaces. The α_1–β_2 contact shows signs of strain, indicating that on dissociation of ligand, various residues are trying to move toward the deoxy conformation. It looks as though entry of Tyr HC_2 into its pocket presses on the indole ring of tryptophan $C_3(37)$ β_2, causing it to tilt over and press on proline $C_2(36)$ β_2, which helps to push the β chain in the required direction.

These results demonstrate that on binding of ligand to the α subunit, helix F moves so as to expel Tyr $HC_2(140)$ from its pocket (a similar movement of helix F occurs in the β subunit). On removal of ligand, residues at the α_1–β_2 contact show strain, moving them toward the deoxy formation even though the required sliding of the contact is inhibited by the BME.

Quaternary structure changes involve small shifts (< 1 Å) at the pair of subunit contacts α_1–β_1 and α_2–β_2, but large shifts (of approximately 7 Å) at contacts α_1–β_2 and α_2–β_1.

Mutations replacing residues at the α_1–β_2 contact diminish heme–heme interaction, but mutations at α_1–β_1 contact do not. Also, nearly all residues at the α_1–β_2 contact are invariant, whereas those at the α_1–β_1 contact vary.

The contact α_1–β_2 is dovetailed so that the CD region of one chain fits into the FG region of the other. During the change of quaternary structure the two subunits rotate relative to one another so that the dovetailing of CDβ with FGα remains much the same but that of CDα with FGβ changes.

In oxyhemoglobin a knob consisting of the side chain of threonine $C_3(38)$ α fits into a notch made up of the main chain of valine $FG_5(98)$ β. In deoxyhemoglobin the same notch is filled by the side chain of Thr $C_6(41)$ α, the one protruding from the next turn of the helix. At the same time the hydrogen bond linking Asp $G_1(94)$ α to Asp $G_4(102)$ β in oxyhemoglobin is replaced by a hydrogen bond between Tyr $C_7(42)$ α and Asp $G_1(99)$ β in the deoxyhemoglobin.

Stereochemically, the existence of an intermediate quaternary structure is unlikely. Spectroscopic evidence favors the existence of an intermediate but not necessarily one with a quaternary structure different from the oxy and deoxy forms. ESR spectra of spin-labeled oxy- and deoxyhemoglobin are markedly different. If the spin label is attached to Cys $F_9(93)$ β, which lies near the tyrosine pocket, and the α_1–β_2 contact, the label senses changes in the conformation of the β and α chains as well as shifts in that contact. If hemoglobin existed in only two alternative conformations, the ESR spectra at successive stages of oxygenation would exhibit common isosbestic points; however, this is *not* so, demonstrating an intermediate structure is involved in the process.

Further evidence for intermediary structures comes from difference Fourier maps of the BME-hemoglobin. These show that constrained subunits take up an intermediate conformation. In a comparison of the difference Fourier with the actual differences between native oxy and deoxy structures, many but *not all* the expected pairs of positive and negative peaks are found, indicating the existence of an intermediate tertiary structure.

On uptake of oxygen the tertiary structures of the individual subunits click to the oxy conformation, but the quaternary structure may remain in the deoxy conformation until several of the subunits have reacted with oxygen. Thus subunits in the oxy conformation are constrained in the quaternary structure of the deoxy form, and vice verse.

Mechanism of the Conformational Change. The distance between the heme groups is too large (25 to 37 Å) for electromagnetic interactions to be effective, suggesting that the trigger for the change is stereochemical. Two possibilities are changes triggered by deformation of the heme group or by the ligand prising the heme pocket apart. In the α subunit, the ligand fits into the heme pocket in both the oxy and deoxy forms; therefore, the second mechanism can be discounted. However, the transition from deoxy to oxy involves movement of the proximal histidine toward the plane of the porphyrin, approximately 0.75 to 0.95 Å. In acid met-hemoglobin a similar movement occurs but is approximately 0.45 to 0.65 Å. Such movements should be sufficient to cause the changes in tertiary structure since the iron atom is rigidly linked to histidine F_8 and the porphyrin is in contact with about 60 atoms of the globin (which is flexible and can change conformation). This can be seen from difference electron density maps of BME deoxy- and met-hemoglobin, showing that helix F moves toward the center of the molecule and expels tyrosine $HC_2(140)$ from its pocket between helices F and H. The expelled TYR must pull arginine $HC_3(141)$ with it, breaking its salt bridges with the opposite α subunit and releasing Bohr protons.

In the β subunit, thermal vibrations provide the activation energy needed to open the heme pocket, allowing the ligand to reach the iron atom. Reaction with the heme then moves the iron into the plane of the porphyrin (again helix F moves as in the α subunit), breaking salt bridges between His $HC_3(146)$ and Asp $FG_1(94)$.

In deoxyhemoglobin the α subunits have room for ligands but the β subunits do not; therefore, the α subunits are likely to react first. Reaction of the iron in α_1 with O_2 causes Tyr $HC_2(140)$ α to be expelled from its pocket, and the links between Arg $HC_3(141)$ α_1 and the α_2 subunit to be broken with release of Bohr protons. The iron in α_2 reacts next, Tyr is expelled and links between C-terminal Arg and the α_1 subunit are ruptured, causing release of further Bohr protons.

At this stage four of the six salt bridges constraining the deoxy tetramer have been broken, with a resulting change in the allosteric equilibrium constant in favor of the quaternary oxy form. At this point (which may, in fact, occur at any stage in the reaction scheme) the $\alpha_1-\beta_2$ and $\alpha_2-\beta_1$ contacts give way and the tetramer clicks to the oxy form, breaking the remaining salt bridges [i.e., those between Lys $C_5(40)$ α and His $HC_3(146)$ β, and the ones between DPG and the 2β subunits]. This leads to the liberation of DPG but does not release Bohr protons.

We now have the intermediate in the quaternary structure of oxyhemoglobin, with the two-fold symmetry of the molecule conserved. The α heme and its immediate environment have the oxy conformation, while those in the β chain have the strained deoxy conformation [i.e., Tyr $HC_2(145)$ β are still in their pockets, keeping the ligand sites obstructed by Val $E_{11}(67)$]. The change in quaternary structure has halved the activation energy needed to expel Tyr $HC_3(146)$ β from its pocket because it has broken the salt bridges between the C-terminal His and the α subunits, leaving only the internal salt bridges between C-terminal His and Asp of the same β chain to hold the tyrosines in place. This results in an increase in oxygen affinity. The iron atoms of the β chain now react with ligand, accompanied by the expulsion of a Tyr and the rupture of one of the salt bridges, leading to the release of the last Bohr protons.

Nature of the Interaction Energy. There are fewer van der Waals interactions at contacts between α and β subunits in deoxyhemoglobin than in oxyhemoglobin, and about the same number of hydrogen bonds, suggesting that neither of these is likely to provide the stabilizing interaction that constrains the subunits in the deoxy form. The salt bridges linking $\alpha_1-\alpha_2$, $\alpha_1-\beta_2$, and $\alpha_2-\beta_1$ are more likely sources of the interaction energy. Enzymatic removal of the four C-terminal residues makes the quaternary deoxy-structure unstable and inhibits heme–heme interaction. The bond energy per salt bridge is about 1 to 2 kcal/mol and therefore for six salt bridges is approximately 6 to 12 kcal/mol (which is the same order as that observed for the interaction energy of 12 kcal/mol). If only salt bridges constrain the deoxy tetramer, part of the energy released by the reaction of the heme groups with oxygen must be expended to break them.

In the genetic variant *Hemoglobin M Iwate*, where His $F_8(87)$ α is changed to Tyr, the α chains are unreactive and the tetramer is locked in the deoxy conformation. The β chains have low affinity for oxygen and the reaction produces *no* heme–heme interactions and *no* release of Bohr protons, indicating that these effects are absent when change of the quaternary structure is not possible. This suggests that the changes in free energy of the subunit contacts accounts for *all* the energy of interaction.

Implications for Allosteric Models. The MWC model assumes that all the subunits in the quaternary T state are in the unreactive conformation, and that all of those in the R state are in the reactive form whether they are liganded or not, and hence allows no intermediate forms. However, the hemoglobin subunits change their tertiary structure in response not to the change in quaternary structure, but to the binding of ligand, as in the sequential models, and intermediate forms have been shown to exist. However, the Koshland model implies that the change in tertiary structure of each subunit directly affects the ligand affinity of its neighbors, and there is *no* evidence for this in hemoglobin. All or most of the interaction energy arises via the step-by-step release of constraints on the unreactive quaternary structure and diminishes the work required to change the tertiary structure of each subunit from the unreactive to the reactive form.

In the sequential models, the change in tertiary structure of the α subunits would be caused by induced fit in response to a change in conformation of the heme on reacting with substrate, while the β subunits change in response to both the change of heme conformation and to steric adaption of the active site to the substrate.

The equations developed for the concerted or the sequential allosteric models both give adequate fits to the experimental data observed for oxygen binding to hemoglobin, illustrating a common problem with fitting experimental data to allosteric models. With sigmoidal saturation curves, Eqs. (21-23) and (21-14) both give adequate fits. As a result, the emphasis on distinguishing between the two types of models relies heavily on the experimental detection of either the preexistent equilibrium between the R and T forms, or on the ability to demonstrate sequential conformational or intermediate forms. In the case of negative homotropic interactions, this is not a problem since only sequential models account for this behavior.

In this chapter we examined the theoretical basis of a variety of models for allosteric interactions in oligomeric proteins. The major emphasis has been on homotropic interactions, since heterotropic effects are simply explained in terms of separate, nonactive ligand binding sites and do not *require* the protein to be oligomeric. We have also considered mechanistic details of the allosteric properties (both homotropic and heterotropic) of hemoglobin. Although it is not an enzyme, an appreciation of these details give insight into the types of mechanisms that must operate in the various enzymes known to exhibit allosteric properties.

We have examined in great detail the conformation of proteins, the interaction of ligands with proteins, and models describing changes in the activity of proteins that are induced by ligand binding. We have also discussed some of the ways in which amino acid side chains and the overall conformation of the active site of an enzyme contribute to its catalytic activity. From these discussions a picture of a biologically active protein has emerged, one in which the properties of the protein are governed by the conformation of the polypeptide chain or chains that make up that protein. The conformation is dynamic, and ligands interacting with the protein shift the distribution of accessible configurations the protein sees. This changes its properties and its overall conformation: binding sites may appear or be lost. Regions of the protein that might be involved in subunit–subunit interactions may change conformations, leading to alterations in the ways in which subunits interact. The local conformation around individual amino acid side chains changes, leading to alterations in the chemical properties of those side chains. These conformational changes form the basis for the regulation of the biological activity of a protein. In this chapter we have looked at some formal models for both homotropic and heterotropic allosteric regulation of enzyme activity. However, we have considered these models with little attention to the mechanisms by which the conformational changes leading to the activity changes were elicited or the biological role of such changes. One of the purposes of this chapter is to examine the mechanisms that biological systems employ to alter the conformation of proteins, and hence regulate their activity.

It is important to remember that one purpose of regulation of activity is to modulate the amount of a particular compound that is present. To achieve this, it really does not matter whether "allosteric" regulation occurs, and in some instances "regulation" is achieved without benefit of ligand-induced changes in the enzyme's activity.

Finally, we must not lose sight of the fact that another purpose of regulation is to allow integration of a vast number of different metabolic reactions, and a variety of ways of achieving this integration are available to the cell. Regulation of particular points in metabolic pathways must be considered in the context of the pathways. It is impossible in a text of this type to consider metabolic regulation in detail. However, some general conclusions about the types of regulation that may operate at particular points can be drawn. One of these involves the interaction of multiple regulatory ligands with a particular enzyme, and later in this chapter we examine this topic in the context of the complex regulatory properties of glutamate dehydrogenase. Other ways in which this regulation of integration can be achieved involves compartmentalization of enzymes or metabolites, and synthesis or breakdown of the enzymes involved in metabolic pathways. The latter may be considered as an adaptive or course-control mechanism. Both of these, although outside the domain of this book, are important in terms of biological control processes. The purpose here is to discuss the protein chemistry and enzymatic consequence of what might be considered to be essentially instantaneous, or fine control, regulatory mechanisms.

ALTERATIONS IN ENZYMATIC ACTIVITY

We come now to the main focus of this discussion, the reversible regulation of an enzyme activity via induced conformational changes. There are two ways in which the activity of an enzyme can be changed. Its maximum velocity can be altered, thus making the enzyme either more or less efficient at a particular degree of saturation with substrate. Alternatively, the degree of substrate saturation at a particular substrate concentration can be altered. Provided that the enzyme is not operating at V_{max} conditions (i.e., $[S] \gg K_m$), the latter mechanism effectively changes the amount of enzyme being utilized at a given substrate concentration. These two mechanisms can be summarized by saying that a regulator must affect V_{max}, K_m, or both to be effective.

By reducing an enzyme mechanism to three phases, we can analyze what types of changes may be involved in regulation. The phases are (1) substrate binding steps, (2) catalytic transformation steps, and (3) product release steps.

Effects on V_{max}

To *increase* the rate of the catalyzed reaction via an effect on V_{max}, the regulator must speed the slowest step in the reaction. To *decrease* the rate via an effect on V_{max}, the regulator can either slow the slowest step or create a new slowest step by inhibiting a previously non-rate-limiting step. Depending on the formal kinetic mechanism, the slowest step in the overall reaction is one in either the catalytic transformation process (usually the case with rapid equilibrium mechanisms) or in the product release phase (often the case with steady-state mechanisms).

In a particular enzyme any one or more of these may contribute to the catalytic transformation phase of the reaction. In some manner all involve the chemical properties of individual amino acid side chains. These properties are governed by their environment, which in turn is governed by the enzyme's tertiary (and quaternary) structure. A change in enzyme conformation produced by regulator binding must change the chemical environment of one or more of the side chains involved in the catalytic mechanism so as to increase the catalytic efficiency of the enzyme if activation of an equilibrium-type mechanism is to occur. Inhibition in an equilibrium mechanism can be brought about only by a conformational change slowing the same step, as discussed previously. Alternatively, a new step in the product release phase of the reaction may become rate limiting; this can result in a change in the enzyme's formal kinetic mechanism. In its simplest manner the product release phase of the reaction can be expressed as $EP \rightarrow E + P$, and to slow this phase, the off-velocity constant for P must be slowed: that is, for $EIP \rightarrow EI + P$, the release of P is slower than previously. This is equivalent to stating that the dissociation constant for P is *smaller* for EIP than it is for EP. In either case the dissociation constant is made up of an "on"-velocity constant, which is unlikely to be affected, and an "off"-velocity constant. An example gives some idea of the range of possible ways in which the regulator may affect the product release phase of a reaction. The enzyme glutamate dehydrogenase catalyzes the reaction

$$\text{glutamate} + \text{NAD(P)} \rightleftharpoons \text{2-oxoglutarate} + \text{NAD(P)H} + \text{ammonia}$$

and involves an enz-(2-OG)-NAD(P)H-ammonia quaternary complex. Release of 2-OG and ammonia is presumed to occur more rapidly than NAD(P)H release, which under some conditions is thought to be rate limiting. ADP and GTP are both allosteric regulators of the enzyme. ADP increases the dissociation constant of the enz-NAD(P)H complex, while GTP decreases it. Under circumstances where release of NAD(P)H is rate limiting, ADP acts as an activator and GTP as an inhibitor, by having effects on product release stages in the reaction. Under optimal conditions, however, NAD(P)H is *not* rate limiting in the absence of either of these regulators. As a result of the slowing of this step in the presence of GTP, it is possible that it becomes rate limiting. However, the kinetic mechanism may not be as simple as involving only those steps required by the minimal formal kinetic mechanism. In Chap. 16 we discussed the effects of abortive complexes. In terms of regulation, they represent an ideal step to elicit either an activatory or an inhibitory response. Under conditions where an abortive complex is formed, it is by definition the rate-limiting step. Activation can be achieved by destabilizing the abortive complex, or inhibition by stabilizing it.

In all these cases the effect is produced by regulator binding producing a conformational change that alters the off velocity constant of the product from the abortive complex.

Effects on K_m

The other step at which a regulator may operate involves the K_m for a substrate. Figure 21-9 illustrates conditions under which such regulation could be effective.

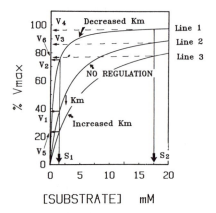

Figure 21-9 Regulation of an enzyme by effects on K_m. In line 1 the K_m of the substrate is decreased relative to the unregulated enzyme (line 2). In line 3 the K_m of the substrate is increased relative to the unregulated enzyme.

Consider how an activator must operate via a K_m influence. At a given substrate concentration, S_1, below the K_m concentration of the unregulated enzyme, a certain velocity, V_1, is achieved. If an activator binds and *decreases* the K_m for the substrate, the effect, without any change in V_{max}, is that a new velocity, V_2, results at the same substrate concentration. If the substrate concentration is originally much greater than the K_m concentration (i.e., at S_2), the effect of such an activator is minimal; the rate changes from V_3 to V_4. Similarly, if the regulator *increases* K_m, an effective inhibition is achieved at the low substrate concentration, S_1; the rate decreases from V_1 to V_5. At the highest substrate concentration, again little effect is found (compare V_3 and V_6). For a regulator to operate via a K_m effect, it is necessary that the enzyme operate at substrate concentrations at or below the K_m concentration. As V_{max} conditions are approached, a regulator operating at the level of K_m is relatively ineffective.

How such effects on K_m can be elicited depends on the formal kinetic mechanism. In rapid-equilibrium mechanisms K_m represents a true substrate dissociation constant, and a change in K_m must be explained at the level of the "off"-velocity constant for the appropriate substrate. For steady-state mechanisms K_m embodies both binding and catalytic steps, and as a result the point of change may involve either type of step. Both changes require a conformational alteration affecting the properties of either a binding site or some amino acid side chains involved in the chemical mechanism of the enzyme.

Heterotropic and Homotropic Regulation

The effects on V_{max} and K_m discussed in the preceding two sections have been examined from the standpoint of heterotropic regulation. As was discussed extensively earlier homotropic regulation exists in a number of enzymes. The concept of homotropic allosteric effects must also be considered at the level of heterotropic regulation. In a number of cases a heterotropic activator or inhibitor of an enzyme's activity binds to the enzyme in a cooperative manner. Although such a ligand is clearly exerting a heterotropic effect on the enzyme activity, it itself has a homotropic binding effect. As with homotropic substrate binding effects, the purpose of this is either to sensitize or desensitize the enzyme to the heterotropic regulatory ligand's effect, as may be necessary.

If this situation seems confusing, it is because *it is confusing*. However, it is also realistic, and only a part of the complexity to which this enzyme and many other regulatory enzymes are subjected. To solve the any remaining mysteries (and there are many) of enzyme regulation, the idea of multimode regulation must be addressed. A clear picture of the regulatory properties of an enzyme requires an understanding of the interactions between different regulatory ligands in terms of their binding effects and elicited effects on the enzymatic properties of their target enzyme. Many regulatory ligands—as in the case of adenine nucleotides with glutamate dehydrogenase—may have different effects at different pH values or at different concentrations of substrates, and it is most important to examine these regulatory interactions at physiologically relevant concentrations of the substrates and the regulatory ligands themselves.

Index